A MODERN ILLUSTRATED MILITARY HISTORY

LAND
POWER

A MODERN ILLUSTRATED MILITARY HISTORY

LAND POWER

EXETER BOOKS · New York

in association with Phoebus

First published in USA 1979 by
Exeter Books
Distributed by Bookthrift
New York, New York 10018
ISBN 0-89673-010-7
Library of Congress Number 79-87557

Made and printed in Great Britain by
Redwood Burn Limited

Imperial War Museum

*A British 9.2-in howitzer in action in an orchard
near Albert, July 1916*

INTRODUCTION

In the first part of this book we present the full story of the weapons of the First World War. Artillery, tanks and infantry weapons, as well as gas, flame and other methods of destruction, are described in detail, and illustrated by a superb combination of action photographs and detailed drawings of the weapons themselves. The authoritative text in Part I is written by high-ranking officers and expert military historians. There is even a brief but illuminating passage by the young Erwin Rommel, a German company commander during operations in the Argonne.

Part II is subdivided into three main sections, the first of which presents an account of the infantry during the Second World War: their tactical role, their organisation, and, particularly, the weapons they used. These ranged from tracked vehicles to automatic pistols, and are illustrated by John Batchelor in his own magnificent style.

The second section takes the reader through the whole story of the German panzers – from the secret training machines of the Weimar era, through the intense period of innovation and experiment, to the Second World War itself, and the greatest tank battles in history. When Germany went to war in 1939, the pride of her army was the Panzer Divisions. Intended to be the armoured spearhead in a war of rapid conquest, they were equipped with a wide range of vehicles designed to function together. Tanks would make the initial breakthrough, self-propelled guns would give fire support, while half-tracks brought up the Panzergrenadiers.

The third section examines Germany's armoured fighting vehicles, and the ideas that have influenced AFV development ever since. As German engineers developed an outstanding range of weapons, so the authors analyse the vehicles model by model and provide a wide range of information on prototypes and technical detail.

JOHN BATCHELOR, after serving in the RAF, worked in the technical publications departments of several British aircraft firms before becoming a freelance artist. His work has appeared in a wide range of books and magazines, and his work for Purnell's History of the World War Specials has established him as one of the leading artists in his field.

CONTENTS

A German 80-cm Gustav rail gun, the largest (though not the largest calibre) gun ever built, in action during the siege of Sebastopol in 1942

JIM

A British 12-in howitzer in action during the Battle of Albert, July 1916

PART I:
THE FIRST WORLD WAR

The outbreak of war in 1914 found the bolt-action rifle at the peak of its development and the machine-gun being used in large numbers for the first time. No infantry charge could hope to succeed in the face of such concentrated firepower, and from Switzerland to the sea a system of trenches appeared which made nonsense of the theories of warfare current in the pre-War years.

The first reaction of the generals was to use longer and heavier artillery barrages in a futile attempt to break the deadlock — with the result that trenches were dug deeper and shell-holes made the ground even more impassable. On other fronts this pattern was repeated, until the advent of the tank which at last brought hope of a solution to the stalemate.

A British 14-in rail gun. The crew have added the name 'Boche Buster' on the mounting

Italian artillerymen with a 280mm mortar, 2000
metres up on the Carnic Alps

4

CONTENTS

Much capital has been made in the half-century following the First World War of the fact that in the period of lively military speculation which preceded its outbreak the only theorist fully to appreciate what its nature was likely to be was a civilian—Bloch, a Polish banker and economist writing in 1897. His thesis was that the development of industrialism had fundamentally altered the character of war. He claimed that 'the outward and visible sign of the end of war was the introduction of the magazine rifle. The soldier, by natural evolution, has so perfected the mechanism of slaughter that he has practically secured his own extinction'. The picture he painted of what actually was to happen in 1914 and 1915 on the Western Front turned out to be startlingly accurate. *At first,* he wrote, *there will be increased slaughter on so terrible a scale as to render it impossible to get troops to push the battle to a decisive issue. They will try to, thinking they are fighting under the old conditions, and they will learn such a lesson that they will abandon the attempt. The war, instead of being a hand-to-hand contest in which the combatants measure their physical and moral superiority, will become a kind of stalemate, in which, neither army being able to get at the other, both armies will be maintained in opposition to each other, threatening each other, but never being able to deliver a final and decisive blow. Everybody will be entrenched in the next war; the spade will be as indispensable to the soldier as his rifle.*

Fifty years later it can be seen that these were the deductions which the General Staffs and defence philosophers of all the belligerents should have drawn from their studies, ardently pursued before 1914, of the American Civil War, the Franco-Prussian War, the South African War and the Russo-Japanese War. The effect of rifle fire at Gettysburg, Gravelotte and Saint Privat indicated that the defence had become the stronger form of war. The two gallant but abortive French cavalry charges at Wörth showed that the day of the 'arme blanche' was over: never again would armies be able to justify the large amount of railway rolling stock their forage absorbed. The great potential of modern artillery was already apparent. The development of railways with the inevitable risks arising from being tied to fixed lines of supply had been shown to be both an advantage and a disadvantage. British experience in the South African War had emphasised the great potential of the skilled marksman firing from cover and consequently that of the machine gunner: it had also brought out the need for more mobile and intelligent infantry exercising their initiative to the full.

These then were the facts of life which the armies were to discover by bitter experience on the Western Front in 1914 and 1915. It is therefore surprising, in view of the emphasis that had been laid on the study of military history, that the French, German and British military leaders should have gone to war in 1914 steeped in doctrines which flouted to a considerable degree the lessons of the immediate past.

In view of their obsession with Napoleon and their tradition since Waterloo of arriving at the wrong conclusion by apparently faultless logic, the French pre-war theories must be considered first. Foch, Grandmaison and Langlois were disciples

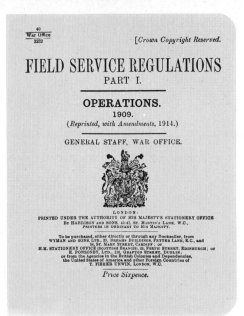

THE NEW WARFARE

As the belligerent powers surveyed the events of 1915, they looked with dismay at the mistakes they had made, the opportunities they had lost. Nevertheless, both sides hoped for a successful conclusion to the war in 1916 *Major-General H. Essame Above:* Sixpenny handbook for the troops at the front

of Clausewitz who taught that the secret of success in war was offensive action and the concentration of superior force in terms of bayonets at the decisive point and time. They were further greatly influenced by Colonel Ardent du Picq's *Études sur le Combat* which stressed that morale was the most important factor in war, which indeed has always been true, but did not necessarily imply that a solid phalanx of heroes in red trousers was the best answer to the bullet and the shell. Foch even went so far as to claim that 'a battle lost is a battle one thinks one has lost; for a battle cannot be lost physically'. He went on to state that future improvements would increase the effectiveness of the attack. Above all, the French envisaged a short and mobile war in which the mystic quality of *élan,* which they believed they possessed to a higher degree than all other nations, combined with the threat of the long bayonet (*Rosalie*), would triumph. The artillery would merely be an auxiliary arm. In the attack they pictured an advance supported by controlled fire to within 400 yards of the enemy; thereafter they considered that aimed fire by the enemy would become impossible and that a mass of infantry in depth charging with the bayonet would carry all before it.

The German theorists prescribed opening the attack with a dense infantry firing line followed by an advance with artillery

support until a position about 600 yards from the enemy was reached. At this stage they thought a struggle for fire superiority would develop which, when won, would enable the infantry to move forward until, on reaching a position about 100 yards from the enemy, they would be able to deliver the final assault on classic lines.

Both nations had studied the battles of Mukden and Port Arthur and had realised the need for increased artillery support for the infantry attack. In the case of the French, this took the form of increased emphasis on the rôle of their excellent 75-mm field gun. The Germans mobilised under Corps control four heavy batteries of 5.9-inch guns per division in addition to the normal divisional artillery.

The British tactical approach to the coming war was understandably influenced by the experience of the South African War. After this war, Roberts had given General Henderson the task of producing a tactical manual. This appeared in 1905 and was entitled *Combined Training.* In it, Henderson embodied not only his own deductions from the South African War but also what were thought to be the lessons of the Russo-Japanese War—the importance of entrenchment, the problems raised by barbed wire obstacles, the need for attack in depth at the enemy's weakest point and the value of indirect artillery fire. *Field Service Regulations (Volume 1),* which superseded this manual in 1909, reproduced much of what Henderson had thought. By this time, however, there had been a good deal of cross-fertilisation of ideas with the French. The results of this liaison are evident in the section entitled *The Decisive Attack,* where concentration of all available artillery and machine gun fire and the building up of a firing line are envisaged. The section goes on to say: *The climax of the infantry attack is the assault which is made possible by superiority of fire to be gained by the artillery, machine guns and infantry. The fact that superiority of fire has been obtained will usually be first observed from the firing line; it will be known by the weakening of the enemy's fire. The impulse for the assault must therefore come from the firing line and it is the duty of any commander in the firing line who sees that the moment for the assault has arrived, to carry it out and for all other commanders to co-operate. Should it be necessary to give the impulse from the rear, all available reinforcements will be thrown into the fight and as they reach the firing line will carry it with them and rush the position.* Its companion volume, *Infantry Training (Volume 1 1914),* was even more explicit: *The action of the infantry in attack must therefore be considered as a constant pressing forward to close with the enemy. When effective ranges are reached there must usually be a fire fight, more or less prolonged according to the circumstances, in order to beat down the fire of the defenders. The leading lines will be reinforced and as the enemy's fire is gradually subdued, further progress will be made by bounds from place to place, the movement getting renewed force at each pause until the enemy can be assaulted with the bayonet.* This manual, the bible of the vast New Armies, goes on to stress the need for close liaison with the supporting artillery and places responsibility for communication with them on the infantry.

Field Service Regulations (Volume 1)

also implied that at the crisis of the battle the chances of successful cavalry intervention *en masse* would increase—something Henderson had never said and probably now inserted as a result of the influence of Haig, Director of Staff Duties at the relevant time, and the coterie behind the *Cavalry Journal,* brought privately into being in 1906 to ensure that the well bred horse occupied a distinguished place on the battlefield—a concept which even managed to survive the First World War.

Thus so far as their approach to battle was concerned, all the belligerents in August 1914 had this much in common: all paid lip service to mobility and thought in terms of a short and highly mobile war; all stressed the importance of offensive action; all gave the impression that if the assault

They fell for what seemed to them the obvious solution— overwhelming bombardment. They would now crush all resistance by sheer weight of shell.

was determined enough it must succeed; all agreed on the importance of close co-operation between the artillery, machine guns and infantry, in order to gain superiority of fire and thus enable the infantry to close with the bayonet. Furthermore, all implied that cavalry had still a future as an assault arm and none fully appreciated the great strength of defensive systems in depth based on entrenchments, wire, interlocking arcs of machine gun fire and artillery defensive fire. As a result of years of brainwashing drill on barrack squares all were wedded to the ideas of units attacking in straight lines thus producing perfect targets for machine guns firing in enfilade.

The first three months of the war were to show that in armament and organisation, although the Germans had the

Above: French St Chamond 400-mm rail gun, the largest land gun used in the First World War. *Total traverse:* 12°. *Elevation:* −15° to +73°. *Maximum range:* 17,500 yards. It had three different types of shells, weighing 1,410 pounds, 1,960 pounds and 1,980 pounds respectively.

Below: French Batignolles 320-mm rail gun. *Muzzle velocity:* 2,630 feet per second. *Maximum range:* 2,950 yards. There were several different types of shells, varying between 790 pounds and 1,100 pounds in weight

7

advantage in heavy howitzers and numbers of machine guns, this was counterbalanced by the French 75s and the British outstanding superiority in rifle marksmanship vividly demonstrated at Mons and Le Cateau. It has been claimed that if French had not allowed his divisions to drift to battle and had made a more vigorous and determined attempt to capture the Chemin des Dames in the Battle of the Aisne he might have ended the war in 1914. This is doubtful; his army had neither the technique nor the equipment.

At the First Battle of Ypres in the following month the rapid and accurate rifle fire of the British army in the opening engagement had virtually the stopping effect of machine gun fire. In this bitter struggle — a 'soldiers' battle' if ever there was one — the fallacy of prewar theory was finally demonstrated. The defence, for the time being at any rate, had shown itself to be much the stronger form of war. A whole generation of British regular officers and men had been destroyed in the process: consequently they would not be available to train the New Armies. For the future, the blind must lead the blind. Virtually a whole generation of young Frenchmen had also sustained such losses that the national morale would never be quite the same again. With the onset of winter, the front finally congealed: trench warfare began. Tactical and strategical stalemate on the Western Front was complete.

The Allies and Germans now contemplated a deadlock. Meanwhile, by trial and error, they would all learn their lessons, item by item, by costly instalments. For the Allies there was no alternative to offensive action: the Germans were on French and Belgian soil and public opinion unanimously demanded their removal with the greatest possible speed.

The battles of the Aisne and Ypres had resulted in a demand for artillery ammunition far exceeding anything anticipated by the British War Office before the outbreak of war and in consequence an acute shortage developed during the winter. This led those preparing for the renewal of the offensive in the spring to think that overwhelming artillery bombardment — to swamp the German machine guns and cut the enemy wire with shrapnel — would provide the key to success. These ideas dominated the meticulous preparations for the battle of Neuve Chapelle in March 1915, the aim of which was to capture Aubers Ridge. In this battle, after a brief but intense preliminary bombardment, lasting only 35 minutes, four infantry brigades advanced to the attack on a total frontage of only 2,000 yards. Complete surprise was achieved, the German front line was overrun and the infantry advanced about 1,200 yards. Thereafter communications between the infantry and the artillery collapsed and between commanders below the divisional level as well. It was not until late in the afternoon that anything in the nature of serious exploitation was attempted — obviously too late. Further attempts on the succeeding days came to nothing. Such success as this battle achieved was due to the surprise effect of the preliminary bombardment lasting only 35 minutes. Obviously the front of attack was too narrow. The battle further accentuated in dramatic form a complex problem — that of communications between the forward infantry and the supporting artillery. Tele-

Above: General Sir William Robertson, the newly appointed Chief of the Imperial General Staff. He forced Kitchener to recognise the differences in their respective rôles, and he also believed that, given sufficient resources, any line could be broken, any offensive won

'The soldier, by natural evolution, has so perfected the mechanism of slaughter that he has secured his own extinction'

phone lines were found to be and continued to be almost hopelessly vulnerable.

The success of the short and sharp preliminary bombardment at Neuve Chapelle encouraged Haig to try a similar technique when he was ordered to make a further attempt to take Aubers Ridge on May 9, in conjunction with a French offensive under Foch north of Arras. This time there was to be a 40-minute preliminary bombardment to cut the wire, swamp the Germans' forward troops and cut off their reinforcements. Haig planned two converging attacks 6,000 yards apart hoping to cut off some six or seven German battalions. To ensure continuous artillery support, specific batteries were detailed to follow the advancing troops and batteries of mountain artillery were placed in close support of attacking battalions.

Unfortunately, the Germans too had learnt something at Neuve Chapelle, notably the need for solidly constructed fortifications in depth. The British bombardment was a failure, much wire remained uncut when the infantry advanced to find the German defences to be immensely strong. Twenty-foot breastworks and bomb-proof shelters protected the garrisons and enabled them to line their parapets with rifles and machine guns the moment the bombardment lifted. According to an officer survivor of the 2nd East Lancashire Regiment: 'the artillery entirely failed to shake the enemy, who maintained heavy rifle fire and machine gun fire throughout the bombardment.'

Casualties among the attacking troops were as high as 75%.

In his next offensive at Festubert a month later Haig, encouraged by the French who had had some success in attacking after four days' intensive shelling, arranged for a 36-hour artillery preparation to be followed by a night attack to gain a footing in the first two German lines. It was then proposed to exploit the penetration next day. Although the Germans were not taken by surprise, this gambit at first had some success but the artillery support proved inadequate, ammunition ran out, no reserves were available when the crisis came and troops already tired were driven forward to assault once more. The attack soon expired in the teeth of greater machine gun fire than had ever previously been encountered. Haig concluded however from the experience gained in the battle that provided the preliminary bombardment was long enough, heavy enough and meticulously prepared an advance to a depth of about 1,400 yards could be guaranteed.

'Miserable offensive'
The British wanted no more major offensives in 1915. The shell shortage was now at its height and it would take the new ammunition factories in the United Kingdom and the United States at least a year to reach peak production. Delay, however, did not suit the French who demanded and finally obtained British assistance in the build-up area about Lens and Loos on the immediate left of Foch's September offensive towards Namur. French and Haig could see little future to operations in this area of coal mines and industrial slums but were overruled in what were thought to be the interests of the Alliance. Haig, now committed to the policy of prolonged artillery preparations, had only sufficient ammunition to support two of the six divisions available for this battle. He therefore decided to make good the deficiency by using gas discharges from cylinders operated by Special Companies of the Royal Engineers. As the German gas attack at Ypres in April had already shown, success depended on the vagaries of the wind: on the day of the attack there was virtually none. Nevertheless the attacking troops did succeed in capturing the German first line and in advancing beyond it. Now, however, there was delay in bringing forward the reserves who were controlled by French himself, and the Germans were able to stabilise the battle. Haig continued this dismal offensive for a further miserable 17 days. It was the first experience of the New Armies and despite their outstanding courage and high grade human material it emphasised not only the lack of training of the troops but also of their commanders and staffs as well. It also showed that the Germans once again were a step ahead: they now catered for a second line of defence out of range of any bombardment the Allies could mount.

At Loos, Haig had been right in asking for control of the reserves and French wrong. French therefore was removed. His replacement in somewhat dubious circumstances by Haig involved no dramatic leap forward in tactical thought. In the planning and conduct of the battle he had failed to find a means of keeping his finger on the pulse of it. After zero hour the battle so far as he was concerned had virtually

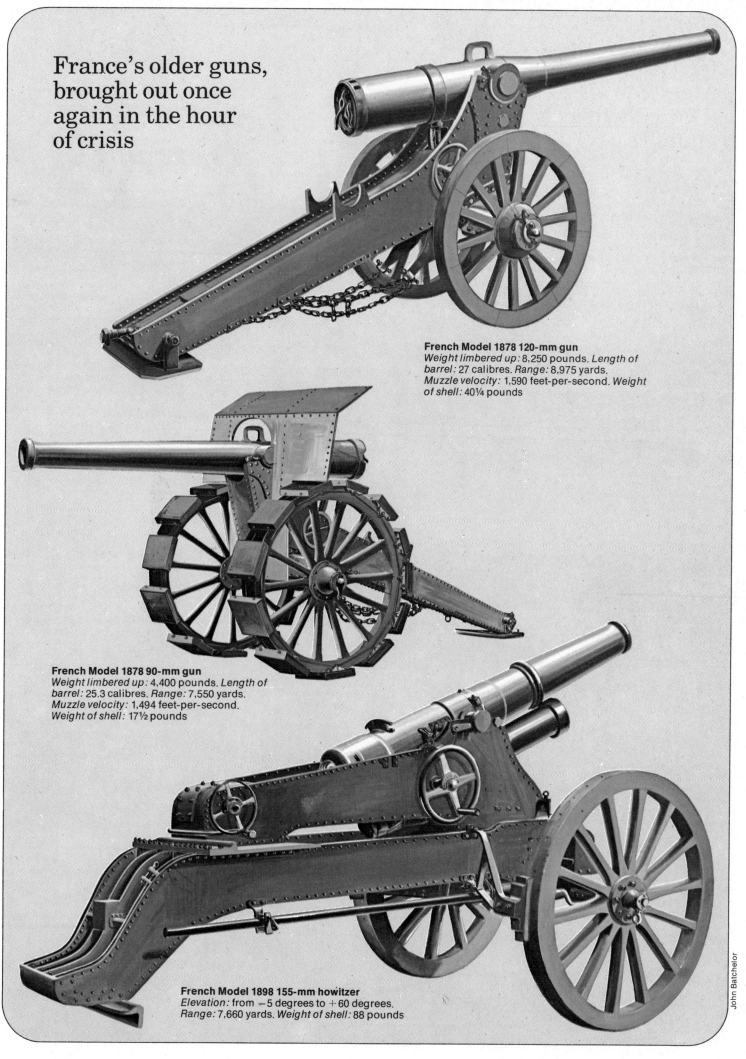

France's older guns, brought out once again in the hour of crisis

French Model 1878 120-mm gun
Weight limbered up: 8,250 pounds. *Length of barrel:* 27 calibres. *Range:* 8,975 yards. *Muzzle velocity:* 1,590 feet-per-second. *Weight of shell:* 40¼ pounds

French Model 1878 90-mm gun
Weight limbered up: 4,400 pounds. *Length of barrel:* 25.3 calibres. *Range:* 7,550 yards. *Muzzle velocity:* 1,494 feet-per-second. *Weight of shell:* 17½ pounds

French Model 1898 155-mm howitzer
Elevation: from −5 degrees to +60 degrees. *Range:* 7,660 yards. *Weight of shell:* 88 pounds

John Batchelor

'Infantry alone
does not possess
effective offensive
power against
obstacles defended
by gunfire'

Right: A German supply party carries billy-cans of food up to the troops on the front line.
Below: French troops take over a former German strongpoint. They are equipped with the Chauchat 8-mm light machine gun with its characteristic moonshaped magazine.

fought itself. Experience gained up to the end of 1915 had convinced him that the heavier the bombardment, the greater the chances of success, even if it inevitably involved the sacrifice of surprise. He proceeded to create senior artillery appointments at every level of command. In the development of tactics, Major General Birch, at first Rawlinson's and later Haig's principal artillery adviser, was henceforth to exert ever increasing influence. He and Haig saw the coming battles as series of step by step advances, each stage covered by an elaborate fire plan. This procedure, they thought, would eventually so demoralise the enemy that the waiting cavalry divisions could gallop through his shattered defences.

On the Eastern Front the operations had on the whole conformed to the mobile type envisaged by Allied and German prewar theory. Here, the cavalry divisions had played a prominent part both in Russia and the Balkans, largely on account of their mobility. The general incompetence of the Russian commanders at all levels, the virtual absence of an efficient logistic system, the acute shortage of munitions, had combined with the vast distances and featureless terrain to give the Germans comparatively easy victories. Politically, the Germans intended to embody Belgium in the *Reich*; it therefore sufficed in 1915 to stand on the defensive on the Western Front and in the process concentrate on economising in manpower by the ever increasing use of wire, machine guns and deep dugouts. The end of active operations in 1914 had left the Germans, unlike their opponents, in tactically sound defensive positions. Throughout 1915 they had pushed ahead with the construction of well-drained and well-revetted trench systems with an industry which put the efforts of the British in this respect to shame. Their prewar theory had stressed the importance of depth in defence and provision for counterattack. Constructed in accordance with these principles, their defensive systems by the end of 1915 were formidable: four to six miles deep, crisscrossed by interlocking arcs of machine gun fire, dead ground covered by artillery defensive fire and protected by ever-increasing belts of wire. By the end of 1915 they had put an amazing amount of work into the construction of shell-proof and weather-proof accommodation, thus enabling them to keep their troops for long periods in the forward defences without undue discomfort and with small loss of life.

Falkenhayn, like Joffre and Haig, saw the coming battles of 1916 in terms of fire rather than of large scale movement, but with greater subtlety. He planned to attack with very light forces after an intense bombardment, thus forcing the French to counterattack straight into the maw of his massed artillery and machine guns. By this means he hoped to bleed them to death.

Weighed down
It is surprising that so little thought was given in the early years of the First World War to the tactics needed once the much desired breach in the German defences had been blasted by the artillery. One consideration, which the prewar regulations had stressed, was that the infantryman would have to show great initiative and be highly mobile. Instead, throughout 1915 his mobility had been continuously reduced. In August 1914 his load had been 59 pounds 11 ounces, by Christmas 1915 this had risen to 66 pounds. Apart from his weapons, he had to carry into the attack a pick or shovel, wire cutters, sandbags, an anti-gas respirator and 170 rounds of ammunition. To this had to be added his personal necessities — a heavy greatcoat, a ground sheet, three pairs of thick socks, a spare woollen undervest and long underpants, a cardigan, a spare shirt and his iron ration of a one pound tin of corned beef, tea and sugar and two packets of cement-hard biscuits. The steel helmets issued in this year were of solid steel and very heavy, often causing headaches. Each man also had a heavy mess tin, a waterbottle, a knife, fork and spoon, an enamel mug and washing and shaving kit including a cut-throat razor. In winter, thigh gum boots and leather jerkins were added for good measure. In an attack the infantryman was expected to carry on for days on hard rations washed down with such tea as he could brew for himself in his mess tin with solidified paraffin. At all times his food included an inordinate amount of cheese — a constipating factor which the almost continuous shell fire did something to mitigate. The infantryman was thus almost always overloaded and inadequately sustained by hot food.

The short Lee-Enfield rifle carried by each man was an accurate and simple weapon capable, in skilled hands, of producing 15 aimed rounds a minute.

Thus to a limited extent platoons and companies, although overloaded, had the means to get forward on their own in a fluid situation once clear of the enemy's main defences. Despite the bias of the prewar manuals to what they called open warfare and the stress on it as the ultimate aim, little or no thought at the higher levels of command was given to it after 1914. As late as spring 1916 the Infantry Officers' Basic Course of one month at the School of Musketry at Hythe was devoting a complete week to firing the rifle in the standing position. Platoon tactics received two hours in the whole course — two sections of instructors advancing by alternate rushes from the 800-yard firing point to the butts. A high proportion of training time was devoted to bayonet fighting. So important was this subject deemed to be that all troops passing through the Base at Étaples were rushed through an institution known as the Bull Ring. Here instructors in red jerseys showed how 'easy' it was. You stabbed your opponent in the stomach, stirred your bayonet round a bit, then pulled it out smartly.

Also available within brigades was the Stokes Mortar which had a maximum range of about 400 yards. It consisted of a plain steel tube with a spike at the bottom end. The crew attached a sporting cartridge to the base of the shell, slid it down the barrel onto the spike and hoped for the best. It was advisable to wait for the departure of one round before inserting the next: increased range could be obtained by adding rings of ballastite to the cartridge. The main drawback to this weapon was its weight, particularly that of the base plate. Nevertheless, it does seem that these drawbacks could have been eliminated quickly if more attention had been given to the problem at GHQ.

More than half a century later it is not easy to explain the shortsightedness of the Allied High Command in tactical matters in 1915. It may have sprung from the fact that most of them had never in their lives had to fight on foot as infantrymen. It may well be that from sheer inexperience and lack of imagination they literally could not understand that it was virtually impossible for a battalion commander in the attack to control his companies and maintain touch with his supporting artillery. Now that the tactical problem by which they were confronted had taken an unmistakable form they fell for what seemed to them the obvious solution — overwhelming bombardment. They would now crush all resistance by sheer weight of shells. Accordingly they proceeded to pile gun on gun, substitute high explosive for shrapnel and bring in ever increasing numbers of heavy artillery batteries with calibres ranging from 6 inches to 18. In almost all their eyes, the developments inspired by Swinton and Churchill which would eventually completely change the character of land warfare had no significance.

The views expressed by Robertson, at the time Chief of the General Staff of the BEF, in his memorandum to Asquith on November 6, 1915, may be taken as epitomising the official attitude. He wrote: *Experience has taught us that, given sufficient guns and ammunition, any front system can be broken. It is the depth of the enemy's defences and the power of bringing intact reserves up quickly to occupy rear lines which make attack difficult on the Western Front.* He later went on to state that the main lessons of the attacks in 1915 had been that *given adequate artillery support there is no difficulty in overwhelming the enemy's forces in front line and support. The principles are that sufficient force should be employed to exhaust the enemy and force him to use up his reserves, and then, and then only, the decisive attack which is to win victory, should be driven home. There are therefore no grounds for considering that the prospects of a successful offensive next spring are anything but good.*

Further Reading
Edmonds, Sir James, *Military Operations France and Belgium* (HMSO)
Falkenhayn, E. von, *General Headquarters and its Critical Decisions* (Hutchinson 1919)
Fuller, Maj-Gen. J. F. C., *Decisive Battles of the Western World* (Eyre and Spottiswoode 1957)
Liddell Hart, Sir Basil, *The Tanks* Vol. 1. (Faber 1959)
Montgomery, Viscount, *A History of Warfare* (Collins 1968)
Sixsmith, E. K. G., *British Generalship in the 20th Century* (Arms and Armour Press)
Weller, J. A. C., *Weapons and Tactics* (Vane 1966)

MAJOR-GENERAL H. ESSAME served as an Infantry Officer from early 1915 to 1949. In the First World War he fought at the battles of the Somme, Third Ypres and Passchendaele and was present at the March retreat of 1918 and the final advance to the Armistice Line. He was awarded the Military Cross. Between the wars, he graduated at the Staff College, Quetta and in the Second World War he commanded the 214th Infantry Brigade from Normandy to the Baltic. Since his retirement he has been active as a radio commentator, military television adviser, lecturer, freelance journalist and as a writer on military historical subjects. His literary works include: *The 43rd Wessex Division at War* (Clowes), *Battle for Germany* (Batsford) and *Normandy Bridgehead* (Ballantine).

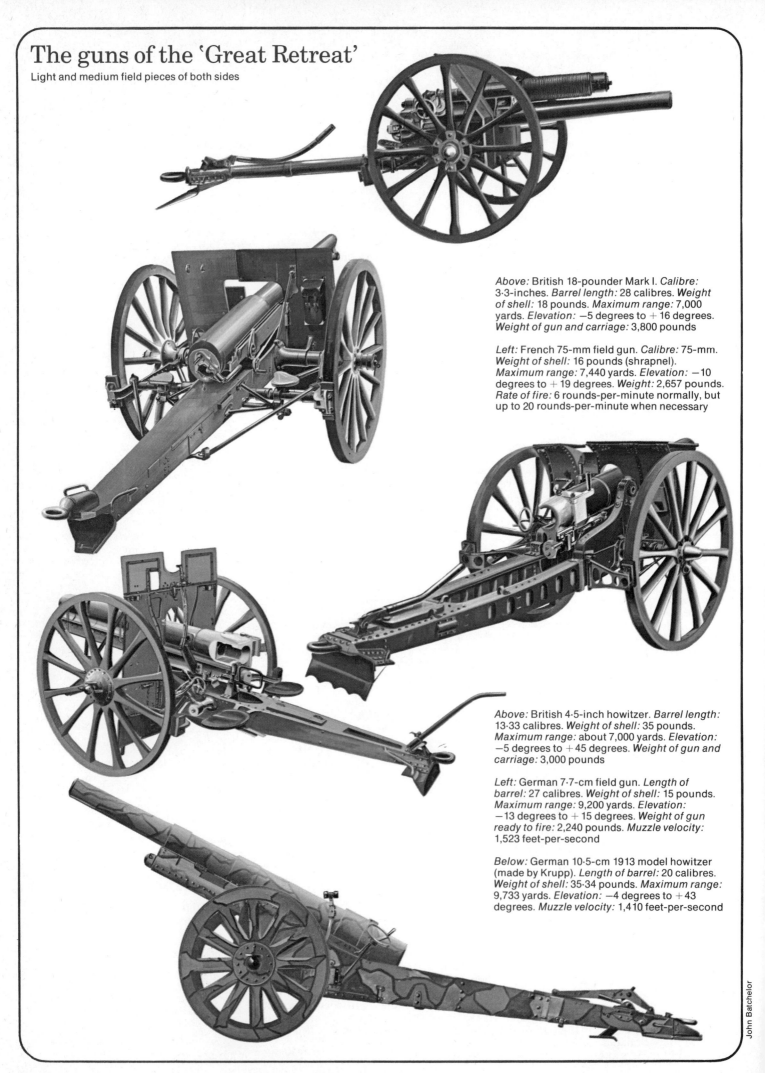

The guns of the 'Great Retreat'

Light and medium field pieces of both sides

Above: British 18-pounder Mark I. *Calibre:* 3·3-inches. *Barrel length:* 28 calibres. *Weight of shell:* 18 pounds. *Maximum range:* 7,000 yards. *Elevation:* −5 degrees to + 16 degrees. *Weight of gun and carriage:* 3,800 pounds

Left: French 75-mm field gun. *Calibre:* 75-mm. *Weight of shell:* 16 pounds (shrapnel). *Maximum range:* 7,440 yards. *Elevation:* −10 degrees to + 19 degrees. *Weight:* 2,657 pounds. *Rate of fire:* 6 rounds-per-minute normally, but up to 20 rounds-per-minute when necessary

Above: British 4·5-inch howitzer. *Barrel length:* 13·33 calibres. *Weight of shell:* 35 pounds. *Maximum range:* about 7,000 yards. *Elevation:* −5 degrees to + 45 degrees. *Weight of gun and carriage:* 3,000 pounds

Left: German 7·7-cm field gun. *Length of barrel:* 27 calibres. *Weight of shell:* 15 pounds. *Maximum range:* 9,200 yards. *Elevation:* −13 degrees to + 15 degrees. *Weight of gun ready to fire:* 2,240 pounds. *Muzzle velocity:* 1,523 feet-per-second

Below: German 10·5-cm 1913 model howitzer (made by Krupp). *Length of barrel:* 20 calibres. *Weight of shell:* 35·34 pounds. *Maximum range:* 9,733 yards. *Elevation:* −4 degrees to + 43 degrees. *Muzzle velocity:* 1,410 feet-per-second

John Batchelor

One of the most heroic actions fought by the BEF during the retreat from Mons to the Marne was that at Néry, a village near Compiègne, where on September 1 a single battery of the Royal Horse Artillery which, with the 1st Cavalry Brigade, was covering the withdrawal of III Corps, held off for several hours the whole of the German *4th Cavalry Division*. The battery, which in 1914 was equipped with six 13-pounders, is now known as 'L' (Néry) Battery, RHA

NÉRY

'L' Battery, Royal Horse Artillery, following in the wake of its Cavalry Brigade to Néry, reached it after the other units had begun to shake down into their quarters. The allotment of the village was as follows: at the northern end were the 5th Dragoon Guards with their horses in the open; the 11th Hussars were billeted on the eastern face and up the east side of the village street, the men and horses being under cover — in houses, yards, barns, sheds, or lean-to's. On the west side of the village street, and in the fields behind the village on this side, were the Queen's Bays, one squadron being in a field further to the south; all their horses were in the open.

'L' Battery on arrival was given a field to the south to bivouac in, and the sugar factory was allotted to it as its headquarters. In the north-west corner of the field were some haystacks.

While the battery was completing its arrangements for the night the Battery Commander proceeded to Cavalry Brigade Headquarters, situated in the main street, to ascertain what protective arrangements had been made to cover the bivouac of his battery.

He received orders that 'L' was merely required to block the two roads which led east and south from the sugar factory. He was also told that the force would continue the march at 0430 hours on September 1. Major Sclater-Booth returned to his battery, and the necessary posts were found by 'L' to cover the southern end of the billeting area.

Gradually the work was finished and, wearied with the day's march under the hot August sun, men and horses settled down to rest. Silence brooded over the little village and the surrounding bivouacs that nestled around it on the western slope and at the bottom of the narrow valley, which was shut in to east and west by its guardian heights. Day broke cool and very misty, and when the march should have been resumed it was quite impossible to see objects more than 150 to 200 yards away. Orders were, therefore, issued that units should stand fast until 0500 hours.

The battery, which was standing halted in mass with the teams hooked in, took advantage of this delay to let down the poles and water the horses by sections at the sugar factory. Generally, it may be said, 'the only desire of our force in Néry at this moment was to get outside an excellent breakfast'. This very natural desire was to be roughly frustrated. The mist was nearly as thick as ever when, just before 0500 hours Major Sclater-Booth, with his

officers, walked down from the sugar factory to the north-west corner of the battery field, where the haystacks stood. Leaving the others here, the Battery Commander walked on up the main street of the little village to Brigade Headquarters in order to get the latest instructions as to the resumption of the march.

Going into the house he found the Brigadier and his Brigade Major. Hardly had he entered when a high-explosive shell burst over the village, and a roar of gun and rifle fire broke out from the heights overlooking the eastern side of Néry.

At the same moment Lieutenant Tailby, who had been sent with a patrol to reconnoitre the high ground north of Néry, reached headquarters and reported that he had ridden into a body of German cavalry in the mist and had been chased back. It was now about 0505 hours and the 1st Cavalry Brigade had been taken completely by surprise.

Despite the disadvantage at which the British Cavalry and Horse Artillery were taken, and despite the heavy artillery, machine gun and rifle fire pouring into the open bivouacs around the village, steps were taken by all units to offer an effective resistance and hold on till assistance arrived from neighbouring troops.

Imperial War Museum

A British crew poses beside their gun — a 13-pounder, the standard equipment of the RHA

As soon as firing broke out the Brigade-Major went out to see that the necessary action was being taken. Major Sclater-Booth also went out into the street with the Brigadier, and then left at once to return to his battery. Suddenly a mob of maddened horses came galloping wildly down the main street. They were the horses of the Bays, stampeded by the enemy's fire. At the same moment a high-explosive shell burst among the surging mass of animals and rendered the road impassable. Crossing over to the western side of the street the Battery Commander ran behind the houses and so came to the field where 'C' Squadron of the Bays had bivouacked during the night. From here the battery field was open to view, and Major Sclater-Booth saw that three guns had been unlimbered and brought into action to answer the fire of the German battery, the flashes of which could be seen stabbing through the slightly thinning mist. Apparently the German guns were in action on the heights to the eastward a short half-mile away. The din was terrific. There was one incessant roar of gun and rifle fire, punctuated by the violent detonations of 'Universal' shells bursting over the battery.

As he ran forward to reach his battery

Bayer. Hauptstaatsarchiv München Abt.

a shell burst immediately in front of him, knocked him down, and put him out of action for the rest of the fight.

An inferno of shells

At the moment the surprise was effected, Captain Bradbury and the other officers of the battery were standing near the haystacks. Suddenly, with no previous warning, a shell burst over the battery, and immediately afterwards the bivouac came under very heavy rifle fire from the ridge. Captain Bradbury shouted out 'Come on! Who's for the guns?' and running out from behind the haystacks, made for them, followed by all the other officers. Meanwhile, in the exposed battery, horses and men were falling fast. Joined by those men who were engaged in steadying the horses in the inferno of bursting shells, the officers got three guns unlimbered and swung round to face the German battery. Captain Bradbury, Sergeant Nelson, and others took one gun; Lieutenant Giffard took another; while Lieutenants Campbell and Mundy were at a third. The ammunition wagons were 20 yards away, and over that death-swept open space the ammunition had to be brought up. Hardly were the three guns in action when one of them,

Flyweight guns at Néry

The heroic action of 'L' Battery at Néry, which resulted in the award of three VCs. *Inset:* Superior in numbers and calibre, the German guns nevertheless suffered heavy casualties

under Lieutenants Campbell and Mundy, was knocked out by a direct hit; the other two guns opened fire on the enemy.

These two guns of 'L' carried on an unequal struggle. A few rounds only had been fired when Lieutenant Giffard, in charge of one of the guns, was severely wounded and all the detachment either killed or wounded. This left only one gun — under Captain Bradbury — still in action.

Lieutenants Campbell and Mundy, when their gun was knocked out, at once ran to the gun where Captain Bradbury and Sergeant Nelson were working, while Gunner Darbyshire and Driver Osborn crossed and recrossed the shell-swept zone behind the gun to bring up the necessary ammunition from the wagons.

Almost immediately after the two subalterns joined Captain Bradbury's detachment Lieutenant Campbell was killed, and the distribution of the duties at the gun became as follows: Lieutenant Mundy in position close to the gun, acted as Section Commander, while Captain Bradbury carried out the duties of layer, and Sergeant Nelson those of range-setter. The gun appeared to bear a charmed life and remained untouched. Also it was clear that its fire was not without result, for the Ger-

man guns were being badly mauled.

When the action began the German guns seem to have been in two groups — one battery in action on the heights, and now busily engaged with 'L' Battery, and two more batteries, unlimbered farther to the north almost opposite the centre of the village and firing on it.

Drawn by the fire kept up by 'L', the Germans now apparently decided to mass all their guns, and the two batteries in action abreast of the centre of the village moved round to join that engaged with 'L'.

The solitary gun of the latter was now opposed to heavy odds; for the hostile guns were under 800 yards away and in a commanding position. The action broke out with renewed fury and the massed German batteries made a determined effort to crush the single undaunted gun. Lieutenant Mundy was now seriously wounded, and the tale of casualties began to mount up, until at last at 0715 hours there remained only Captain Bradbury, still unhit, and Sergeant Nelson, who had been severely wounded. They kept up the best rate of fire they could, but naturally it became very desultory. A reinforcement now reached the little detachment, in the person of Battery-Sergeant-Major Dorrell, and on

his arrival Captain Bradbury, knowing that the ammunition up with the gun was running low, went back to fetch up more from the wagons. As he left the gun he was hit by a shell and mortally wounded. There now remained only the Battery-Sergeant-Major and the wounded Sergeant Nelson. With these two to serve it, the gun fired its last remaining rounds and was silent. The end had come.

But it had not been fought in vain, for, as its last discharge boomed and echoed, reinforcements of all arms reached the field and the result it had fought so hard to attain was achieved.

'L' Battery's casualties amounted to 45 officers and men killed and wounded out of a strength of 170. Among the killed was Captain Bradbury, who was awarded posthumously the Victoria Cross: it was also awarded to Sergeant Nelson and Battery-Sergeant-Major Dorrell. The German cavalry division lost more heavily, was driven into the surrounding forests, did not emerge from hiding until late next day and was still unfit to move on September 4.

[*From the* Royal Artillery War Commemoration Book 1914-1918, *reprinted by permission of the Royal Artillery Institution.*]

BOLIMOW AND THE FIRST GAS ATTACK

Since the beginning of the war, the idea of using poison gas as a weapon had been regarded with suspicion by soldiers of the front line. But Germany's scientists had nevertheless continued their research into various forms of offensive chemicals and by January 1915 a new gas was ready for use. *Ian Hogg*

Christmas 1914 found the armies on the Eastern Front locked solidly together, entrenched in the ice, snow and bitter cold of the Polish winter, as static as the opposing forces on the Western Front. In an endeavour to break this deadlock and get the German army rolling once more on the road to Warsaw and beyond, Hindenburg planned an attack in the central sector. If it succeeded, the momentum would carry the Germans on to Warsaw; if it failed, at least it would have caused a big enough diversion to occupy large numbers of Russian defenders and keep them occupied while his northern armies deployed from their positions in East Prussia in a flanking attack towards Bialystok. The problem was to overcome the inertia at the beginning of the attack – the same problem which confronted the commanders on both sides in Flanders. Some new approach, some new technique, some new weapon was called for.

Although the war was only five months old, there had already been talk of poison gas as a weapon. The subject had recently been given a public airing in 1908, when *The Panmure Papers* were published. These were a volume of reminiscences by Lord Panmure, and revealed in it were copies of letters between Panmure, Palmerston and Admiral Lord Dundonald which showed that the admiral had, during the Crimean War, suggested the use of sulphur fumes to overcome the defenders of Sebastopol. The authorities of the time, while admitting the idea showed promise, turned it down on the grounds that 'no honourable combatant would use such means'.

At about the same time as the publication of these papers was causing a mild stir, the French army developed a rifle grenade charged with a small quantity of tear-gas. This was intended for the attack of casemates and fortifications by firing the grenade through the loopholes and so incapacitating the defenders. These grenades were brought into the public view when they were used in 1912 by the military and police authorities when the notorious Bonnot Gang – the Motor Bandits – was trapped in a house at Choisy-le-Roi.

After the war had begun, the German army was alert to the possibility of these grenades being used against them, though there appears to be no record of their use. Any suspicious smell was treated with respect; cases of asphyxiation after artillery bombardment were at first thought to be due to the use of gas, and chemists were sent to make checks in the front line. It was found that the casualties were due to carbon monoxide poisoning, the gas being generated by incomplete detonation of high explosive shells of faulty design in confined spaces such as dug-outs. Then in late September 1914 the Allied press featured stories of a new French shell filled with a mysterious substance called Turpinite, and once more the gas bogey arose in German minds. But Turpinite was no more than a liquid explosive of doubtful efficiency which the French army employed for a short time to overcome shortages of the more usual shell fillings, and which had suffered a change in the over-heated prose of the newspapermen.

These incidents, minor in themselves, served to keep the poison gas idea alive in some minds, particularly in German minds. They were ever-alert for signs that gas was being used against them.

Towards the end of October 1914, the Kaiser Wilhelm Institute for the Advancement of Science, in Berlin, was approached by the German army with an unusual request. Due to the shortage of steel, a number of artillery shells had been made from cast iron; because of the relative weakness of the metal, the shell walls had to be much thicker than those of a steel shell in order to withstand the stress of firing. This meant that the space within the shell for high explosive was so small that the effect of the shell's detonation was merely to split the shell into one or two large chunks, instead of the shower of fine, lethal fragments desired. The Institute was asked to examine these shells and see if it could suggest some form of chemical filling which would give them back their lethal effect. After a violent explosion in a laboratory, which killed one professor and severely injured another, this particular investigation was quietly dropped, but the 'chemical warfare' idea had taken root in the Institute.

The first serious suggestion for using some form of offensive chemical agent came from a Professor Nernst, who proposed placing an irritant substance among the bullets of shrapnel shells. His idea was accepted, and he was given the task of supervising the production of a batch of these shells. In October 1914 some were used at the front, in an attack near Neuve Chapelle. It speaks volumes for their lack of effect that nobody on the Allied side knew that such a weapon had been used, and it was not until after the war that the facts became known. It is said that Professor Nernst's shells were abandoned after General Falkenhayn's son had won a case of champagne by betting he could remain inside a cloud of this 'irritant substance' for five minutes, unprotected and unharmed.

The next move was by a Dr von Tappen, also of the Kaiser Wilhelm Institute. His brother was Operations Chief at OHL, and through this convenient channel, Tappen suggested the use of a shell carrying a charge of xylyl bromide; this was a lachrymatory (tear producing) compound which Tappen had studied extensively for a research degree. Professor Haber, head of the Institute, also interested himself in this proposal, to such an extent that he eventually became director of all German chemical warfare research. Tappen's shell was tested on the artillery ranges at Kummersdorf in December 1914 with reasonable success, and opinion began to form that here was the sought-for weapon which could break the deadlock on the fronts, flush the enemy out of his trenches, and get the war mobile and fluid once again. After the trial, General von Tappen was placed in charge of a programme to produce quantities of 15-cm howitzer shells charged with high explosive and xylyl bromide; Haber suggested using powerful mortars to fling large canisters of gas; and someone, never identified, suggested putting gas in cylinders and simply turning the tap when required, an idea which was seized upon and given to Professor Haber to turn into a workable proposition.

It is worth looking closer at Tappen's shell and its contents; this was a pioneering effort, based on nothing but faith and a certain knowledge of chemistry. The shell itself was the standard 15-cm high explosive type, filled from the base – a common German design feature – with TNT. Into a cavity in this TNT was placed a lead canister containing the xylyl bromide, so that when the shell detonated on striking the ground, the blast of the explosion would break open the lead canister and disperse the chemical in a fine mist. It would then vaporise and give rise to severe irritation of the eyes and nose of anyone unfortunate enough to breathe it. Indeed, xylyl bromide – or *T-Stoff* as it was code-named – could be lethal if inhaled in a heavy enough concentration, but Tappen appreciated that such concentrations were out of the question with artillery shells as the delivery system, and was quite satisfied that the irritant effect would sufficiently incapacitate the Allied troops to allow the German attackers to overrun them. By way of protection, the German troops were to be provided with a simple face mask, a pad of cotton waste tied in front of the nose and mouth with tapes, and moistened with a chemical solution carried in a bottle and reapplied when necessary. In view of the known tear-producing effect of xylyl bromide, such a mask seems a trifle optimistic.

While manufacture was getting under way, a more difficult job had to be done – the soldiers had to be persuaded that the proposal was worth consideration. This was to be the first example of a difficulty which was to persist throughout the war in all the armies. While the scientists were full of ideas, particularly in the chemical warfare field, the front-line troops were more than a little opposed to their proposals. Very little was known about the effect of chemical agents, while a great deal had been discovered about the effects of shrapnel shells, machine guns, wire and bayonets. Gas was a mystery, and nobody liked the idea of going forward with nothing but a mystery to sustain him.

Eventually Hindenburg was persuaded to use the new *T-Stoff* shells in his forthcoming attack on the Eastern Front, and supplies were shipped to ammunition dumps near Lódź. Some sources aver that large quantities of explosive rifle bullets were also supplied for this attack, but this is open to considerable doubt. After the war, the Germans took great pains to point out that the wording of the Hague Convention—*d'employer du poison ou des armes empoisonnées* (to use poison or poisoned weapons)—was never intended to apply to war gases; that gas did not cause undue suffering by comparison with other weapons; that their use was not excluded by the convention's phrase *d'employer des armes, des projectiles ou des matières propres à causer des maux superplus* (to use weapons, projectiles and materials likely to cause the utmost injury); and that since the *T-Stoff* shells carried a large charge of TNT they were not contrary to the convention. In view of all this it is difficult to believe that they would deliberately employ explosive small arms bullets whose use was most definitely proscribed by the same convention paragraphs. It seems likely that these reports were simply atrocity propaganda of the type both sides engaged in.

By the end of January 1915, Hindenburg's position in the centre of the Eastern Front straddled the road and rail lines from Łódź to Warsaw. Just north of the railway is the village of Bolimow, and the line in this sector rested its left flank on the River Bzura and ran southwards along the line of the Riva Rawka, through forests and farmland. The opposing Russian line rested its northern flank on the village of Sochaczew and ran south through the villages of Borzymow, Humin and Wola-Szydlow to a point on the Rawka where it loops to the east below the railway line. It was here that Hindenburg proposed to demonstrate his new weapon.

On the high ground west of the Rawka, Mackensen's army artillery was massed, upwards of 600 guns across the seven-mile front, a gun density of one gun for every 20 yards or so. This was, for January 1915, a considerable concentration. The majority of these guns were of 7.7-cm and 10-cm, the usual field pieces for the close support infantry, but there was a considerable leavening of 15-cm heavy howitzers spread across the front, and it was these guns which received the supplies of *T-Stoff* shells. On Sunday January 3, the bombardment of the Russian positions began with a rain of high explosive and shrapnel shells together with the new threat of lachrymatory gas.

While the bombardment thundered on, seven divisions, almost 100,000 men, were concentrated ready for the attack. Their three main objectives were the Russian positions round the three villages of Borzymow, Humin and Zduńska Wola. One of these three must crack, and the breach thus made would allow the main weight to pour through, turn, and take the remaining defensive positions in the rear. Contemporary newspaper accounts spoke of these troops as 'Divisions of Death', a highly-coloured phrase coined in an attempt to convey their determination and contempt for casualties. The seven divisions had been carefully assembled from selected regiments of high morale and proven fighting ability, since a great deal hung on the result of this attack, and the physical condition of the battlefield was sufficient to deter all but the hardiest; waist-deep snow, intense cold in which tree branches snapped like glass and unprotected hands froze to bare metal, harsh east winds and the numbing dry cold of the Polish plains.

Early on Monday the attack began, after an intensified artillery bombardment

VHU, Prague

which included more gas. There are no records of what the climate did to the primitive liquid-soaked respirators of the German troops, if indeed any were worn. In that climate, such a mask would freeze in seconds and effectively suffocate the wearer. In the event, the attackers had no trouble from the after-effects of their own gas bombardment. But having fought across the icy Rawka and through the snow to the Russian defensive line, it seemed to have had no effect on the defenders either. There were no crowds of weeping men beside their arms, no defensive positions in which men were so bemused by tears and sneezing that they were unable to man their weapons. The reception was harsh and deadly, and it was a long and bitter struggle before the Russians yielded their positions, fell back into the forests behind them, and enabled the Germans to take the three villages almost simultaneously.

Falling back, the Russian defence was still cohesive and gave the Germans no chance to make the breakthrough their plans had envisaged. But by falling back into the forest, and widening the front, the Russian move caused the German attack to lose formation, and the attacking divisions were split into smaller groups as they took up the pursuit. As the Russians vanished into the forests, they filtered through a mass of field guns and machine guns disposed ready for just this situation, and as the dispersed German groups reached the forest edge, the trap was sprung.

A withering fire of shrapnel and machine gun bullets tore the heart from the attacking force, and dispersed the groups into smaller and smaller parties, more interested in staying alive than forcing the forest barrier. Rough defensive positions were formed and a new line of battle struck as the German armies regrouped and prepared for a fresh assault. Artillery moved forward, observers were deployed, and a fierce artillery duel now broke out while the infantry slowly hammered their way into the forests. The battle now degenerated into a frozen slogging-match, every yard disputed and every tree a strong-point.

On the southern end of the attacking line, the Germans crossed the railway and managed to advance as far as Bednary, some six miles from the start line, by Wednesday. The remainder of the line barely reached the River Sucha, a matter of three or four miles of ground taken. The Russian resistance was so fierce that no sector could be stripped of troops to reinforce the southern push in the hopes of making a break-through at Bednary, and the advance got no further. On Thursday the issue was still in doubt, but on Friday the doubt was removed by the appearance of reserves of Siberian troops on the Russian side. By this time the Germans were depleted and exhausted, and in no condition to withstand the onslaught of fresh troops, who, moreover, were more accustomed to the climate conditions than were the attackers. With increasing speed, and

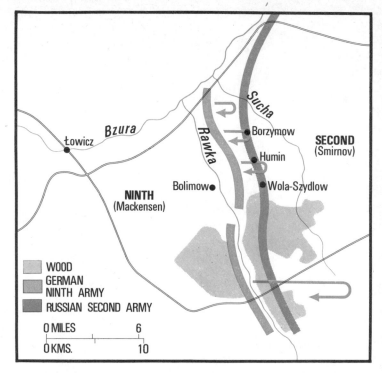

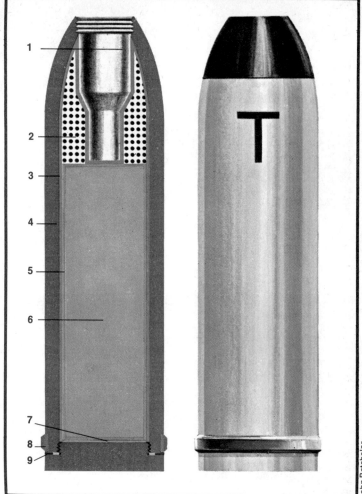

Above: The site of Hindenburg's demonstration of his new weapon—xylyl bromide gas. The German attack on Russian positions near Bolimow relied on a massive concentration of artillery: 600 guns positioned across a seven-mile front. *Left:* The system of releasing gas in artillery shells used at Bolimow was not the only method of dispensing it. Here, Austrian soldiers prepare chlorine gas in steel bottles. *Below right:* A French contemporary cartoon depicts a 'true to life portrait of the pioneers of civilization and culture'

Right: The new German 15-cm *T-Stoff* gas shell
1 Fuse pocket
2 3·3 lbs explosive
3 Paraffin wax
4 Shell
5 Lead container
6 7·0 lbs liquid gas
7 Felt wad
8 Copper driving band
9 Lead washer

John Batchelor

mounting casualties, the attack was stopped and then rolled back. By Friday the Russian troops had recovered the three villages and had ejected the German advanced troops in Bednary. By Saturday they had driven the whole German line back across the Rawka to their starting point, and by Sunday both sides were licking their wounds. The battle was over. After six days of intense effort and sacrifice, the Divisions of Death were back where they had started, leaving over 20,000 dead behind them. Hindenburg's long awaited demonstration had failed as an attack, but, at a cost, had succeeded in occupying the defenders while the attack in the north got under way.

And what of the gas shells? What effect had the new weapon which had promised to flush out the enemy from his defensive positions at no cost to the attacker? In this, their first battle, they were utterly useless. It is possible to appreciate the desperate attitude which led to their employment on this particular occasion; the Russians were a stubborn enemy, and anything which promised to ease the attackers' task was worth trying. It is also possible that the Eastern Front was used for the first employment of gas so as to preserve some element of surprise for the Western Front and still be able to try the new weapon in combat. But it is impossible to avoid pointing a finger and saying that somewhere, someone did not do his homework. The simple fact which seems to have escaped those responsible for the proposal was that xylyl bromide was unsuited to the climate; the object of the shell was to disperse the liquid in fine droplets by the explosion of the TNT charge, so that it

Bapty

could evaporate and give rise to the irritant effect. But due to the sub-zero temperature at the time of the battle, the liquid was thick and turgid, failing to break down into fine enough drops, and when distributed, froze instead of evaporating. Indeed, the whole effect could be called negative, since the space in the shells taken up by the gas canisters was wasted, where it might have been used to carry more TNT and thus do more damage in the conventional fashion. It was at least realised by the Russians that gas had been used in the action, contrary to the British experience at Neuve Chapelle. But con-

temporary accounts failed to mention it, and it was not until a few months later that it came to light in Western reports. Even then, it was credited with being chlorine gas; since this was the cloud gas used in Flanders, it can be accepted that the writers simply assumed that no other war gas had been discovered at that time. If it did nothing else, the employment of gas at Bolimow alerted the Russians and enabled them to begin provision of rudimentary gas-masks before the Germans had perfected the technique. But as far as its tactical effect on the battle is concerned, gas at Bolimow was a failure.

Further Reading

Fries and West, *Chemical Warfare* (New York: McGraw-Hill, 1921)
Foulkes, C. H., *'Gas!'; the Story of the Special Brigade* (Blackwood 1934)
Hanslian, *Der Chemische Krieg* (Berlin: Mittler, 1939)
Lefebure, V., *The Riddle of the Rhine* (New York: Chemical Foundation, 1923)
Meyer, J., *Der Gaskampf und die Chemischen Kampstoffe* (Leipzig: Hirzel, 1926)

IAN HOGG was born in 1926 in Durham City, enlisted in the Regular Army during the war, and is now a Master Gunner in the Royal Artillery. After serving in Europe and the Far East, including duty with a Field Regiment during the Korean War, he became a member of the Instructional Staff of the School of Artillery, Larkhill. He is at present employed in the Ammunition Branch of the Royal Military College of Science as an Assistant Instructor. He has made a study of the history and development of modern artillery equipment for several years, with particular emphasis on ammunition.

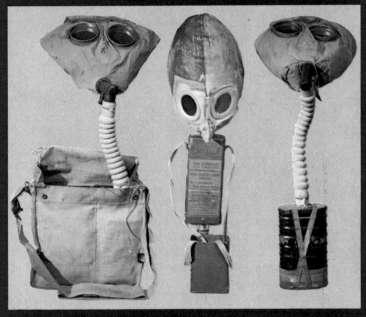

Top left: The German Model 1917 gas mask, made of heavily oiled leather. It was not proof against high concentrations of gas, however, and was stuffy, as the wearer had to breath both in and out through the drum containing the chemicals. *Above right:* The Italian Modello Piccolo No. 2 mask. *Above:* The German Model 1916 mask, made of rubberised fabric. It was again stuffy, and the eye pieces were prone to misting up

GAS

Gas was introduced to the arsenal of the First World War in 1915, and quickly adopted by all the other belligerents. There were two methods of launching a gas attack, from cylinders or by projectiles. The cylinder attack entailed bringing a large number of cylinders of compressed gas up to the line, laying pipes from these as far forward into No-Man's Land as possible and releasing the gas in a dense cloud when the wind was from the right quarter to blow the cloud of gas down onto the enemy's front line. The need for the right wind was a distinct disadvantage for the Germans on the Western Front,

Left: The British 1917 pattern Small Box Respirator. This had splinter-less eye pieces and a breath-outlet valve, and was therefore more comfortable to wear than its German counterpart. It was the best mask of the war, proof against all gases except, of course, mustard gas. *Centre:* The Russian box respirator, with the box attached to the nose piece. *Right:* The British 1916 Respirator, with an extra filter on the box

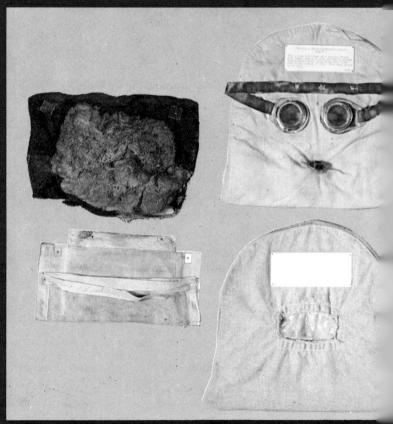

Above: A bellows-operated resuscitator. There was always the chance that if uncontaminated air could be given quickly to a man who had been only slightly gassed, he might make a far better recovery

Top row left: The British 1915 Veil Respirator, an impregnated mouth pad. *Centre:* The British 1916/17 Phenate-Hexamine-Goggle helmet. *Right:* The British tank-crew respirator and goggles (left) and anti-lachrymator gas

THE ODOUR OF DEATH

where the prevailing wind was a westerly one. The other disadvantages were the labour involved in installing the equipment and the vulnerability of the cylinders to shell-fire. The main gases used were phosgene and chlorine. The other method of gas attack was by means of projectiles fired from guns or trench mortars. The advantages of this system were independence of the wind, simplicity and the ability to attack rear areas. The disadvantages were the need for heavy bombardments, as each shell carried so little gas, and the need for windless conditions to keep the gas from dispersing

British gas casualties in France, 1915–1918
from British official sources

	🧍	🧍
APRIL AND MAY 1915 Cloud attacks with chlorine	7 000 *	350 *
MAY 1915–JULY 1916 Lachrymator gas shell	?	—
DECEMBER 1915–AUGUST 1916 Cloud attacks	4 207	1 013
JULY 1916–JULY 1917 'Lethal' gas shell	8 806	532
JULY 1917–NOVEMBER 1918 (Mustard gas period) Gas shell	160 526	4 086
DECEMBER 1917–MAY 1918 PROJECTOR ATTACKS	444	81
TOTAL	180 983	6 062

German cloud gas attack April 30 1916

Lindenhoek, Messines, Neuve Eglise, Bailleul, Ploegsteert

FRONT LINES
BRITISH
GERMAN
GAS STRONGLY FELT, HELMETS WORN
GAS SLIGHTLY FELT
PLACES WHERE CATTLE WERE KILLED
0   400 YARDS
🧍 CASUALTIES
🧍 DEATHS
* Approx figures for casualties admitted to medical units only; many died on the field or were taken prisoner.

Gas casualties in the First World War

	Non-fatal injuries 🧍	Deaths 🧍	Total
BRITISH EMPIRE	180 597	8 109	188 706
FRANCE	182 000	8 000	190 000
UNITED STATES	71 345	1 462	72 807
ITALY	55 373	4 627	60 000
RUSSIA	419 340	56 000	475 340
GERMANY	191 000	9 000	200 000
AUSTRIA-HUNGARY	97 000	3 000	100 000
OTHERS	9 000	1 000	10 000
TOTAL	1 205 655	91 198	1 296 853

Note: These figures, apart from the American ones, are only approximate. In many cases they differ from the official published figures, which, since they do not include men who died on the battlefield, tend to be far too low.

Production of poison gases during the First World War (in tons)

	Lachrymators	Acute Lung Irritants	Vesicants	Sternutators	Totals
BRITAIN	1 800	23 335	500	100	25 735
FRANCE	800	34 000	2 140	15	36 955
UNITED STATES	5	5 500	170	—	6 215
ITALY	100	4 000	—	—	4 100
RUSSIA	150	3 500	—	—	3 650
GERMANY	2 900	48 000	10 000	7 200	68 100
AUSTRIA-HUNGARY	245	5 000	—	—	5 245
TOTALS	6 000	123 335	13 350	7 315	150 000

Poison Gases used during the First World War

	date of introduction	approx concentration to incapacitate a man in a few secs. (in parts per 10m)	approx concentration which, if breathed for more than 1 or 2 mins, would cause death	used by: British, French, Germans, Austrians
ACUTE LUNG IRRITANTS				
Chlorine	1915	1 000	1 000	
Phosgene	1915	100	200	
Chlormethyl-chloroformate	1915	100	1 000	
Trichlormethyl-chloroformate	1916	50	200	
Chloropicrin	1916	50	200 ▲	
Stannic chloride	1916	—	1 000	
Phenyl-carbylamine-chloride	1917	50	1 000	
Cyanogen bromide	1918	—	300	
Dichlor-methyl-ether	1918	1 000	1 000	
LACHRYMATORS (tear producers)				
Benzyl bromide	1915	5	—	
Xylyl bromide	1915	5	—	
Ethyl-iodoacetate	1916	5 to 2	200	
Bromacetone	1916	5	1 000	
Monobrom-methyl-ethyl-ketone	1916	2	2 000	
Dibrom-methyl-ethyl-ketone	1916	2	2 000	
Acrolein	1916	—	—	
Methyl-chlorsulphonate	1915	—	—	
PARALYSANTS				
Hydrocyanic acid	1916	5 000●	5 000	
Sulphuretted hydrogen	1916	10 000●	1 000■	
STERNUTATORS (sensory irritants of eyes, nose and chest)				
Diphenyl-chlorarsine	1917	1	200	
Diphenyl-cyanarsine	1918	1	200	
Ethyl-dichlor-arsine	1918	20	500	
Ethyl-dibrom-arsine	1918	—	—	
N.ethyl carbazol	1918	—	—	
VESICANTS (blister producers)				
Dichlor-ethyl-sulphide (Mustard gas)	1917	—	10○	

▲ Cumulative   ● Immediately fatal   ■ Affects eyes and lungs   ○ With 60-minute exposure

Chris Barker.

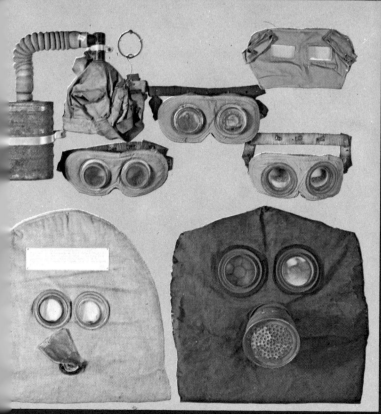

...les (three on right). *Bottom row left:* The Maw civilian mask of 1917. ...re left: The British 1915 Hypo Helmet. *Centre right:* The British 1916 ...ate-Hexamine helmet. *Right:* An unrubberised German fabric helmet

Above: Gas casualties, production and types in the First World War. It is worth noting that the most effective type was not one of the so-called lethal types, but one of the incapacitators, mustard gas

AUSTRIA ON THE DEFENSIVE

Along the Isonzo valley, the Austrians held excellent natural defensive fortresses. Thrown time and time again against these, the Italian infantry was cut down in its tens of thousands — as much for political as strategic reasons.

Below: An Austrian crew prepares to fire a 15-cm howitzer

Far left: An Austrian tunnelling machine in action on the Galician Front.
Right: A Krupp-made Russian 15-cm howitzer Model 1880. Little is known about this gun except that it had a crew of 5 and a rate of fire of 3 rounds-per-minute.
Centre left: An Austrian 22.5-cm mortar in travelling position. Little is known about this gun also except that it had a crew of 4 and a rate of fire of 2 rounds-per-minute.

Centre right: A Russian 25-cm coastal gun. Little is known about this gun too except that it had a crew of 22 and a rate of fire of 1 round every five minutes.
Bottom left: An Austrian 10-cm light field howitzer, Model 1899. *Weight:* 2,250 lb. *Range:* 6,015 yards. *Rate of fire:* 5 rounds-per-minute. *Crew:* 6.
Bottom right: A German 25-cm heavy minenwerfer. *Weight:* 1,362 lb. *Range:* 992 yards

John Batchelor

25

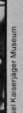

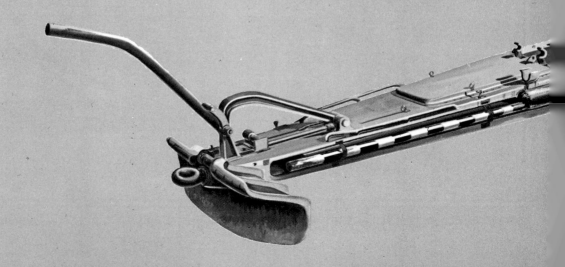

Berg Isel Kaiserjäger Museum

Berg Isel Kaiserjäger Museum

Top: Loading an Austrian 12-cm trench mortar.

Above: The breech swung back into place: the mortar ready for firing. In the trenches in Italy as in those on the Western Front, trench mortars came to play an increasingly important rôle in defence as they were handy weapons and could be used right up in the front line.

△ Austrian 15-cm heavy field howitzer M 99. *Range:* 6,780 yards. *Weight in action:* Two tons 8.5 cwt. *Weight limbered:* Two tons 14.6 cwt. *Weight of shell:* 156 lbs. *Muzzle velocity:* 960 feet-per-second. *Elevation:* −2 degrees/+45 degrees. *Crew:* Eight. *Rate of fire:* Two rounds-per-minute

◁ Austrian 15-cm heavy field howitzer M 14. *Range:* 8,858 yards. *Weight in action:* Two tons nine cwt. *Weight of shell:* 28 lbs. *Muzzle velocity:* 1,120 feet-per-second. *Elevation:* −5 degrees/+70 degrees. *Crew:* Eight. Motor transport was frequently employed

▽ Austrian 10.4-cm field gun. *Range:* 13,670 yards. *Weight in action:* Two tons five cwt. *Muzzle velocity:* 2,230 feet-per-second. *Elevation:* −10 degrees/+30 degrees. The gun was made of nickel steel, and the shrapnel shell had 70 bullets

▽ The *feld grau* uniform of the Austrian infantry, introduced in late 1915. Shown is the Austrian-style steel helmet with a ventilation hole in the top, although usually a cap rather than a steel helmet was worn. He is carrying a *tornister* (calf-hide full pack), a bread bag on his hip, and the M 1895 Mannlicher 8-mm rifle

Malcolm McGregor

28

Below left: An enormous flash as an Austrian 30.5-cm howitzer is fired. *Right top:* Culled from every available source, Austrian reserves rush to the front near Doberdo. *Right centre:* A few of the droves of Italian infantry cut down in the almost hopeless battles along the Isonzo. *Right bottom:* An Italian 75-mm mountain gun in the Alps

ROMMEL IN THE ARGONNE

In the summer of 1915, a figure later to play a rôle of the greatest importance in the Second World War was a captain, commanding an infantry company in the Argonne in France. This was the future *Feldmarschall* Erwin Rommel, who described his varied career in the First World War in a book entitled *Infantry Attacks*. *Left:* The young Erwin Rommel

Although General Joffre's summer offensives were usually concentrated on selected sectors of the Western Front in the hope that their impact would achieve far-reaching strategic results, this in no way inhibited his army commanders projecting strong attacks elsewhere to absorb German resources. But to be successful these subsidiary attacks had to be directed at some place where a breakthrough—no matter how unlikely—would threaten some vital German artery. The southern Argonne was one such place, for this heavily forested and hilly terrain lay on the flank of the crucial fortress zone of Verdun and of the bitterly contested St Mihiel salient, and was held on many important heights by the Germans, and here the French persisted in trying to repel the Germans, while developing their own counter threat towards Metz and along the Moselle into Germany.

Here, where the Germans could apply their defensive strategy in a most economical manner, they could also momentarily afford to concentrate their relatively meagre local artillery reserves for a counteroffensive, and it was towards the end of June that they provoked a struggle which was to bring only marginal changes in possession, at a totally unreasonable loss of life on both sides. Yet from amid the squalor of this fighting there came an enlightening account by a young professional German officer who was to achieve great fame in later years—an account which accurately described the battle techniques as well as catching the horrific atmosphere. The officer's name was Erwin Rommel (then a company commander in the 124th Infantry Regiment), who, having recovered from a wound received in 1914, had rejoined his unit in January 1915. Thereafter he had partaken in fierce fighting in the Argonne. In May his unit moved into a fresh sector.

At Bagatelle there were few indications that the Argonne was a dense forest, for the French artillery fire had cleared the trees thoroughly, and for miles all that remained visible were stumps.

As was our custom, we deepened the trench at once and built dugouts for ourselves. Sudden and violent bursts of French artillery and mortar fire, accompanied by hand grenade fights all along the line, kept us from having dull moments. In the warm weather the frightful stench of corpses wafted into the position. Many French dead still lay in front of and between our positions, but we could not bury them because of the French fire.

The nights were really exciting. Hand grenade battles went on for hours along a broad front and became so confused that we never knew whether or not the enemy had broken through at some place or had worked his way behind our front line. Added to this, various French batteries chimed in from the flanks. This was repeated several times nightly, and we soon found it a strain on our nerves.

Below: The devastating and bloody attacks on the French Central I and Central II positions. The close quarter fighting, combined with the use of large catapulted bombs, entailed heavy casualties and very great destruction. *Above right:* Some of Rommel's crack troops. Rommel had nothing but praise for the outstanding bravery and initiative of his men at this time

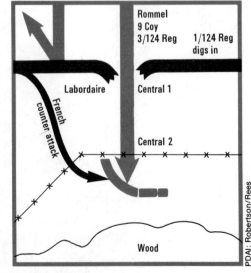

Rommel 9 Coy 3/124 Reg / 1/124 Reg digs in

Labordaire

French counter attack

Central 1

Central 2

Wood

The heat was unbearable during those days. One day Ensign Möricke, an especially fine soldier, visited me. I was down in my dugout, and we had to talk to each other through the shaft because there was not room for two in my warren. I told Möricke I was convinced that we were not safe from the damned flies even when we were 12-feet underground. Möricke said it was no wonder since the edge of the trench was simply black with them. He got a pick and started to dig there, and at the first swing the half-decayed, blackened arm of a Frenchman came to light. We threw chloride of lime on it and left the dead man.

We managed to weather the ten days, and on our return to our regimental sector we were shoved back into the front line. We found that every effort had been made to render trench warfare more unpleasant, for the French had added mining to an increased volume of artillery and trench mortar fire. The opposing outposts were only a few yards apart in half-covered sap trenches heavily reinforced with wire, and the night was full of lively hand grenade battles which from time to time brought the entire garrison to its feet. Each side tried to destroy the other's advance tunnels and positions and hardly a day passed without an explosion.

Swept by machine guns

Our attempts to capture the nearby French sentry posts usually ended in considerable losses. These posts and the sections of trench leading to them were completely enclosed with barbed wire. At the slightest noise, the French in the blockhouses would sweep the obstacles with machine gun fire. These conditions soon became exasperating and we hoped to remedy them by storming 'Central'.

Following a three and a half hour artillery and trench mortar preparation, we were to seize the French strongpoints of Labordaire, Central, Cimetière and Bagatelle. Close behind the front line in shell-proof emplacements, medium and heavy mortars had been sited. Day and night,

Bapty

reserve companies brought up supplies, dismantled mortars and ammunition through the narrow communication trenches. The French harassing fire had increased in violence, and many a carrying party had been hit. Toward the end of June and after a few days in the rest camp, the *9th Company* headed back into the line. We were amazed to see a large quantity of medium and heavy artillery emplaced and camouflaged in the vicinity of Binarville. It was pleasant to note that adequate ammunition seemed to be available. This time we moved up into position in the best of spirits.

The regiment prepared detailed plans for the five assault companies. During the preparation, my platoon remained in reserve about two-thirds of a mile north of Central. Shortly before the jump-off, we were to move up behind the line of departure, follow the assault echelon closely, and keep it supplied with grenades, ammunition, and entrenching equipment.

At 0515 hours on June 30, the artillery opened up with everything it had including 8.3- and 12-inch mortars. The effect of the shells was unbelievable. Earthen geysers shot into the air, craters suddenly appeared before us. The strong French earthworks were smashed apart as if hit by gigantic trip hammers. Men, timbers, roots, obstacles and sandbags flew into the air. We wondered how the French felt, for we had never seen such heavy fire before.

One hour before the assault, the medium and heavy mortars opened up on the blockhouses, wire entanglements and breastworks. The French massed their artillery fire to break up the assault, but their efforts were futile. Our forward line was thinly held and was too close to the main hostile position. The French guns ploughed up the dirt farther to the rear.

I kept looking at my watch. We had 15 minutes to go. A blue-grey pall of smoke from the bombardment obscured vision as both sides increased their volume of fire.

The communication trench assigned us was exposed to strong hostile fire, and I decided to deviate from my orders in dis-placing to the side for about 100 yards. We ran for our lives across the open stretch of ground and found shelter down in the hollow; then, through the communication trench, we rushed into the front line with French standing barrages bursting all around. The storm troops were lying side by side, and across from us the last gun and mortar shells were bursting.

0845 hours, and our assault moved forward over a wide front. French machine guns poured out their fire; the men jumped around craters, over obstacles and into the enemy position. The assault echelon of our company was hit by machine gun fire from the right and a few fell, but the bulk rushed on, disappearing in craters and behind embankments. My platoon followed. Each man had his load, either several spades or else sacks filled with grenades or ammunition. The French machine gun to the right was still hammering away. We passed through its field of fire and climbed over the walls on which the *9th Company* stood on January 29. The position was a mass of rubble. Dead and wounded Frenchmen lay scattered through the tangle of revetments, timbers and up-rooted trees.

To the right and in front of us hand grenade fights were in progress and from rearward positions French machine guns swept the battlefield in all directions and forced us to take cover. The sun was hot. We moved off to the left and then, hard on the heels of our company's assault echelon, pressed on to a communication trench leading to the second position.

Our artillery had shifted its fire to the second French line (Central II), located 160 yards to the south, which was to be captured on July 1 only after renewed artillery and mortar bombardment. The assault echelons of the regiment not engaged in mopping up Central I were pushing on toward Central II.

About 30 yards ahead of us a violent hand grenade fight raged, and beyond that we could see the outlines of Central II some 90 yards distant. French machine gun fire made it impossible to move outside the communication trench, and our own assault group up forward seemed to be stalled. Its young leader, Ensign Möricke, was lying in the trench severely wounded with a bullet in his pelvis! I wanted to carry him back, but he said that we should not worry about him. Stretcher bearers took charge. One last hand-clasp and I assumed the command up front. The ensign died the next day in the hospital.

The French flee

We engaged the garrison of Central II. Our own artillery had ceased firing. A few salvoes of hand grenades followed by a charge, and we were in Central II. Part of the garrison ran down the trench, others fled across the open fields, and the rest surrendered. While part of the unit worked to widen the trench, the bulk of the assault force kept pressing south. We moved through a ten-foot-deep communication trench and had the luck to surprise and capture a French battalion commander with his entire staff. About 100 yards down the line the trench opened on a large clearing. In front of us the terrain dropped sharply down to the valley at Vienne-le-Château, which was obscured from view by the woods. We lost contact on both flanks and to the right. On the edge of the forest some 200 yards away we saw a considerable number of the French. We opened fire and after a sharp fire fight they withdrew into the woods. While this was going on I established contact on my left with elements of the *1st Battalion* who had advanced this far; and I also reorganised my command, which by this time included elements of all units of the *3rd Battalion,* and deployed it in a defensive position some 350 yards south of Central II. Because of our exposed right flank and since we still heard the sounds of bitter fighting behind us, I felt it inadvisable to continue the advances to the south.

A reconnaissance detachment reported that the unit on our right had been unable to clear Central I. This meant we must prepare blocking positions to protect our newly-won positions against attacks from the west. To bolster the defence I put my best veterans in line and was glad of it, for during the next few hours the French launched a series of violent counterattacks to regain their lost positions. I kept the battalion commander informed of developments.

On the left some companies of the *1st Battalion* had advanced down the valley as far as Houyette gorge. Combat outposts reported strong French forces in the woods on the slope 330 yards ahead. I discussed the situation with Captain Ullerich, the commander of the *1st Battalion*, and he decided to have the *1st Battalion* dig in on the left of the *9th Company*.

We wasted no time in getting to work. I kept one platoon in reserve and used it to bring up ammunition and hand grenades and for work on the flank position in Central II. French reconnaissance detachments were probing our front, but we drove them off without trouble.

Digging was easy, and in a short time our trench was more than three feet deep. Since the start of the attack the French artillery had been rather quiet, but now it opened up on Central II with every weapon at its disposal. The French appar-

ently believed we occupied the place in force, for their ammunition expenditure was very heavy. The net result was to smash up their old position and cut our communications to the rear. Supplies ran a gauntlet of fire, and our lone telephone wire was soon knocked out. We did succeed in emplacing one heavy machine gun platoon in the company sector.

By evening our trench was five feet deep, and the French artillery was still falling behind us. Suddenly bugle calls rang out in the woods; and the French, in their usual massed formation, rushed us from the woods an eighth of a mile away. However, our fire soon drove them to earth. There was a slight fold in the ground, and as a result we picked up targets from our trench only when they came to within 90 yards of our position. Perhaps a position farther back near Central II would have been better. We certainly would have had a good field of fire; but, on the other hand, the French artillery would have punished us severely. The French attacked with vigour, and hand grenade fights continued all over the place even after it became quite dark. Our supply of grenades was limited and we did most of our fighting with rifles and the heavy machine gun. The night was black and the smoke from

Strong trenches and heavy guns—defence and aggression in the wooded Argonne

Left: Trench construction in the summer of 1915. These troops do not seem to be unduly concerned with their task, but on their efforts and their thoroughness at this time would depend their lives if and when the French attacked, particularly if heavy artillery or the much-feared flying bombs were used.
Right: It was essential for the troops to get all the rest they could, even when they were at the front, and so the trenches criss-crossing Europe saw the proliferation of all sorts of locally devised methods of making the men more comfortable, such as this hut in the German line ,The superscription over the door reads 'Into the Lions' Den'. *Below left:* A French 155-mm gun. Against guns of this calibre, good trenches stood some sort of chance, but Rommel's men had to contend with these and much heavier artillery, as well as flying bombs, which fell in a more vertical trajectory.
Below: A German 21-cm howitzer. These heavy howitzers, together with even heavier ones, wrought great destruction on the French trenches before Rommel's attack went in. Their shells produced 'great earthen geysers'

Südd Verlag

Kriegsarchiv, Vienna

Imperial War Museum

bursting grenades reduced the effectiveness of our rocket flares. Grenades burst all around us, for the French were within 50 yards of our position. The fight waxed and waned all night, and we beat off all attacks. At daybreak we made out a sandbag wall some 50 yards away and all sounds indicated that the French were busily engaged in digging in behind their improvised shelter. The French infantry kept us on the go during the night and their artillery then took over the morning shift. Fortunately the bulk of artillery fire landed in Central I and II with only a small fraction striking close to our position, and a hit in the front line was a great rarity. So we felt comparatively secure and did not envy the carrying parties who had to move rations and other supplies up through the shell-plastered ammunition trenches.

We spent the succeeding days in improving the position. The trench was soon six feet deep; we constructed small wood-lined dugouts, installed armoured shields, and sandbagged firing points. The front line losses from artillery fire were few, but we lost men daily in the shell-torn communication trenches leading to the rear.

The strong artillery concentrated for the June 30 offensive moved to another front, and our weak organic artillery lacked sufficient ammunition to give us worthwhile support. However, an artillery observer was always up front, and we in the infantry appreciated this very much.

During the early days of July the French began to wreck our trenches daily with flying mines delivered from positions that permitted a certain degree of enfilade fire. These flying-mine projectors are of simple construction and provided but little lateral dispersion; consequently they obtained a high percentage of direct hits, and it was not always possible to evacuate the dangerous localities in time to avoid losses. Our casualties from these mines were considerable, seven men being killed by the blast effect of the 220-pound bomb alone.

During July I began a five weeks' assignment as substitute of the commander of the *10th Company* in a sector where relief was furnished by the *4th* and *5th Companies*. We company commanders worked on a unified plan for construction of shell-proof

Positions entirely flattened by the guns

The French pounded the German trenches in the Argonne with 220-pound flying mines

multi-entrance dugouts 26 feet under ground. This work went on day and night with several parties working on one dugout from different directions. The officers pitched in, and we found that sharing the work helped morale.

Frequently an entire position was flattened by artillery inside an hour. When this happened we saw our lightly timbered dugouts collapse like cardboard boxes. Fortunately the French bombardment followed a rigid plan. Usually they began on the left and traversed right. To remain under heavy fire proved too expensive; so whenever it started I evacuated the trench and waited until they moved their fire laterally or lifted it to our rear area. If French infantry had followed their artillery and attacked us, then we would have thrown them out with a counterattack. This did not bother us because, man to man, we felt ourselves to be superior.

(The battles, which took place in the Argonne between June 20 and July 14, cost the French over 32,000 casualties: they also cost General Sarraill, the Commander of Third Army, his job. And nothing of value had been achieved.)

[Reprinted from *Infantry Attack* by Erwin Rommel.]

Mortars firing —
a painting by the German war artist Frost

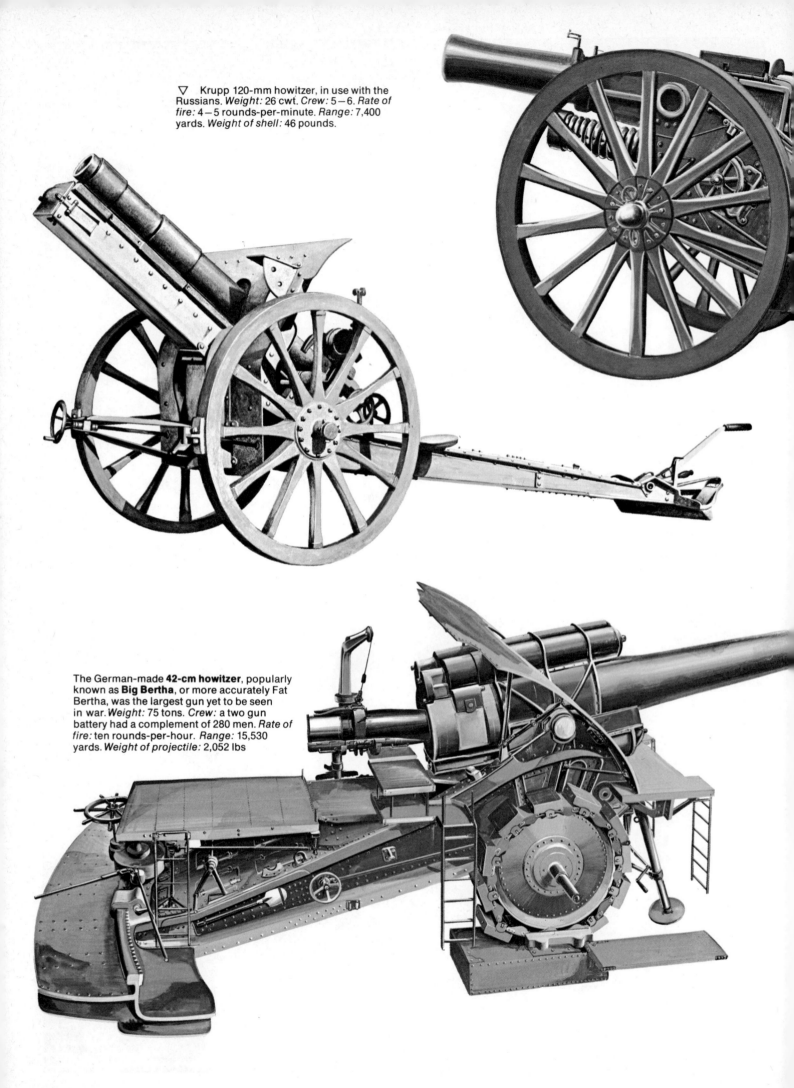

▽ Krupp 120-mm howitzer, in use with the Russians. *Weight:* 26 cwt. *Crew:* 5—6. *Rate of fire:* 4—5 rounds-per-minute. *Range:* 7,400 yards. *Weight of shell:* 46 pounds.

The German-made **42-cm howitzer**, popularly known as **Big Bertha**, or more accurately Fat Bertha, was the largest gun yet to be seen in war. *Weight:* 75 tons. *Crew:* a two gun battery had a complement of 280 men. *Rate of fire:* ten rounds-per-hour. *Range:* 15,530 yards. *Weight of projectile:* 2,052 lbs

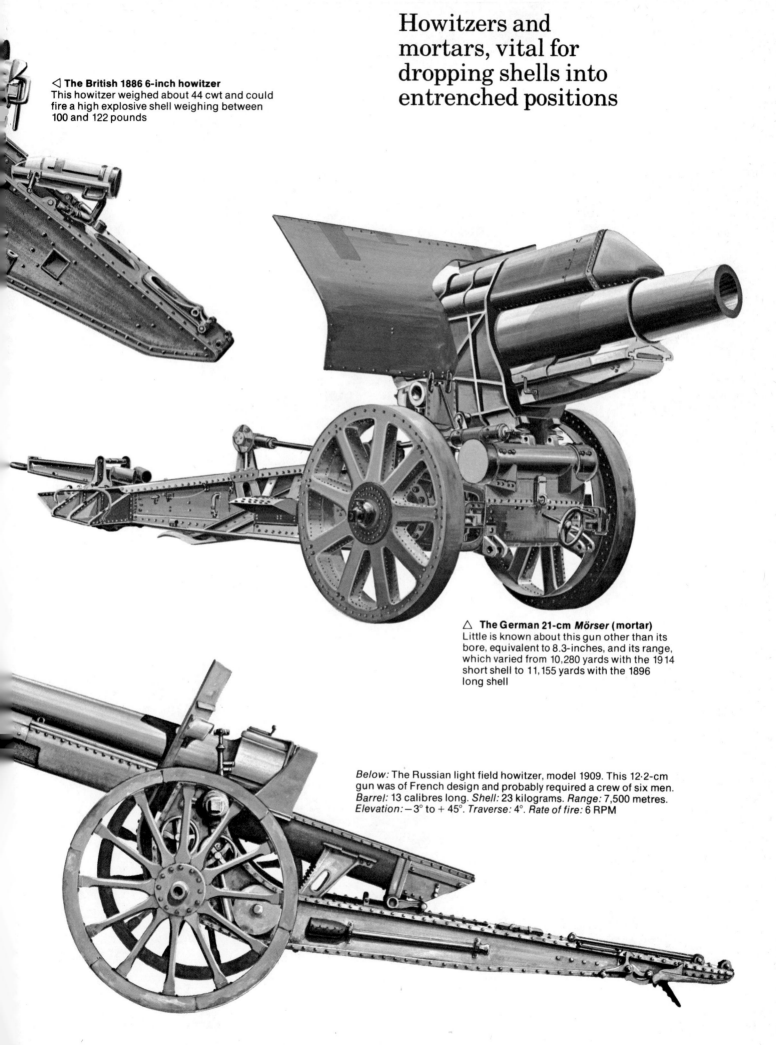

◁ **The British 1886 6-inch howitzer**
This howitzer weighed about 44 cwt and could fire a high explosive shell weighing between 100 and 122 pounds

△ **The German 21-cm *Mörser* (mortar)**
Little is known about this gun other than its bore, equivalent to 8.3-inches, and its range, which varied from 10,280 yards with the 1914 short shell to 11,155 yards with the 1896 long shell

Below: The Russian light field howitzer, model 1909. This 12·2-cm gun was of French design and probably required a crew of six men. *Barrel:* 13 calibres long. *Shell:* 23 kilograms. *Range:* 7,500 metres. *Elevation:* −3° to + 45°. *Traverse:* 4°. *Rate of fire:* 6 RPM

Far left: 21-cm howitzers *(top)* firing from the German trenches. The black smoke produced by the exploding shell earned them the nickname 'Jack Johnsons' after the prewar negro prizefighter. German 25-cm trench mortars *(bottom)* maintained a crippling bombardment of British front lines. *Left:* A German 15-cm L/40 navy gun, originally designed as a secondary armament for pre-dreadnought battleships. *Total weight:* 25,400 lb. *Weight of barrel:* 10,300 lb. *Length of barrel:* 19¾ ft. *Elevation:* −8° to +32°. *Traverse:* 27° left and right. There were two types of shells, with maximum ranges of 14,000 yds and 19,000 yds

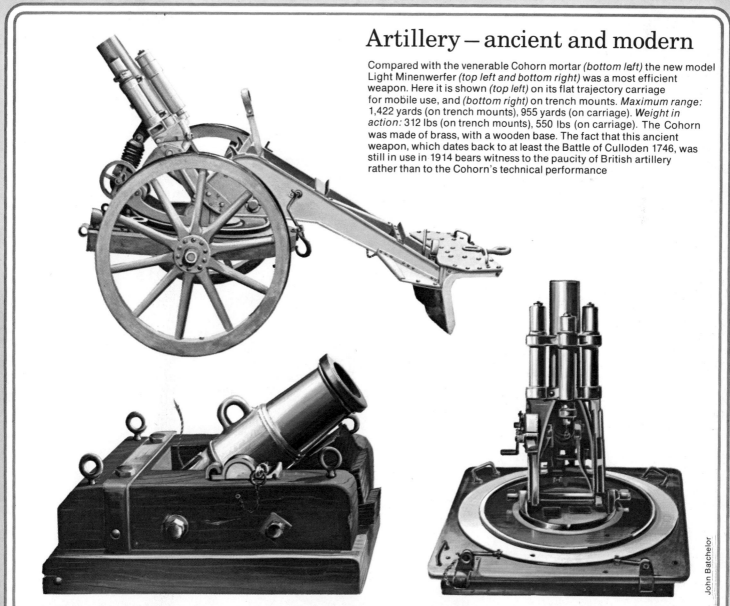

Artillery — ancient and modern

Compared with the venerable Cohorn mortar *(bottom left)* the new model Light Minenwerfer *(top left and bottom right)* was a most efficient weapon. Here it is shown *(top left)* on its flat trajectory carriage for mobile use, and *(bottom right)* on trench mounts. *Maximum range:* 1,422 yards (on trench mounts), 955 yards (on carriage). *Weight in action:* 312 lbs (on trench mounts), 550 lbs (on carriage). The Cohorn was made of brass, with a wooden base. The fact that this ancient weapon, which dates back to at least the Battle of Culloden 1746, was still in use in 1914 bears witness to the paucity of British artillery rather than to the Cohorn's technical performance

John Batchelor

39

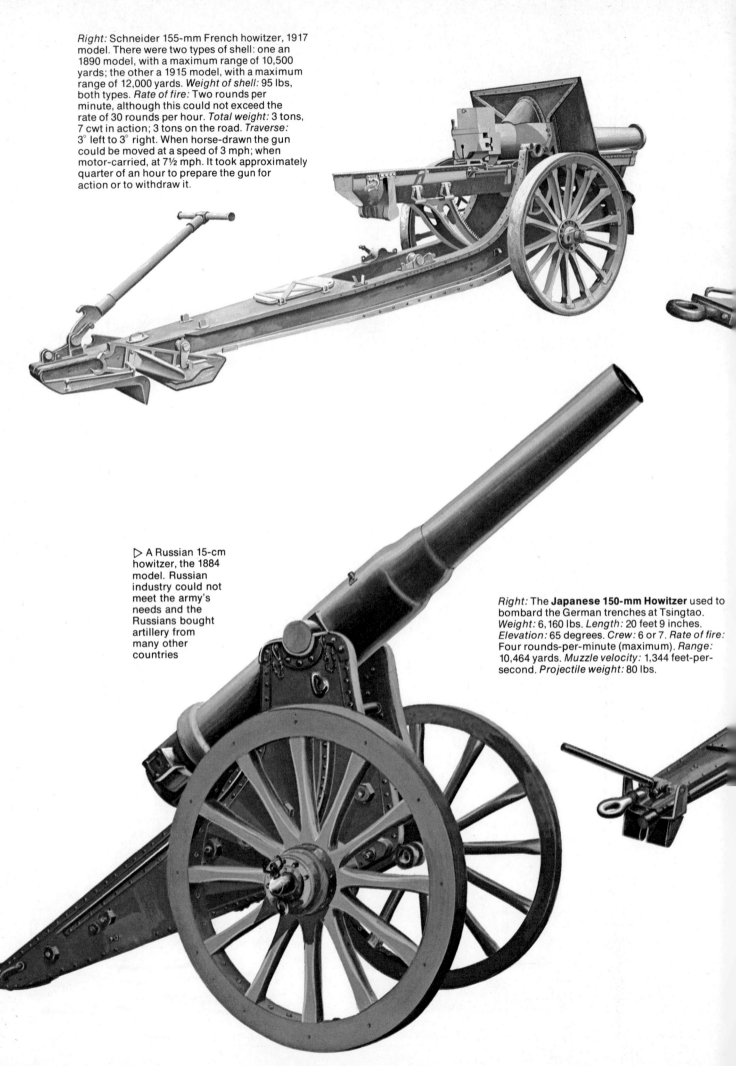

Right: Schneider 155-mm French howitzer, 1917 model. There were two types of shell: one an 1890 model, with a maximum range of 10,500 yards; the other a 1915 model, with a maximum range of 12,000 yards. *Weight of shell:* 95 lbs, both types. *Rate of fire:* Two rounds per minute, although this could not exceed the rate of 30 rounds per hour. *Total weight:* 3 tons, 7 cwt in action; 3 tons on the road. *Traverse:* 3° left to 3° right. When horse-drawn the gun could be moved at a speed of 3 mph; when motor-carried, at 7½ mph. It took approximately quarter of an hour to prepare the gun for action or to withdraw it.

▷ A Russian 15-cm howitzer, the 1884 model. Russian industry could not meet the army's needs and the Russians bought artillery from many other countries

Right: The **Japanese 150-mm Howitzer** used to bombard the German trenches at Tsingtao. *Weight:* 6,160 lbs. *Length:* 20 feet 9 inches. *Elevation:* 65 degrees. *Crew:* 6 or 7. *Rate of fire:* Four rounds-per-minute (maximum). *Range:* 10,464 yards. *Muzzle velocity:* 1,344 feet-per-second. *Projectile weight:* 80 lbs.

▽ The Rimailho 155-mm howitzer, the French equivalent of the German 5.9-inch German howitzer, and the only modern piece of heavy artillery in the French army at the outbreak of the war. It saw much use in the artillery duel at Verdun

The Austrian-made **Skoda 30.5-cm howitzer** played a vital rôle in crushing the Liège forts. *Weight:* 28 tons. *Crew:* variable, but usually 12. *Rate of fire:* ten rounds-per-hour. *Range:* 13,124 yards. *Projectile weight:* 846 lbs

THE AMMUNITION SCANDAL

The blindness of those in command to
the real demands of the war machine
led, during the first year of the war,
to a critical shortage of ammunition.
As the factories in Britain struggled
to meet the new demands, the press
laid the blame for the crisis at
Kitchener's door
Major Henry Harris

It is generally considered beyond question that at the outbreak of war in 1914 none of the combatant nations foresaw either the lengthy duration of the war or the form it would take.

Field-Marshal Lord Kitchener, the lonely, formidable, Irish-born Englishman, who was brought into the Liberal government to fill the vacant post of Secretary of State for War (the first serving soldier to hold the job since Monk), had been one of the few to predict that the war would be a prolonged one. But not even he had foreseen that the lines would become static and that the artillery, the standard counter, would be used so extensively and would go on firing continuously day and night, using up enormous quantities of ammunition.

In anticipation of besieging Belgian and French fortresses, Germany was at first better prepared for heavy ammunition expenditure than her foes, but by the spring of 1915 she too was deficient in the types of guns and ammunition which experience was showing to be necessary.

And the pattern of warfare which had emerged now called for an unprecedented number of guns and enormous reserves of ammunition, not only of existing calibres and types, but also of new weapons more suited to trench conditions. Of these, high explosive shells for all calibres of guns were to be the most critical. By early October 1914 there was general concern among the Staff and the Ordnance in the BEF about the ammunition shortage. So meagre was the supply that every round had to be rationed. At the First Battle of Ypres, the allocation for 18-pound shell had to be reduced first from 20 to ten rounds per gun per day and towards the end of that month's fighting to two rounds per gun per day; the 4.5-inch howitzers were on two rounds and the 6-inch howitzers on six rounds per gun per day. Reports of daily stocks at railheads were telephoned to a senior Ordnance Officer at GHQ who prepared a tabulated statement for the QMG, BEF. From this, allocations on a priority basis were made. At times this meant not more than six rounds per gun per day, or less for some if the situation was worse than usual in another sector. To provide some flexibility, the Ordnance kept certain types of ammunition on special trains, which could be diverted from one army railhead to another in accordance with operational needs. During the 1915 battles nothing was left in depots and ammunition was being brought up to the guns straight from the ships. At Ypres in 1915 the British Second Army demands for howitzer ammunition had to be refused.

Shared responsibility
In 1914 the British War Office's Army Council included two military and one civilian member concerned with supply — the Quartermaster-General (QMG), the Master-General of the Ordnance (MGO) and the Finance Member.

The QMG and the MGO shared the chief responsibility, the latter being concerned with artillery supplies and technical munitions. The MGO's principal officer for material was the Director of Artillery whose branches dealt with all types of field and fixed guns, their pattern, design, manufacture, inspection and employment. The total staff of this Directorate was only 40. The MGO also had various technical bodies, such as the Ordnance Board and

Above: Kitchener visits France for talks during the ammunition crisis. The Press, in their campaign against him, underestimated his popularity and found themselves boycotted by major institutions. Kitchener survived the crisis and public faith in him remained unshaken. *Opposite:* Men and women at work in a shell filling factory. The system of tenders governing shell production was largely inefficient and led to inflated prices

the Small Arms Committee, as well as the production establishments of the Royal Ordnance Factories (ROFs) at Woolwich Arsenal, Waltham Abbey and Enfield Lock. The QMG had a principal officer, the Director of Equipment and Ordnance Stores (DEOS), whose duties included receiving, storing, issuing and repairing artillery equipment and ammunition.

Under the Finance Member was the Director of Army Contracts, in charge of the Contracts Branch born out of the British troubles in the Crimea. This was responsible for purchasing the total requirements of all departments of the army, but could order nothing without the concurrence of the military branch concerned. It had a staff of eight in August 1914. The size of the army being determined by Parliament and scales of equipment approved, the ordering followed a well-defined pattern through two main channels.

A large amount of artillery equipment and ammunition was obtained from the ROFs under direct MGO control. ROF contracts were regulated so as to keep them busy, but when necessary outside orders were placed to prevent civilian firms from closing through lack of work. Shells were ordered from 'the trade' so as to give the firms concerned experience of such work in the event of war. ROF work also acted as a check on contractors' prices.

To obtain such work, civilian firms had first to get on the List of Contractors. There were rigid rules on eligibility for this list — fair wages, no sub-letting, cutting out agents and middlemen, financial and quality control criteria and inspection of plant — all to be satisfied before a manufacturer could be put on the list. There were arrangements and facilities for firms

to inspect 'sealed patterns' and obtain specifications of equipment which was highly standardised.

Supplies were invited by tender. Tendering was a formal and leisurely business; the tenders came in by a fixed date in sealed envelopes and were dropped through a slit in a metal tender box, the key of which was held by a specially nominated officer. At an appointed hour this box was ceremonially opened and the tenders listed, signed and witnessed by those attending. Tenders were afterwards tabulated and considered, the basis being the most favourable price, though some note would be taken of the competence of individual firms. The procedure which laid down that 'The established principle of public purchase is competition and the acceptance of the lowest offer' was designed to ensure fairness between rivals and to avoid suspicions of favouritism or collusion. In normal times the system ensured reasonable prices for standardised articles.

But the supply requirement in August 1914 was far from normal. The supply departments realised this, but with no precedents to guide them, their steps to cope with the new situation were slow and hesitant, based on the existing machinery. Had the demands escalated immediately, the fundamental reorganisation on a national rather than a localised basis might have taken place earlier. But the gradual build-up obscured the fact that the current organisation for supply was only capable of limited expansion.

The defects were not immediately apparent, even to Kitchener who considered the Liberal government guilty of the collective sin of irresponsibility, and himself took on the burden of raising and equipping new armies of unprecedented size, and supervising the conduct of British military strategy throughout the world. Apart from lack of shells, shortages of all kinds were immediately apparent within a couple of weeks of the outbreak of war. Some requirements, such as clothing and footwear, harness and saddlery, could be met by large scale purchases of certain non-standard patterns.

When shortages became crucial and it was widely known that, contrary to his orders, many recruits were without boots and other necessities, Kitchener dismissed the Director of Army Contracts.

But guns and ammunition could not be obtained by speedy resort to trade patterns. No action by Kitchener — or anyone else — could redress for many months the prewar blindness and indifference of the British people and their leaders. But the demand for more heavy guns had to be met. In September 1914 a Siege Committee of officers from the War Office and GHQ, BEF recommended the provision of 140 heavy howitzers 'for use against fortresses on the Rhine', as well as anti-aircraft guns and weapons for use in trenches. Kitchener immediately authorised these but when the MGO (Lieutenant-General von Donop) demurred about the size and cost of the order which ran to £3,000,000 Kitchener had it checked with the French War Ministry. When they replied that they thought these were essential, he overruled von Donop and gave peremptory instructions for their purchase.

In the early days of the war, shells were put to tender to the approved firms, and the ROFs put on full capacity production.

These first orders, totalling 500,000 rounds (mainly shrapnel), were contracted for by separate components and assembled at Woolwich Arsenal. But unplaced demands began to accumulate as early as August 12. Kitchener invited US manufacturers to send representatives over to discuss production and the War Office placed some orders in Canada and India. Troubles and misunderstandings were to hold up deliveries of these overseas orders; there were variations in meanings between US and British technical terms, and the British were ignorant of US methods and vice versa. So keen were US firms to get orders that one company signed a contract a month before they knew what they were going to make! Another firm quoted for an entirely different article from that which they were expected to supply. Canadian production was to be particularly disappointing as a result of the difficulties in organising manufacture among a large number of inexperienced firms.

One of the first relaxations of the supply procedure was to permit sub-contracting—as this was recognised as giving training to firms inexperienced in munitions work—but the main contractors were held fully responsible. But lack of Inspecting Officers to visit the new firms led many to ask for and receive orders they were unable to fulfil. The most that could be done by the overworked inspection staff was to show samples and give advice to would-be contractors who came to Woolwich.

'To get them and hold them'

Unnoticed and unrealised by the government, a grave shortage of skilled labour developed. The initial and temporary breakdown of trade impelled many men to enlist rather than face the miseries of unemployment. The engineering trade alone lost 12% by October 1914 (this was to go up to 20% by July 1915). In the scramble for orders, firms began to poach each other's labour, inducing tradesmen who were already on war work with other employers to join them. Some firms, wishing to give their men on war work some recognition, suggested special badges. Churchill agreed to this and ordered one for Admiralty Service but it was initially vetoed by the Treasury. Full peacetime Treasury control (prior approval of all expenditure) was not completely relaxed for munitions until February 1915.

Employers made various other proposals; the Chief Superintendent of the Ordnance Factories suggested that suitable skilled men be given two to three year contracts of employment—'the only way to get them and hold them'. Messrs Vickers proposed bringing over Belgian workers and implementing large scale employment of women. It was also suggested that the French system of industrial mobilisation should be adopted, under which workmen were conscripted by the government into their jobs and moved about as munitions production required. This was fiercely resisted by the trade unions. Some of the drain off into the army was checked by telling Recruiting Officers not to enlist men from certain firms without their employers' consent. But despite these measures, by the end of the year the munitions programme was in deep trouble.

The tendering system created artificial scarcities and inflated prices. All the tendering firms went after the same raw materials at the same time, the net demands multiplying several times over in the market. Two hundred firms would put in for orders which involved the use of 2,000 tons of raw materials. Only 20 firms would get the orders but all 200 would have obtained options on 2,000 tons each. This caused chaos in the markets and forced options up to fictitious and unwarranted levels. Calling for tenders also led to the War Office revealing its total requirement to everyone in the trade and the relation of such demands to probable supply was quickly and accurately gauged by most. The War Office, being in a hurry, accepted high tenders in the hope that disappointed firms would reduce their prices next time.

The abandonment of many standard patterns and specifications for 'trade pattern' substitutes made inspection for quality control and other checks difficult, with innumerable queries arising. Most of this work was done physically after delivery, in the labyrinthine Peninsula-period accommodation of Woolwich Arsenal. Delays occurred often, with masses of items lying around in confusion until decisions could be made.

The most frequent excuse for failing to keep to delivery dates was shortage of machine tools. The UK output was insufficient and US orders were not delivered on time. One firm working on shells, Cammell Laird, in two months had only obtained 26 machines out of over 200 ordered, and had to put off their planned expansion. Gun contractors were also badly affected, and by the end of 1914, though their production had only just begun, they were finding it impossible to live up to their promises.

The greatest amount of delivery failures occurred among sub-contractors, who, despite their initial experiences, continued to give too optimistic a picture of their powers of production. They were harder hit by labour shortages than the big firms and were also hampered by delays in obtaining samples, drawings, specifications and of course in delivery of machine tools.

The War Office machinery for dealing with this situation was itself in poor shape. There was confusion in the offices as a result of over-crowding, the staffs' long hours (some worked 14 hours a day) and the ordinary civil servants' ignorance of commercial matters clashing with the contempt shown by the businessmen brought in to assist. Despite the addition of these 'expert buyers and advisers' in the Army Contracts Department, the old-fashioned purchasing procedure was little altered, some 'expert buyers' often being subjects of suspicion on grounds of gaining unfair advantage over their competitors.

In October the government set up a Cabinet Shells Committee. It was presided over by Kitchener and met six times up to December 1914. On December 21 there was a 'Shells Conference' which debated all the munitions troubles (in this month the government had considered it imperative to ship munitions to Russia to avert a military collapse there) and on December 23 the Cabinet agreed to co-ordinate the supply of labour in the following ways:
● substitute Belgian for British workers;
● divert labour from less urgent industries 'such as the railways'; and
● put pressure on firms not on munitions work to release their men.

To strengthen the War Office, an additional civilian member was appointed to the Army Council, and a general officer added to the MGO's Branch to visit and report on the flood of untried firms offering their services (as a result of an appeal by the Board of Trade) and whose production capacity was unknown.

From this point on, the responsibilities of the War Office were shared by other government agencies, and certain interests in and outside Parliament began to feel uneasy about the way munitions production was being directed. But the government still maintained that no drastic revision of policy was necessary except for machinery to deal with the delays in deliveries. The Board of Trade's efforts included attempts to obtain relaxation of trade union restrictive practices, and a drive against drinking by workers.

Unfair load

A large section of the engineering industry was dissatisfied with the War Office's methods of contracting, feeling that more direct contracting would help to reduce the loss of workers. Cammell Laird and Vickers were two firms to put this view when revising downwards their delivery dates. Demands for direct contracts by sub-contracting firms were stimulated by the steady loss of their work force. But it was not easy, as shown above, to give orders direct to untried firms and the policy of organising supply through firms already on the Contractors List had much in its favour. In 1914-15 these were the only firms with experience of such work; with little or no higher organisation to instruct and supervise, ordinary firms were quite incapable of going on to such work at short notice. To a great extent, the 'education' firms received as sub-contractors enabled them to organise themselves better in the summer of 1915. But apart from delays, deliveries in early 1915 were often defective. The ROFs were compelled to carry an undue share of the load at the expense of more specialised manufacture and every branch of the Arsenal was congested with work in hand and for inspection. Kitchener tried to conceal these troubles from his Cabinet colleagues, although he informed Asquith, who told him that *any* form of conscription (including industrial conscription, which Bonar Law was advocating) would wreck the government. Kitchener therefore addressed himself anew to enlisting private firms into making munitions.

But by March it was no longer possible to conceal that, as far as the immediate future was concerned, adequate supply was not assured. Kitchener abandoned his usual reticence in a speech in the House of Lords on March 15, 1915. He admitted that 'progress in equipping the new armies and forces in field is hampered by failure to obtain sufficient labour and by delays in production of plant, largely due to enormous demands', and went on to say: 'Labour has a right to say that their patriotic work should not be used to inflate the profits of directors and shareholders and we are arranging a system under which the important firms come under Government control. Men working long hours in the shops by day and night, week in and out, are doing their duty for King and Country the same as those on active service in the field.'

By now, Lloyd George and others had decided that a radical reorganisation was vital but Kitchener opposed this. He did, however, accept the Munitions of War Committee which was set up on April 12 under Lloyd George and he was represented on it by the MGO. But he refused to delegate any War Office responsibility for munitions, and also, on security grounds, to provide the committee with information about the numbers of men being put into the line by particular dates. Lloyd George claimed they were unable, as a result, to prepare accurate estimates of quantities required by specific dates.

Britain's other enemy

Lloyd George, though Chancellor of the Exchequer, was active on many other matters at this time in addition to munitions supply. He was giving advice on strategy (with Churchill he had strongly advocated an attack on Turkey), industrial unrest and the problems of drink. He was particularly keen to deal with 'the lure of the drink' which he felt was detrimental to good work and the cause of many serious production delays. He announced that: 'If we are to settle with Germany we must first of all settle with drink.' He induced the King to set an example by banning all alcoholic liquor from Royal establishments and sought a law to control or close public houses near where war work was being done. He claimed that it was 'proved quite clearly that excessive drinking in the works [sic] connected with (munitions) operations is seriously interfering with output'. During March the government negotiated an agreement with trade union leaders with recommendations to workers including eschewing of strikes, the use of negotiating machinery and temporary relaxation of demarcation and dilution rules. Most moderate trade unionists welcomed this and readily co-operated with local munitions committees in 'delivering the goods' and 'doing their bit'. But Lloyd George was not satisfied with these steps and felt he had a duty to play a more direct part in munitions supply. He considered it proper to invite the megalomaniac Lord Northcliffe to take up the munitions question in the press and so compel Kitchener and the government to act. Northcliffe, who considered Kitchener an impediment to victory over the Germans, naturally favoured a campaign against this overburdened titan, based on information obtained from Sir John French.

At this time things were not going very well for the British on the Western Front and Sir John French was induced to lend his name to the campaign. He appears to have done so willingly after his defeat at Neuve Chapelle, claiming that a shortage of ammunition robbed him of victory. But there was in fact no shortage during that four-day battle; ammunition was fired off at an unprecedented rate, described by Kitchener as 'irresponsible'. Shortages had hampered French on other occasions, mainly because of his refusal to accept the logistic situation and curtail offensive operations until supplies were adequate for his plans. Such appreciations were not well understood by that generation of Commanders and Staffs. However, on April 14, Kitchener told Asquith that 'French says that with the present supply of ammunition he will have as much as his troops will be able to use in the next for-

ward movement.' The 'next forward movement' was Festubert, a week before which French told Kitchener that 'the ammunition will be all right'. But on the first day of the battle, Kitchener had had to reserve 20,000 shells on paper for Gallipoli – an earmark which was cancelled the next day but it was enough to upset Sir John French and spark off the crisis.

In the preceding weeks, press comments on munitions had become more pointed. On April 7, *The Times* (a Northcliffe paper), writing about Kitchener's Armaments Output Committee, criticised the 'extraordinary failure of the Government to take in hand in business-like fashion the provision of full and adequate supply of munitions. The War Office has sought to do too much and been jealous of civilian aid. It should chiefly devote itself to the organisation of the armies and should state its supply requirements and leave to others the far more complex task of organising industry.' On April 10, *The Times* stated that *the primary reason why Sir John French is unduly short of munitions is not drink at all. It is that in our previous wars the War Office has been accustomed to rely for all such supplies on the MGO as a sort of Universal Provider. In this unprecedented war the Government ought to have insisted on instant organisation of the whole of our national resources, leaving the War Office to state its requirements and raise its armies.* But Lloyd George was saying that it was *obviously better to get your men under the direct supervision and control of those who for years have been undertaking this kind of work, obviously better than going to those without experience. The failure [of the Board of Trade to get men through Labour Exchanges] drove us to other courses – to introduce the Defence of the Realm Act to equip War Office and Admiralty with powers to take over engineering works.*

In mid-April the Prime Minister set up the Munitions of War or Treasury Committee under Lloyd George. He announced that the decision to do so had been taken a month earlier but the ground had had to be prepared for its activities. Kitchener was not a member of it and by its terms of reference it superseded his Armaments Output Committee. One of its schemes was for National Shell Factories and it was in effect the embryo of the Munitions Ministry.

The Battle of Aubers Ridge of May 9 resulted in heavy British casualties and on the 14th *The Times* printed a telegram from its correspondent in Northern France, stating that the 'attack had failed because of want of unlimited supply of HE (High Explosive)'. This correspondent was Colonel Repington, a retired officer, who had been staying at GHQ as French's personal guest. Kitchener had cautioned French about having Repington with him as it was a breach of orders, but French replied: 'Repington is an old friend and stayed for a day or two in an entirely private capacity – I really have no time to attend to these matters.' French gave Repington access to the papers on munitions and said all his efforts were crippled by lack of shells. He also sent copies of the papers to the opposition leaders, unknown to Asquith and Kitchener.

A week after *The Times* reported Repington's telegram, the *Daily Mail* (another Northcliffe paper) opened a direct attack intended to hound Kitchener from office

with the banner headline 'The Shells Scandal – Lord Kitchener's Tragic Blunder'. But Northcliffe had underestimated the nation's belief in its idol and the ceremonial burning of the paper containing the headline outside the Stock Exchange was but one reaction. Clubs and other institutions cancelled subscriptions to the *Daily Mail* and the general public indignantly boycotted it to such an extent that sales became seriously affected. After a few weeks, Northcliffe dropped the campaign, ruefully admitting his failure to a fellow press lord – Beaverbrook.

The press campaign, described by Churchill as 'odious and calculated' and by Haig as 'reptilian', had, if anything, increased public faith in Kitchener, and the King made him a Knight of the Garter. In the reconstruction of the government on a coalition basis, Asquith therefore had no choice but to retain him, but his position was weakened. In the new coalition National Government formed on May 26, Kitchener remained as secretary of State for War and Lloyd George went from the Exchequer to the new Ministry of Munitions. In the latter's words, this was *from first to last a businessman's organisation. Its most distinctive feature was the appointments I made to chief executive posts of successful businessmen to whom I gave authority and personal support that enabled them to break through much of the aloofness and routine which characterised the normal administration of Government Departments.* These businessmen had powers to take over any land or buildings they required, to engage such labour as they needed and to make conditions of work binding for the war period. Strikes and lockouts were forbidden, profits were limited, but workers were encouraged to earn as much as they could by piece work. But despite all these powers (and an increase of staff from 137 to 2,350 in five months) the time needed from creation of new capacity to delivery of bulk outputs differed little from that experienced by the War Office. The shortage of shells – the prime reason for handing over supply to a new department – was brought about by arrears of deliveries, not lack of orders, and these could be remedied by time alone.

Further Reading
Magnus, Sir Philip, *Kitchener, Portrait of an Imperialist* (Murray 1958)
Maurice, *Life of Haldane* (Faber & Faber)
Maurice, *British Strategy* (Constable)
Official History of the Great War Vols. I to III (Macmillan)

MAJOR HENRY HARRIS, a retired British Army Officer, is well known in historical circles for his lectures to the Military History Society of Ireland, of which he is a founder member. He has contributed articles on military matters to the *Irish Times*, the *Irish Army Defence Journal*, the *Army Quarterly* and *Brassey's Annual*. He has written two histories of the British Ordnance and has contributed 50 monthly articles on the 'Lost Generation' of 1914-1918 to the *British Ordnance Gazette*, chiefly on logistic and administrative aspects of the First World War. His *Model Soldiers*, based on his collection to be seen in Curragh, Kinsale, London and Camberley, has appeared in four languages. He is working on four more books, including a dictionary of Irish battles and one on the Anglo-Norman invasion of Ireland in 1169.

The fire power that French industry laboured to produce; what could not be made at home had to be imported

Below: The French 270-mm mortar, model 1885. *Length of barrel:* 6 calibres. *Elevation:* 30°. *Range:* 5 miles. *Projectile weight:* 204-384 lbs. *Crew:* 11 men. **1** The French 8-mm Hotchkiss machine gun, the principal machine gun of the French army during the First World War. Although it was rather heavy (gun 55 lbs, tripod 60 lbs) it was very reliable. **2** The French 8-mm Chauchat light machine gun, model 1915. This gun was reputedly unreliable and the standard of its manufacture was poor, nevertheless it was used extensively during the First World War. *Opposite page, top:* The French 220-mm mortar, model 1901 (modified). *Elevation:* 40°. *Rate of fire:* 1 round per minute. *Range:* 5 miles (approx). *Projectile weight:* 200 lbs. *Crew:* 4-5. **3** *Top:* The Lebel

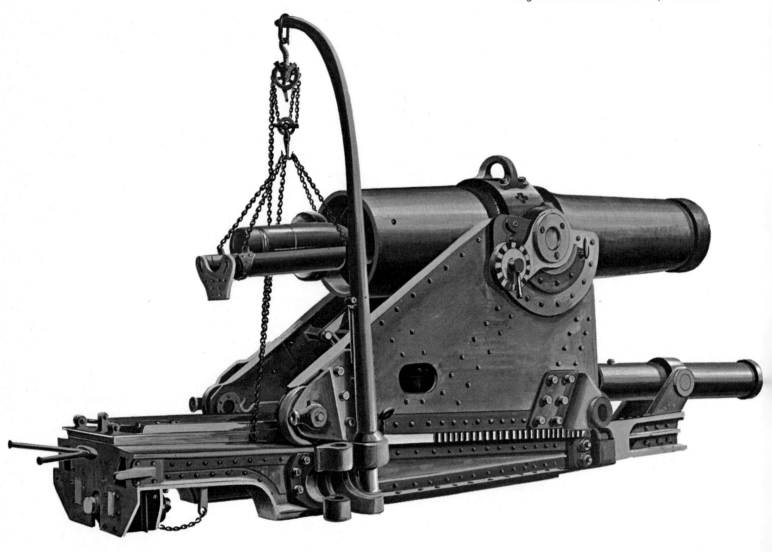

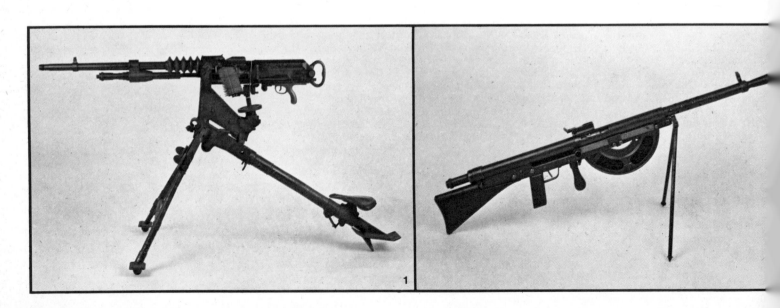

8-mm rifle, model 1886. *Centre:* Dandeteau 6.5-mm rifle, model 1895 (not widely used). *Bottom:* Single shot rolling block Remington rifle, chambered for the 8-mm Lebel cartridge. Many of these obsolete rifles were purchased by France during the early part of the war. **4** *Left:* 8-mm Modèle d'Ordonnance 1892 revolver. *Centre:* The 11-mm revolver, model 1873. *Right:* A Spanish-made pistol poorly made but typical of many of the pistols purchased by the French during the war. **5** Flare pistols — the top one is a modification of the 1886 model R35 carbine. **6** *Left to right:* 'Hair brush grenade', grenade projector. *Top:* 'Vivien Bessier' grenade, assault grenade. *Bottom:* 37-mm cannon shell, bomb for pneumatic mortar

John Batchelor

Jack Pia

47

THE FIRST FLAME ATTACKS

After an ominous silence in the early hours of the morning the Germans released the terror of a new weapon: from the nozzles of six flame throwers they launched jet upon jet of liquid fire at the British trenches at Hooge. Although it was a very minor action, the new weapon achieved a spectacular success. *Michael Dewar*

The necessary basis of this first flame attack was proximity of the trench lines, but this basis vanished immediately the Germans advanced

The idea of the modern flamethrower was first conceived by one Richard Fiedler, a Berlin engineer, in 1900. Fire had, of course, been employed as a means of warfare for hundreds of years, but it was Richard Fiedler who converted this art into a science. In 1901, the German army tested two models, both submitted by Fiedler. The smaller *Flammenwerfer*, which was sufficiently light to be carried by one man, used gas pressure to send forth a stream of flaming oil for a distance of about 20 yards. The larger version, which was more cumbersome to transport and operate, had a range of some 40 yards, and fuel enough for 40 seconds continuous firing.

It is the smaller version with which we are chiefly concerned. It consisted of a steel cylinder resembling a milk churn in shape and filled with an inflammable liquid. The interior of the cylinder was divided into two parts, the lower containing a compressed gas to provide the pressure, and the upper, the inflammable liquid. To one side of the cylinder was fitted a rubber hose, six feet in length, with a long steel nozzle at the end. This whole apparatus was attached to the back of the operator by padded metal arms.

The principle upon which the *Flammenwerfer* worked was extremely simple: a valve released the gas in the lower chamber which pushed the liquid in the upper chamber into the rubber pipe. Two other valves held the fluid in check before it reached the device for igniting it at the nozzle. This device consisted of a small tube containing a spring, a detonator, some gun cotton, and a wick soaked in paraffin. When the gas pressed the fluid against the spring, the wick ignited, and a jet of flame projected from the nozzle for 20 yards or more. To add to the effect, volumes of black smoke were also produced. The flame had a total duration of approximately two minutes. However, should bursts of a shorter duration be required, a firing tube had to be fixed into the end of the steel nozzle for each separate ignition.

Numbers *23, 24* and *25 Pioneer Battalions* were issued with this equipment in 1911. In 1912 the first *Flammenwerfer Regiment* was formed, namely the *3rd Guard Pioneer Regiment*, consisting of 12 flamethrower companies. This unit was commanded by *Major der Landwehr* Reddermann, and was subsequently to see action on both the Western and Eastern Fronts, and from the *Flammenwerfer Regiments* were·formed the *Sturmbatallions*.

An ominous silence
The Official French History of the First World War records a flame attack by the Germans as early as October 1914: 'In Malancourt wood, between the Argonne and the Meuse, the enemy sprayed one of our trenches with burning liquid, so that it had to be abandoned. The occupants were badly burned.' However, it is not clear whether this was a *Flammenwerfer* attack, or simply a spray of pure petrol, later ignited by incendiary bombs, another technique used by the Germans. *Flammenwerfer Apparate,* however, were definitely used against the French on February 26, 1915, near Verdun. *Major* Reddermann of the *3rd Guard Pioneer Regiment* commanded this particular flame attack. In retrospect, it would seem that this attack was a dress rehearsal.

The larger version of the German flame thrower. The smaller version could be carried by one man

The tactical situation at Hooge in July 1915 was as follows. Two actions had taken place recently in the area, but on July 18 the Germans still held the line of brick heaps which had once been Hooge Château, while the stables remained in the hands of the British. From that point the British line ran westwards, crossing and recrossing the Ypres-Menin road through the ruins of Hooge village. On July 19 a mine was exploded by the British and the resulting

Stupefying in its effect, the 'flame' was like a thick smoke incandescent at the centre and stretching for approximately 20 yards

crater which was 120 feet wide and 20 feet deep, was occupied. Hereabouts, No-Man's Land was some 70 to 150 yards wide, but in the vicinity of the Hooge crater it was as little as 15 yards. In fact at one point what had become a German communication trench led from the German line right into the British line: it was barricaded at the British end, but by means of a periscopic arrangement the German sentry could be seen on the other side of the barricade precisely five yards away. On July 22 two further attempts, involving, in each case, two platoons, supported by bombers and an RE detachment, were made to seize parts of the German lines near Hooge. Both failed, breaking down under heavy German fire.

The Hooge sector was held by the 41st Brigade of the 14th Division. On the night of July 29 the 8th Battalion of the Rifle Brigade and the 7th Battalion of the King's Royal Rifle Corps replaced the 7th Battalion of the Rifle Brigade, and the 8th Battalion of the King's Royal Rifle Corps. The Rifle Brigade held the front at the Hooge crater, and the King's Royal Rifle Corps was situated on their right. They were total strangers to the front.

The Germans knew exactly what was taking place. They knew that the new troops were comparatively inexperienced, and that they found the crater and mining operations difficult. They almost certainly knew not only the identity of the division and the fact that it belonged to the New Army, but also the composition of the brigades, and even the names of the commanding officers, gaining this information by means of 'listening sets', which enabled them to intercept all British telephone messages. Consequently they could act with full prior knowledge of British movements, and previous British attacks had failed for precisely this reason.

An ominous silence pervaded the Hooge sector of the front on the night of July 29-30. A few bombs, thrown into the Ger-

49

man trenches, provided no reply. Then, at 0315 hours on July 30, came the carefully planned German stroke. From the nozzles of six *Flammenwerfer Apparate* placed unobtrusively over the parapets of their trenches, they launched jet upon jet of liquid fire, simultaneously dropping a three-minute intense artillery bombardment on the British line. A sudden hissing sound was heard by 2nd Platoon, A Company of the Rifle Brigade who were situated on the left of the crater and a platoon of C Company on the right of the crater. Seconds later they were hit by a deluge of flame. Intense small arms fire swept the 300 yards of open ground between the British front and support lines in Sanctuary and Zouave woods. Then the Germans attacked in force.

The Rifle Brigade dispositions were too tightly packed, and also lacked depth. They had inherited an unsatisfactory position, which they had had neither the time nor opportunity to improve. The unoccupied Hooge crater dominated the centre of their position and was a positive invitation to disaster. The German attack broke through at the crater, and fanned outwards left and right, bombing along the trenches. 'Exactly what took place,' says Reginald Berkeley, 'will never be known, for there is no one alive to speak.' One can only surmise that the Germans occupied the trenches on either side of the Hooge crater with comparative ease as there can have been no real opposition remaining. Brigadier-General Oliver Nugent, commander of the 41st Brigade at Hooge, says of the attack: 'Those that were on the flank of the flame attack speak of the great heat generated by the flame, and their evidence tends to indicate that it was in the nature of thick smoke, incandescent in the centre and up to about 20 to 25 yards from the nozzles of the projectors rather than an inflammable gas.'

'A sudden hissing sound'
Perhaps the most accurate firsthand account is provided by Lieutenant G. V. Carey, who was in A Company at Hooge. He gives the following extremely interesting account: *There was a sudden hissing sound, and a bright crimson glare over the crater turned the whole scene red. As I looked I saw three or four distinct jets of flame, like a line of powerful fire hoses spraying fire instead of water, shoot across my fire trench. How long this lasted it is impossible to say, probably not more than a minute, but the effect was so stupefying that I was utterly unable for some moments to think correctly. About a dozen men of Number 2 platoon were all that I could find. Those who faced the flame attack were never seen again.*

Most of the 8th Battalion of the Rifle Brigade were overwhelmed and fell back. The Germans did not follow, but consolidated in the trenches either side of the Hooge crater. All but a small sector of the King's Royal Rifle Corps' trenches were also lost. The Germans brought up their *Flammenwerfer* again, but they were unable to use them since there were no targets sufficiently close. Moreover, rapid fire was directed upon the crews by the remainder of A Company, who had now recovered sufficiently from their initial surprise to offer organised resistance. By 0900 hours all that remained of the 8th Battalion, reinforced by one company of

the King's Royal Rifle Corps, held a line along the northern edge of Zouave wood.

The Germans had achieved complete surprise, and the employment of flamethrowers was not only totally effective within the limited area in which they were used, but also terrorized the troops in the peripheral area of the attack. Colonel Carey, the only witness to the attack surviving today, states that the numbers actually killed by the flame were comparatively small. Most were able to duck the flame, but the German infantry were so close on its heels that they were able to bayonet the British troops while they were still sheltering from the effects of the flame. He says, 'If the flame is being discharged from 15 yards range, there is every possibility of someone with a bayonet jumping on top of you before you have time to get up. No doubt this happened at Hooge.'

The events in the vicinity of Hooge crater, and the German success in that area, formed only part of what was a much larger attack. German troops were launched against the whole front held by the 41st Brigade. But the only sector in which they succeeded in taking any substantial area of ground was either side of the Hooge crater, and there they drove the British back to Zouave and Sanctuary woods. The trenches lost by the British were some hundreds of yards in length, for the Germans had worked their way into part of the line held by the King's Royal Rifle Corps on the right. They had succeeded in gaining a footing on a commanding ridge, and the division decided that the ground must be retaken without delay, otherwise the position in Zouave wood might become untenable.

It was determined that the assault should be made by the 8th Battalion of the Rifle Brigade attacking from Zouave wood, and by the 9th Battalion of the King's Royal Rifle Corps from Sanctuary wood, with the 7th and 9th Battalions of the Rifle Brigade in support. The objective was Hooge and the trenches in its neighbourhood.

Against the advice of the brigade commander, the attack took place that afternoon. Artillery preparation was limited to three-quarters of an hour's bombardment. At 1500 hours the four battalions duly went over the top and were swept out of existence by an enemy whose machine guns there had been no time to locate, and on whom the meagre artillery bombardment had made no impression. Many were caught on the British wire, and none got more than 50 yards beyond the edge of Zouave wood. The utilisation, in the forefront, of the 8th Battalion of the Rifle Brigade, a spent force, was a serious error of judgement. Consequently, the casualties on July 30 were extremely heavy.

Unwarranted criticism
The flame attack at Hooge was in itself only a small part of an attack which was aimed at the front of a whole brigade. Thus it was an episode of limited importance, but because of the nature of the attack, Hooge has assumed a special place in the history of the First World War. Hooge serves as an example of the ingenuity of the German military mind, and of the tenacity and courage of these representatives of the New Army. Despite the obvious disadvantage of facing a new and terrifying weapon, the British troops at Hooge recovered from

their initial surprise remarkably quickly, and offered a spirited resistance. A certain degree of unwarranted criticism was levelled at the British troops at Hooge. It was pointed out that a man using a *Flammenwerfer* which carries only about 20-30 yards, is bound to be a vulnerable target, and that a rifle or a machine gun brought up on a flank will make short work of him. This is true provided, firstly, that the trench of the flamethrower is more than 30 yards away and secondly, that his opponent has flanks which can be utilised. In this case the Germans had no need to leave their trenches in order to discharge the flame, and A Company had no flanks for offensive purposes, for the right flank was left 'in the air' by the crater, and on the left flank the trench bent back towards the support line. The same facts applied to C Company's sector.

Hooge was, in many ways, the ideal place to prove the effectiveness of the *Flammenwerfer* in battle. The apparatus had been used before, but with little effect, and a success was needed if the weapon was to be retained. The *Flammenwerfer* could only be used effectively in certain limited conditions. It was cumbersome, and therefore could only be operated safely from the security of a trench. Since the German trench systems were seldom less than 40 yards from those of the French and British, there were few places where the *Flammenwerfer* had any potential use. But at Hooge, not only was the British position weak, but the two front lines were sufficiently close to be within the limited range of the *Flammenwerfer*. There was every chance that an attack with *Flammenwerfer* might yield the success which the weapon so sorely needed. For these very simple reasons, the Germans specifically chose Hooge for their experiment. The success at Hooge undoubtedly explains the retention of the weapon for future operations. But the same combination of favourable circumstances was seldom presented to the Germans, which explains why *Flammenwerfer* failed to achieve a success as spectacular as Hooge on the Western Front again. The British and French, for their part, ensured that their trench systems remained, where possible, beyond the range of the German *Flammenwerfer*.

Further Reading
Das Ehrenbuch der Deutschen Pionere (Germany 1923)
Flammenwerfer und Sturmtruppen (Germany 1921)
Hare, Steuart, *Annals of the King's Royal Rifle Corps*
King's Royal Rifle Corps Chronicle, The KRRC Club Ltd (London 1915)
Military Operations, France and Belgium 1915, Vol II, Official History of the War (Macmillan & Co 1928)
The Rifle Brigade 1914-1918, The Rifle Brigade Club Ltd (London 1927)
Roberts, A. A., *The Poison War*

MICHAEL DEWAR was born at Fulmer in Buckinghamshire in 1941, and was educated at Downside, RMA Sandhurst and Pembroke College, Oxford, where he obtained an Honours Degree in History and specialised in Military History. He saw active service with his regiment, the 3rd Royal Green Jackets, during the Malaysian-Indonesian confrontation in Borneo in 1965-66. He now instructs at the RMA Sandhurst and writes for military history publications in his spare time.

Mainstay of the Infantry: The Bolt-Action Rifle

Of the millions of men who fought in the First World War, the majority were infantrymen whose personal weapon was a rifle. The First World War was fought at a time when the bolt-action magazine rifle had reached a peak of development upon which no subsequent improvements have been made. From the comparatively crude Dreyse Needle Gun of the Austro-Prussian war, the modern rifle gradually took shape, but no improvements would have been possible without the one outstanding feature of the Needle Gun: the combination of projectile, propellant and detonator. From this improvement all others sprang.

The principle of the bolt action is as follows:

The bolt, in outline resembling an ordinary door bolt, contains a long, spring-loaded, firing pin. On sliding the bolt towards the breech, the bullet is inserted into the breech, the spring, making contact with the trigger mechanism via the cocking piece, is compressed, thus withdrawing the firing pin, and the bolt is locked into the breech mechanism. The most effective of the locking systems is by dual lugs which revolve into recesses in the breech; these can be either vertical or horizontal. Other locking systems are of a

wedge type in the breech or by a rear lug. This last system, which is employed in the British Short Magazine Lee-Enfield, has been subjected to criticism because it allows too much 'play' in the bolt and, in general, is not as strong as the front locking system. Various methods of triggering the mechanism to release the firing pin have been developed. Essentially, however, the process is the same — by squeezing the trigger the resistance to the cocking piece is released and the firing pin makes contact with the percussion cap. With a turn-bolt, the locking mechanism is engaged by turning the bolt handle down when the bolt has been fully pushed home. The straight-pull bolt has an internal spiral system which automatically revolves the lugs into place as the bolt slides forward.

The box magazine, variations of which were employed by the majority of rifles of the period, is an extremely simple mechanism; it consists of a metal box incorporating a spring-loaded platform. The magazine is situated, forward of the trigger guard, below the mechanism. The capacity of the various designs varied from 3 to 10 rounds. For the most part, the magazines were an integral part of the rifle but there were several removable designs. In some instances,

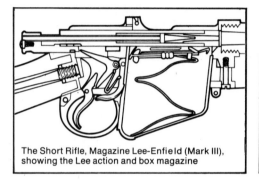

The Short Rifle, Magazine Lee-Enfield (Mark III), showing the Lee action and box magazine

The Russian M1891 Moisin-Nagant rifle with Moisin action and the Belgian Nagant magazine

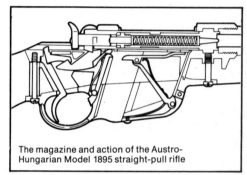

The magazine and action of the Austro-Hungarian Model 1895 straight-pull rifle

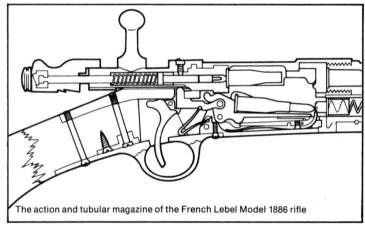

The action and tubular magazine of the French Lebel Model 1886 rifle

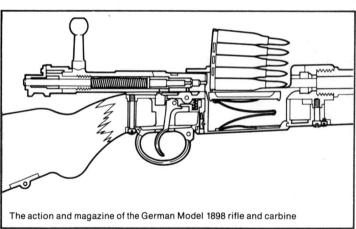

The action and magazine of the German Model 1898 rifle and carbine

the cartridges were staggered within the box, in others they were in-line. Cartridges were injected into the magazine by means of a clip. Clips assumed a variety of different shapes, but the principle was the same in all. They consisted of open-ended slides or cases within which a number of cartridges, 3, 5, or 6, were gripped whether by the spring metal of the case or a spring incorporated in the base. When the bolt was withdrawn, the clip was positioned between the bolt-head and the breech, and the cartridges were usually pressed down into the magazine by the thumb. The subsequent forward movement of the bolt ejected the clip and pushed the first round into the breech. The major deviation from this method was that adopted by Mannlicher for most of his designs. The Mannlicher system involved placing the clip with the cartridges into the magazine, a spring then pushing the cartridges up *within* the clip. When the last round was in the breech, the clip would fall through an aperture in the bottom of the magazine.

Without doubt the German contribution to small arms design has been the greatest. In particular the rifles of Peter Paul Mauser demonstrated a remarkable ingenuity and have influenced rifle design throughout the world. The Mauser Gewehr 98 was probably the most successful rifle of its kind ever designed.

Second only to Mauser was Mannlicher. A frequently underestimated inventor, he was often badly used. It is significant, however, that his designs and elements of his designs have been incorporated in the weapons of many different countries.

The United States claims, with some justification, much credit for advances in small arms design. Most of these advances, however, applied to weapons of eras before and after that which concerns us here. The biggest single contribution from America was that of the Lee box magazine and Lee, although he worked in America, was born in Scotland.

Britain's contribution was in the field of rifling. Before the introduction of the smokeless 'cordite' propellant, the Metford system of rifling was general; the new propellant, however, necessitated a different system and the Enfield system was almost universally adopted and is still used. Illustrated overleaf are the genealogies of the major design features of the First World War's rifles and carbines, with the colour coding showing the provenance of each weapon's cartridge, bolt system and magazine. A few further points about some of the weapons follow. The Italian Mannlicher-Carcano is sometimes called the Mauser-Paravicino and has been described, erroneously, as dangerous to fire. The Mauser Gewehr 88 was Germany's answer to the French introduction of smokeless powder with the 1886 Lebel rifle. The Gewehr 98 was a development of this, and can be claimed to be the world's most successful bolt-action rifle ever. The Mauser M1889 was Belgium's first modern rifle, and was the first rifle to utilise the Mauser magazine charger. The Turkish M1893 rifle was basically the same as the Spanish 7-mm weapon of the same date. Both introduced the Mauser staggered-row magazine. The Springfield M1903 was an adaptation of the Mauser ▷

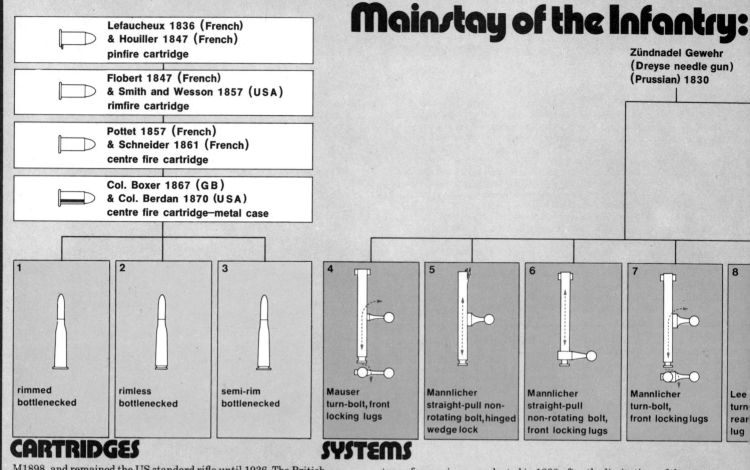

Mainstay of the Infantry:

| Lefaucheux 1836 (French) & Houiller 1847 (French) pinfire cartridge |
| Flobert 1847 (French) & Smith and Wesson 1857 (USA) rimfire cartridge |
| Pottet 1857 (French) & Schneider 1861 (French) centre fire cartridge |
| Col. Boxer 1867 (GB) & Col. Berdan 1870 (USA) centre fire cartridge—metal case |

Zündnadel Gewehr (Dreyse needle gun) (Prussian) 1830

1 rimmed bottlenecked

2 rimless bottlenecked

3 semi-rim bottlenecked

4 Mauser turn-bolt, front locking lugs

5 Mannlicher straight-pull non-rotating bolt, hinged wedge lock

6 Mannlicher straight-pull non-rotating bolt, front locking lugs

7 Mannlicher turn-bolt, front locking lugs

8 Lee turn rear lug

CARTRIDGES   SYSTEMS

M1898, and remained the US standard rifle until 1936. The British Pattern 14 was a modified Mauser design made in the United States. It was used mostly for sniping. The Japanese Arisaka Type 38 was introduced in 1905, and had a sliding bolt cover to keep the action clean. It was usually removed in action, however, as it made a great deal of noise. The Austrian Mannlicher M88/90 was introduced in 1890 to take advantage of the new and more powerful smokeless powder. Otherwise the rifle was identical to the Model 1886. An unusual feature was the fact that the magazine charger clip was ejected from the bottom of the magazine when the latter was empty. The M90 carbine introduced the rotating bolt to the Austrian army. The M95 was the main Austrian rifle of the First World War and was used also by the Italians, Bulgarians, Serbs and, to a lesser degree, by the Greeks. The Mannlicher-Berthier rifles were, with the Lebel, France's standard rifle until well into the 1930s. The Mannlicher

type of magazine was adopted in 1890 after the limitations of the Lebel tubular magazine had been realised. The Rumanian Mannlicher M1893 was identical to the Austrian Mannlicher, but chambered to take a Dutch round. The British Short Magazine Lee-Enfield No 1 Mark III was introduced in January 1907 and was the main infantry weapon used by the British in the First World War. The French Lebel was the world's first modern, mass-produced magazine rifle, and also introduced smokeless powder to the armies of the world. The Lebel's round was, however, of a design unsuitable for automatic weapons, and future French rifles reverted to a more conventional design of magazine and round. Finally, the Russian Moisin-Nagant, which was introduced in 1891. It was designed for a conical-nosed bullet, but the adoption of the Spitzer pointed round in 1908 necessitated no more than a change in the rear sight to accommodate the new ballistics. *Owen Wood*

6·5-mm Mannlicher–Carcano M1891 Italy
| 2 | 4 | 14 |

7·92-mm Mauser Gewehr 88 Germany
| 2 | 4 | 14 |

7·92-mm Mauser Gewehr 98 Germany
| 2 | 4 | 11 |

7·65-mm Mauser M1889 Belgium
| 2 | 4 | 11 |

7·65-mm Mauser M1893 Turkey
| 2 | 4 | 13 |

·30-in Springfield 1903 USA
| 2 | 4 | 13 |

·303-in Pattern 14 Britain
| 1 | 4 | 13 |

6·5-mm Arisaka Type 38 Japan
| 3 | 4 | 13 |

The Bolt-Action Rifle

Lee box
magazine 1879

Chassepot rifle
(French) 1866

Gras rifle 1874

tubular magazine
of a type used in
many early
repeating rifles

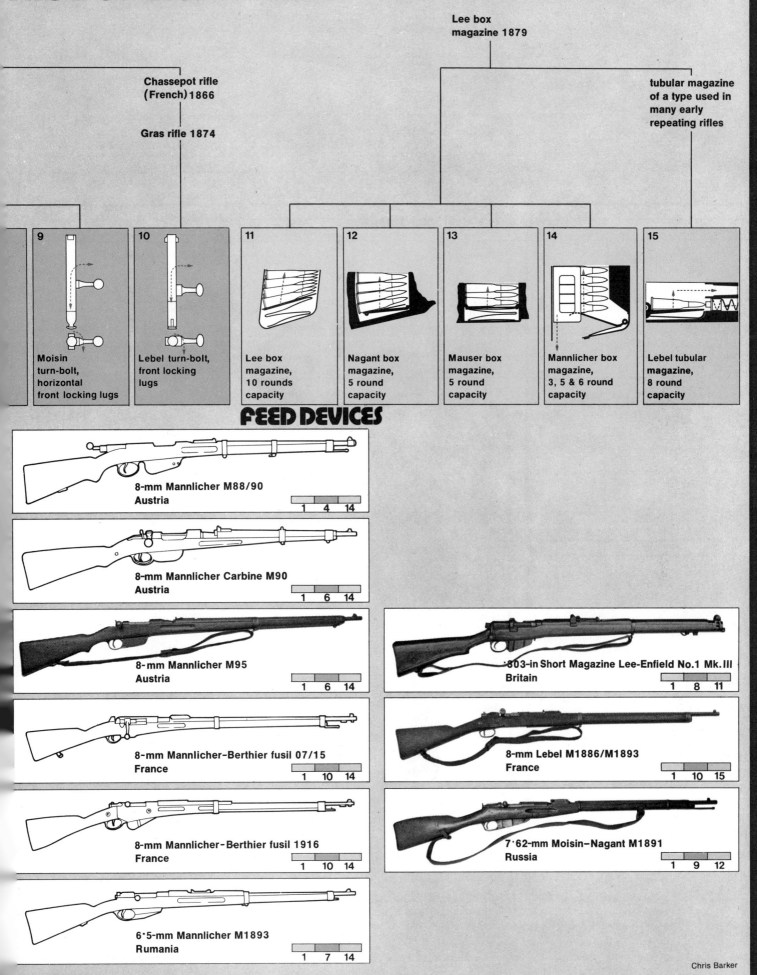

9
Moisin
turn-bolt,
horizontal
front locking lugs

10
Lebel turn-bolt,
front locking
lugs

11
Lee box
magazine,
10 rounds
capacity

12
Nagant box
magazine,
5 round
capacity

13
Mauser box
magazine,
5 round
capacity

14
Mannlicher box
magazine,
3, 5 & 6 round
capacity

15
Lebel tubular
magazine,
8 round
capacity

FEED DEVICES

8-mm Mannlicher M88/90
Austria
1 4 14

8-mm Mannlicher Carbine M90
Austria
1 6 14

8-mm Mannlicher M95
Austria
1 6 14

8-mm Mannlicher–Berthier fusil 07/15
France
1 10 14

8-mm Mannlicher–Berthier fusil 1916
France
1 10 14

6·5-mm Mannlicher M1893
Rumania
1 7 14

·303-in Short Magazine Lee-Enfield No.1 Mk.III
Britain
1 8 11

8-mm Lebel M1886/M1893
France
1 10 15

7·62-mm Moisin–Nagant M1891
Russia
1 9 12

Chris Barker

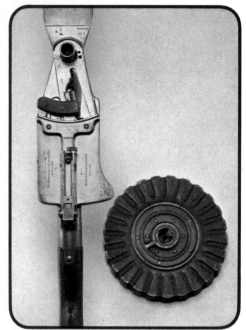

Top, centre and above: The Lewis gun. An air-cooled, American designed .303-inch light machine gun. The ammunition was drum fed, each drum containing 47 rounds. The Lewis gun was a useful infantry weapon, but it was prone to jamming, especially in wet or muddy conditions. *Above right:* Hand and rifle grenades of 1914/15. Column one, a British No 2 Mark I hand grenade. Column two, a German ball grenade *(top)*, a German oyster grenade *(centre)*, a French pear-shaped grenade *(bottom)*. Column three, a British No. 3 Mark I rifle grenade. Column four, a British Malta grenade *(top)* — this was made exclusively in Malta for the Gallipoli Campaign, a British No 12 hand grenade *(bottom)*. Column five, a French Bessozi hand grenade *(top)*, a British No 5 Mark I Mills grenade *(centre)*, a British No 1 Mark II hand grenade *(bottom)*. Last column, a German 1914 rifle grenade

The first sub-machine gun to be used in military service

Top left: **1** The Vetterli Vitali M1871/87 10.4-mm. **2** The Vetterli Vitali M1871/87 10.4-mm converted to 6.5-mm. These two rifles were still in fairly extensive use at the beginning of the Italian campaign. **3** The Mannlicher Carcano M1891 6.5-mm. **4** The Mannlicher Carcano 91TS Carbine 6.5-mm. **5** The Mannlicher Carcano M1891 Carbine 6.5-mm with folding bayonet. The Mannlicher Carcano has a Mauser-type bolt action with a Mannlicher six-round magazine

Top right: **6** The 10.35-mm Glisenti M1891 revolver with folding trigger. **7** The 10.35-mm Glisenti revolver. **8** The 7.65-mm Beretta M1915 pistol. **9** The 9-mm Glisenti M1910 automatic pistol. **10** Flare pistol. **11** The 9-mm Beretta M1915 Glisenti pistol

Left, below left and below right: The Villar Perosa 9-mm sub-machine gun, the first actual sub-machine gun used in military service. Originally designed for aircraft, its light weight (14 lb 4 oz) made it useful for infantry fighting in mountainous regions. Each magazine held 25 rounds, and its two barrels could be fired individually. Here it is shown in a back pack with a container for cartridges (left), modified for infantry use with a bipod (below left), and with a mounting for aircraft or bicycle use (below right)

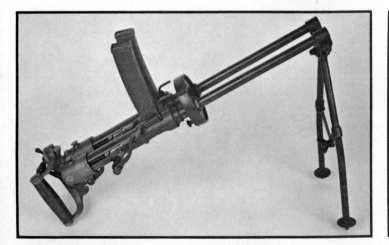

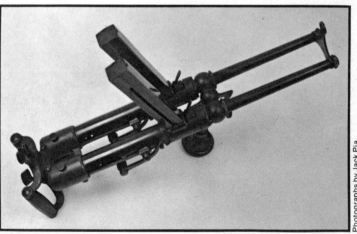

Photographs by Jack Pia

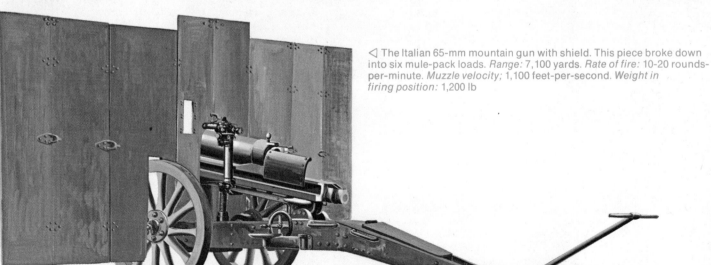

◁ The Italian 65-mm mountain gun with shield. This piece broke down into six mule-pack loads. *Range:* 7,100 yards. *Rate of fire:* 10-20 rounds-per-minute. *Muzzle velocity;* 1,100 feet-per-second. *Weight in firing position:* 1,200 lb

John Batchelor

VERD

UN THE STORM BREAKS

At dawn on February 21 the German bombardment of Verdun and its defences began, a bombardment so massive that it was heard as a steady rumbling 100 miles away. The Germans, superior in the air, in artillery and in numbers of fighting men, seemed poised for a stunning victory, and the fall of Fort Douaumont, the largest of the forts guarding Verdun, was seen as the beginning of a major French catastrophe. Yet the Germans let their opportunity slip, largely owing to caution from above. Fighting desperately under the inspiring leadership of Pétain, the French began to slow the German advance. *Below:* The effect of the German bombardment — the devastated entrance of Fort Souville, one of the forts guarding Verdun *Alistair Horne.*

It was chance—even bad luck—that in the early morning of February 12, 1916, the day fixed for the German assault on Verdun, there was a ferocious blizzard. At dawn deep snow lay everywhere, the storm still raged and mist lay across the whole countryside, concealing all the meticulously pin-pointed targets for the German artillery, on whose accuracy success depended absolutely. So their opening bombardment had to be postponed.

This initial postponement did in fact prevent immediate disaster for the French. On that morning, two newly-arrived French divisions, the 51st and the 14th, were not yet in position. If the Germans had attacked then, with their blanketing artillery and their battalions of chosen assault troops, they would have found the French in half-completed positions—the French 72nd Division alone facing the six German attacking divisions, and the other four French divisions, including the two newly-arrived, spread about through the citadel and the rest of the Verdun salient. Some of the French troops holding the front line were elderly Territorials.

While the Germans waited for the stormy weather to abate, the French settled into their new dispositions. There were now three French divisions—making up General Chrétien's XXX Corps—facing the Germans between the Meuse and the railway to Etain. Of these, the 72nd and 51st Divisions, with 20 battalions, manned the north-eastern curve of the salient, which was to be the sector of the main German attack; and there were 14 further battalions on the French flanks or in close reserve. An elderly gunnery officer, General Herr, was in immediate command of the French troops in Verdun and the salient; but having lost a large part of his fortress artillery to Joffre's recent purges of fixed emplacements, Herr felt himself at a loss in a world of trenches, infantry, wire and chronic lethargy. As a fortress commander, Herr was directly responsible, not to a field army headquarters, but to General de Langle de Cary, commanding the Central Group of Armies.

Opposite these 20 French battalions, the main German attacking force was made up of 72 battalions, or six divisions. On the German right, from the Meuse at Consenvoye eastwards to Flabas, was *VII Reserve Corps* under General von Zwehl—Westphalians. Next, on a short sector opposite the Bois des Caures, was *XVIII Corps* under General von Schenck—mostly Hessians. On the left, from Ville to Azannes, was *III Corps*: these were the famous Brandenburgers, and their commander, General von Lochow, had already gained a reputation for aggressive action—in particular, against Pétain's thrust towards Vimy Ridge in the previous spring. Supporting troops of *V Reserve Corps* and *XV Corps* were on the German left.

But the principal feature of the German line-up was the concentration of guns—over 1,200 for an assault frontage of barely eight miles. There were 654 'heavies', including 13 of the 42-cm mortars, the 'secret weapon' of 1914 that had shattered the Belgian forts, and two 38-cm long-range naval guns earmarked to interrupt communications behind Verdun. There were 21-cm howitzers to pulverise the French front lines, and 15's to seek out and destroy the French batteries. Two and a half million shells had been

Left: A German 21-cm short howitzer. A sudden thaw on February 28, following the long spell of cold weather, turned the surface into a sea of mud and made it difficult for the German artillery to advance

Above: A French soldier is hurled violently back by the impact of a German bullet. By the end of March, 81,607 Germans and 89,000 French had been killed—not only France, but Germany too, was being bled white

brought up to supply this greatest concentration of artillery yet seen on any battlefield.

To preserve secrecy, the Germans had deliberately restricted their front and not included the eastern side of the salient for their attack; for the Woëvres Plain on that side was overlooked by the French-held Meuse Heights. Of much greater significance, however, was the German decision not to attack at the same time on the left, or western, bank of the Meuse. This limitation, and the later German efforts to rectify it, were to be among the terrible hall-marks of the Battle of Verdun.

For nine days bad weather held up the opening of the German bombardment which was to precede the attack of their assault troops. There was snow again; a thaw with fog, rain and gales; more rain and gales; wind and snow squalls; mist and cold. The French troops at least were accustomed to a rough life in poorly finished trenches and improvised dug-outs, with shelter for some in the forts, now largely deprived of their guns, which ringed the inner fortress of Verdun. The Germans probably suffered more in these harsh days of waiting, keyed up as they were for the attack. The concrete *Stollen* in which they assembled expectantly each day were not meant as a permanent shelter for these large bodies of troops; so most of them had a seven-mile march each night and morning to and from their billets through freezing sleet or snow.

On February 19 and 20 the weather improved. So at dawn on the 21st, the massive German bombardment started—heralded by a poor shot from one of the two long-range, long-barrelled 38-cm (15-inch) naval guns which, instead of destroying the Meuse Bridge in Verdun, exploded in the courtyard of the Archbishop's Palace, knocking off a corner of the cathedral. The other naval gun, aiming at Verdun Station, had more success. The ordeal of the salient had begun.

The bombardment of the French lines and rear quickly rose to a shattering peak of intensity, and continued without abatement for nine hours. In all the great artillery barrages of 1915, nothing compared with this concentration of high-explosive shells, with this appalling weight of flying, disintegrating metal. From left to right, from front to rear and back again in the narrow sector of the impending German infantry attack, everything was methodically bombarded. Gas and tear-gas shells were included too, with the aim of incapacitating the French artillery. The French trenches, poorly prepared, were quickly obliterated and many of the troops manning them were buried.

Lieutenant-Colonel Driant, whose warning voice had first drawn attention to the deplorable state of the Verdun defences, was now in the front line in the Bois des Caures, with his *Chasseurs,* as a detachment of 72nd Division—in the very centre of the German attack sector, opposite *XVIII Corps.* They experienced the full intensity and horror of the shelling. It seemed to them as if the wood was being swept by 'a storm, a hurricane, a tempest growing ever stronger, where it was raining nothing but paving stones'. Through the din of the explosions came the splintering crash as the great oaks and beeches of the forest were split or uprooted by the shells. The bombardment was heard nearly 100 miles away on the Vosges front as a steady rumbling.

Unpreparedness played as much a part as the German gas-shells in upsetting the French artillery. Their counterbattery work was generally ineffective from the start: their fire was spread rather aimlessly, and ceased to count as visibility decreased and contact with the infantry broke down. All telephone communication with the French front line had been cut by the shelling within the first hour, and runners often found it impossible to get through the avalanche of shells. As a result, effective command rapidly broke down. There was no question of sending up reinforcements; and in Verdun itself the shelling from the 38-cm naval guns had already dislocated the unloading of trains.

About mid-day there was a sudden pause: the shelling stopped and the French troops emerged from the debris of their line, ready to face the expected onslaught of the German infantry. But it was a trick. The Germans quickly took note of those parts of the French line which were still manned, and then at once took up the bombardment again, concentrating their short-range heavy mortars on the sections where the French had shown themselves, and shelling every part with the same unrelenting fury. To the German infantry, after long days in and out of the flooded *Stollen,* the sight of the French trenches disintegrating was intoxicating.

Even so, the Germans were taking no risks. When the bombardment at last finished at 1600 hours, they cautiously sent forward powerful fighting patrols who, making skilful use of the ground, probed for the sectors of least resistance. True to Falkenhayn's directive for a controlled but insistent advance, the Germans were here evolving a new method of approach and attack, which was quickly to become characteristic of the fighting at Verdun on both sides; and, paradoxically, by prolonging individual survival, these tactics also prolonged the duration and suffering of this inhuman battle.

On this first day at Verdun, however, German caution from above, and their unreadiness to follow up the probing, infiltrating patrols at once with assault troops, lost them their opportunity. For once, the massed waves of attacking infantry, which until now had been the ultimate weapon all along the Western Front, incurring always suicidal losses — for once, these disciplined lines of infantry, coming over immediately while the French were still stunned and shattered from the immensity of the bombardment, could have given the Germans a quick breakthrough and a clear road to Verdun. As it was, two of the corps' commanders stuck rigidly to orders and held back their main bodies of infantry until the next morning, leaving their patrols to do no more than explore French weaknesses. Only Zwehl, commanding *VII Reserve Corps* on the German right, obstinately improved on his orders and sent in his first wave of storm-troops hard on the heels of the patrols, and at once achieved an important success. They man-

aged to occupy the whole of the Bois d'Haumont before dusk and so effect the first breach of the main French defences.

Elsewhere on the French 72nd Division front, and to the east on the 51st Division front, the German patrols, even without support, did damage enough to the French line, exploiting the gaps made by the shell-fire and using their new weapon of horror, the flame-thrower, with terrifying effect, causing panic through the French ranks. There still remained stubborn and heroic pockets of French resistance, flaring out even into disconcerting counterattacks.

None of this, however, was the easy success the Germans had expected. The concentration of shell-fire from the unprecedented opening bombardment had been planned to destroy all life in the French front-line trenches and block-houses. But somehow men had survived. Driant's sector, for example, in the Bois des Caures had consisted of a complex network of redoubts and small strongholds, quite unlike the usual continuous trench lines. Here and elsewhere the German troops were therefore disconcerted to meet intense machine gun and rifle fire from defenders who ought to have been buried in the wreckage of the uprooted forest.

Full-scale attack
By nightfall, Knobelsdorf, Prince Wilhelm's Chief-of-Staff, had become impatient of this cautious progress, and ordered a full-scale attack for the morning after another softening-up bombardment. It was the French, however, who surprisingly took the initiative at dawn on the 22nd, attempting several counterattacks with

△ One of the *Stollen* (concrete bunkers) which the Germans built near the French line to permit first-wave troops to concentrate safely
▽ Douaumont after the German bombardment

Südd Verlag

A German shell explodes and the 'mincing machine' grinds on. Making skilful use of the ground, the Germans sent out powerful fighting patrols probing for the sectors of least resistance

the spirit and dash typical of their army's tradition. But the means were not there, nor the organisation and communications: many attacks failed to get going, others were soon swamped.

Even though Knobelsdorf now placed no limits on corps' objectives the *XVIII Corps*, opposite the Bois des Caures, was still a model of caution, faced as it was with the verve and flexibility of Driant's *Chasseurs*, which masked the absurdly small size of the French opposition; and it took the Germans all day, with the help of their comrades on either flank, to silence Driant's valiant resistance. Zwehl's *VII Reserve Corps* had captured the village of Haumont on Driant's left, after a hard fight lasting into the afternoon, and now the Brandenburgers of *III Corps* came round on Driant's right through the Bois de Ville, split up by the cross-fire from the surviving French machine guns, but relentless with their flame-throwers and overwhelming in their numbers. They surrounded the French positions and picked off their strong-points one by one. Driant himself and a handful of men tried to hold out in the last strong-point, in vain; and as they broke out to the rear, they were mown down by enfilading fire.

The German losses in silencing the Bois des Caures had been unexpectedly heavy, and the confidence of their assault troops was shaken. Above all, their offensive had been held up during one whole day by Driant's resistance in the very centre of the German attack. Nor was Driant alone in this. The anchor position on the left of the French line at Brabant on the Meuse was still held by other elements of 72nd Division; and on the right, 51st Division continued to fight bitterly in Herbebois, holding up the advance of the Branden-

burgers. These delays, though seeming small in the overall picture, robbed the German assault of its required momentum.

Earlier on this day, the French artillery had improved their shooting, and had been reinforced. Fire from French batteries on the left (western) bank of the Meuse onto the Germans' flank was starting to cause trouble. Falkenhayn's parsimony in keeping back reserves for Verdun had deprived the German *VI Reserve Corps*, holding the line on the left bank, of adequate artillery, and this weakness prevented them from silencing the French guns on this side. But on the right bank, behind the battlefield itself, the French field guns were being knocked out one by one by the fire of the long-barrelled 15's.

Once more on the next day, February 23, the Germans made surprisingly little progress, even though they were still held only by the remnants of 72nd and 51st Divisions, whose losses continued at a terrible rate. Brabant had been evacuated, but mixed elements managed to hold out all day in Beaumont, on a rise behind the Bois des Caures. Infantry of the German *XVIII Corps* pressed forward against this position in dense formations, one wave pushing in closely behind another, in the established style of the Western Front, to be scythed down in turn by the French machine guns. At the same time, the French heavy guns under XXX Corps' command had found the range of the advancing Germans, who were suffering too from their own shells falling short. Official German records speak of this as a 'day of horror'.

The French infantry were also starting to get the measure of the dreaded flame-throwers. They were learning to pick off the German pioneers who wielded them, as they moved clumsily under the heavy bur-

den of their equipment, before they came within flame-throwing range. So 51st Division on the right were still able to cling tenaciously to Herbebois. On the other side of the sector, 72nd Division managed to hold up Zwehl's corps at last in front of Samogneux on an 'intermediate line', which de Castelnau had ordered to be constructed in January. When this line was broken, a battalion still held out in Samogneux itself till late into the night, when it was suddenly wiped out by a barrage of French 155's, firing across from the left bank in the belief that the village had already fallen to the Germans.

With this final tragedy, 72nd Division virtually ceased to exist. The two first-line French divisions, 72nd and 51st, with an original combined establishment of 26,523, had together lost 16,224 officers and men.

The Germans, despite their rebuffs, felt that Verdun was now theirs for the taking. So it was. They overran the whole French second position, inadequately prepared to start with, and now pounded out of existence, in three hours on February 24. But following the unexpected and savage opposition which they had experienced in the previous two days and the serious losses they had suffered, the Germans were no longer poised to exploit the situation fully, and their cautiousness had increased. Once more, they missed the opportunity of breaking right through.

The gap in the French lines left by the decimation of 72nd Division was now filled by 37th African Division—the Zouaves and Moroccan *Tirailleurs*. But the bitter, freezing weather, which had returned once the German offensive was launched, and was causing suffering enough to European troops, had cowed the temperamental North Africans. Furthermore, in an attempt to plug the disastrous holes in the French defences, they were being split up into small units under strange officers—survivors of the 72nd and 51st. The incessant bombardment, the horror of which was beyond their comprehension, was the last straw. When the North Africans saw the great grey carpet of German infantry rolling out towards them, some of them lost their nerve and fled.

Crumbling morale

By the night of February 24, French morale was crumbling seriously. Their guns were silent, their clearing stations overflowing with casualties, untended, their wounds often frozen by the intense cold. Only a fraction could be got out: the German 38's, firing with 'diabolical precision', had cut the one full-gauge railway out of the Verdun salient. Chrétien's XXX Corps was finished. Balfourier's XX Corps from the Lorraine front was to relieve them, but only the vanguard had reached Verdun, cold, hungry, and exhausted. Despite every protest, Chrétien insisted on throwing them into the battle at once; but it seemed doubtful if they would hold till the corps' main body, en route in a desperate forced march, could arrive.

It was on the following day that the French Command took decisions which committed them finally to the defence of Verdun, whatever the cost.

On the evening of February 24, de Langle de Cary obtained permission from Joffre to shorten the line in front of Verdun drastically on the still unthreatened Woëvres Plain side of the salient, and to withdraw

The guns were the masters of the battlefield, their thunder was 'like a gigantic forge that never stopped, day or night'

Below: The German 17-cm naval rail gun. *Maximum range:* 25,700 yards. *Barrel:* 22′ 6″. *Traverse:* 13° left to 13° right. *Weight of gun and carriage:* 25 tons. *Weight of shell:* 141 lbs. *Rate of fire:* 1 round-per-minute. *Crew:* 12 men.
Bottom right: The French 370-mm mortar, one of the big guns that helped to press the French advantage at Verdun. It had a range of over five miles, using a 250-pound shell. It weighed 30 tons, complete with platform, could fire one round every two minutes and required a crew of 16 men

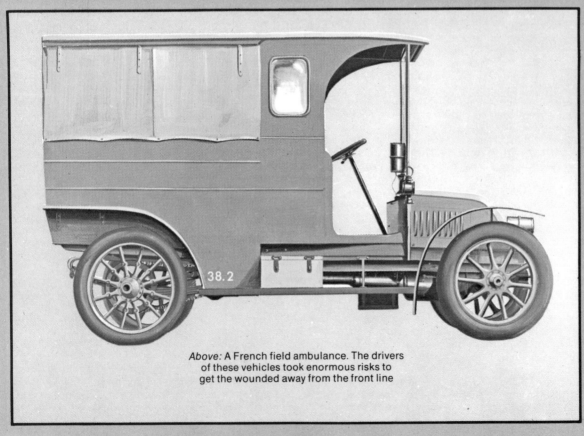

Above: A French field ambulance. The drivers of these vehicles took enormous risks to get the wounded away from the front line

E.C.P.-Armées

Right: Rain-filled craters surround the ruined outer walls of Fort Douaumont

John Batchelor

from the Woëvres Plain to the Meuse Heights east and south-east of the city. Later that night, with the news desperate, Joffre, against all orders and precedent, was rudely wakened from his routine allocation of sleep. One authority says that it was de Castelnau who insisted on waking him; another that it was the Prime Minister himself, Briand, who had driven over from Paris to General Headquarters at Chantilly; a third merely states that Joffre rose to the occasion. The consequence, in any case, was of supreme importance. Joffre sent de Castelnau off to Verdun that night with full powers to do whatever was needed. De Castelnau, as the recently-appointed Chief-of-Staff at French General Headquarters and Joffre's closest colleague (and rival), was the obvious choice for such a critical mission. But the very choice of de Castelnau made certain that the decision over Verdun would be a firm one. He was still very much a 'fighting general', as well as being outstandingly intelligent, quick-witted and flexible; and he had a way of inspiring staff officers and front-line soldiers alike.

De Castelnau reached Verdun at breakfast-time on February 25. He found Herr, who was still in command of Verdun and its salient, 'depressed' and 'a little tired'. He then went straight to the battle-torn right bank, made rapid and accurate appreciations, and effected miracles in reanimating the defence, with his experience and insight and his boundless energy: 'wherever he went, decision and order followed him'. He saw that Verdun could be saved, that an effective defence could be maintained on the remaining cross ridges of the right bank. He telephoned his conclusions to Joffre's headquarters that afternoon, and without waiting for agreement gave orders that the right bank of the Meuse should be held at all costs, and that there must be no retreat to the left bank.

On the previous evening, Joffre had already agreed that General Pétain, the tenacious commander of the French Second Army, should be brought over to defend the left bank at Verdun. Now de Castelnau recommended that Pétain should be put immediately in total command at Verdun, of both sides of the river, and despatched the necessary order without waiting for Joffre's approval. Pétain enjoyed a considerable reputation as a master of the defensive: he had been one of the first to recognise the full implications of firepower. He was also a very able commander, meticulous in his preparations for action, with a real concern for his troops, who therefore trusted him. He had shown his mettle in attack at Vimy Ridge and, less successfully, in the Champagne. Austere, aloof, distrustful of politicians, he was a dedicated soldier.

De Castelnau, by insisting on the defence of the right bank, had accepted the challenge of the German assault on Verdun: without knowing it, he was acting as Falkenhayn wished and expected, allowing the French army to be drawn inexorably into the Verdun salient, to be bled there by the remorseless German pressure. And in appointing Pétain, de Castelnau was ensuring that his decision to defend Verdun would be carried out.

The French, in theory, need not have taken up the challenge. They could have withdrawn to the left bank, eliminating the dangerous salient and shortening their line considerably. Verdun and its ring of forts were no longer of practical defensive value: their guns had by now mostly been withdrawn to augment the French artillery barrages for Joffre's hopeful offensives, and the forts stood there imposing and toothless. In the hilly country behind the Meuse, the French could have conducted a fighting withdrawal with advantage, exhausting the attacking Germans and overstraining their lines of supply. As Winston Churchill wrote at the time: 'Meeting an artillery attack is like catching a cricket ball. Shock is dissipated by drawing back the hands. A little "give", a little suppleness and the violence of the impact is vastly reduced.' Yet de Castelnau's decision to hold was based not only on his own aggressive spirit, but on an intuition of a greater truth: that perhaps, after 18 months without success, the French army was not capable of a controlled, fighting retreat in the face of superior forces. Nor, as Briand was aware, would the French nation or his government have easily survived the shock of abandoning a fortress with the immense symbolic value which Verdun held for the French, a factor of which Falkenhayn in laying his plans was conscious.

Brutal shell-fire

On the ridges before Verdun, whose great forests were being rapidly torn down by the brutal weight of shell-fire, there was still some French resistance on the left, facing *VII Reserve Corps*. But elsewhere in the salient on the morning of February 25 — while de Castelnau was appraising the terrain with his quick, experienced eye — the Germans had before them only the forts which ringed Verdun, sinister guardian monsters which the Germans thought were still fully manned and armed. Fort Douaumont was the first that lay in their path, and the strongest. From every angle its great tortoise hump stood out, stark, menacing, fascinating. It was, in fact, occupied by only a small detachment of territorial gunners, manning the 155-mm turret gun which had survived Joffre's purge; its approaches were inadequately covered by the French infantry. Small bodies of Brandenburgers, heroically defying a powerful garrison existing only in their imagination, managed to edge into the fort unobserved during the afternoon of the 25th and to capture its elderly occupants.

The reaction to the fall of Fort Douaumont was electric. In Germany church bells rang; a newspaper declared, 'Victory of Verdun. The Collapse of France'. In Verdun itself morale went to pieces. An officer ran through the streets crying *'Sauve qui peut'*, and civilians poured out of the city, jamming the vital roads. Worse still, the commander of the 37th African Division, on hearing the news, blew up the unthreatened Meuse bridge at Bras and withdrew his battered troops, quite needlessly, back to the last ridge before the city itself.

Pétain took over command at midnight and set up his headquarters in the village of Souilly, on the main road into Verdun from the south-west. His journey in the rigours of winter weather had been long and tiring, and the last stage, through the distracted rear of the French army in defeat, harrowing. He received his instructions briefly from de Castelnau, who then left. Pétain contacted his corps' commanders by telephone, and spent the night huddled

Risking the shell-fire, French stretcher bearers attempt to reach a fallen comrade

Südd Verlag

in an armchair in an unheated room. Next morning he had contracted double pneumonia; but the secret was strictly guarded – with reason, for the mere knowledge of his arrival, that he was now in command at Verdun, had stiffened the French resistance and given new heart to every man, from General Balfourier down to the exhausted troops holding out in the front line, so high was Pétain's reputation in the army.

In the defence lines themselves the turning-point came at the same time with the arrival of the main body of Balfourier's seasoned XX Corps, the 'Iron Corps'. All troops in the salient were now regrouped under four corps' commanders, responsible to Pétain and his Second Army staff. I Corps was coming in from reserve, and others were on their way. Haig too had at last agreed to relieve the French Tenth Army, which was sandwiched between the British First and Third in the area of Arras.

Pétain, on his sick-bed, took up the threads of command with remarkable vigour. He realised that the loss of Fort Douaumont did not mean final disaster, that the other forts remained, and could be manned and linked into a formidable defence perimeter. He issued orders accordingly: from this line there must be no withdrawal. With his profound understanding of the defensive power which guns could provide, he now had the artillery round Verdun reorganised into one concentrated and effective weapon. From this moment, says the German official history, 'began the flanking fire on the ravines and roads north of Douaumont that was to cause us such severe casualties'.

It was the precarious supply route into Verdun which was Pétain's next concern. The full-gauge railway had been rendered useless by the German long-range shelling. The only substitute was motor transport on the one road in, the narrow second-class road from Bar-le-Duc, just 21 feet wide. Pétain and his able transport officer, Major Richard, threw all their energies and organising ability into making this road a life-line for Verdun. Richard went out requisitioning lorries throughout France, and gangs of Territorials were detailed to devote themselves entirely to the upkeep of the road. It was their tireless labour which kept the road intact and the lorries moving when a sudden thaw on February 28, following the long spell of hard weather, turned the surface to a sea of mud. The road was reserved strictly for the ceaseless flow of motor traffic in both directions. Any lorry that broke down was heaved aside into the ditch. With their solid tyres and top-heavy weight, many skidded on the bad surface; others caught fire. Drivers were inexperienced, and always fatigued.

At night, the procession of dimly lit vehicles resembled 'the folds of some gigantic and luminous serpent which never stopped and never ended'. It is astonishing that the Germans never thought of bombing the road, which would so easily have been blocked: they had the aeroplanes to do it – C-class aircraft – and Zeppelins. Eventually vehicles were passing along the road at a daily average of one every 14 seconds, and for hours on end at a rate of one every five seconds; and the equivalent of more than a division of soldiers was kept mending it. The road was the artery through which the life-blood of France was pumped into Verdun. Maurice Barrès gave it the immortal name of the *Voie Sacrée*.

The tide, for the moment, had turned. By February 28, as the German official history says, 'the German attack on Verdun was virtually brought to a standstill'. The French now were fresher and more determined. Their original divisions had been replaced and the numbers greatly reinforced. The German formations, on the other hand, had not been relieved at all and their troops were feeling the strain of a week's intensive fighting: promised reserves were not forthcoming.

Behind this shifting balance between attacker and defender can be seen a change in the balance of artillery; and it was the artillery's positioning and fire-power which were to rule the battlefield – no longer as a prelude, an opening barrage to clear the path, but as the constant and most demanding element of the battle. Now the German guns were flagging: they were having extreme difficulty in moving forward over the ground which they themselves had so violently cratered; and the sudden thaw turned the clay to deep mud. Worse still, the French had increased their heavy guns in the salient from 164 to more than 500, and they were shelling the German infantry with continuous and effective flanking fire from the left bank of the Meuse, in particular from the Bois Bourrus ridge.

The left bank had therefore become an immediate problem to the Germans, and pressure to open an attack on it intensified in the Crown Prince's *Fifth Army*. The commanding height in this western sector, le Mort Homme, was the obvious objective; for its capture would also neutralise the Bois Bourrus ridge, behind which much of Pétain's artillery was dug in. As late as February 26, with the right bank apparently within the Germans' grasp, Falkenhayn had again refused permission for an attack

on the left bank; but the next few days changed his mind, and he let Knobelsdorf have his agreement on February 29, at the same time sending up reinforcements behind *VI Reserve Corps* holding the left bank trenches – *X Reserve Corps*, two further divisions and 21 batteries of heavies.

Prince Rupprecht of Bavaria, one of the most responsible of German commanders, noted at his headquarters in northern France: 'I hear that at Verdun the Left Bank of the Meuse is to be attacked now, too. It should have been done at once; now the moment of surprise is lost.' These German reserves – quite apart from those held immobilised opposite the currently inactive British front in the north – would have ensured a German breakthrough on the right bank to Verdun itself.

Before the new German offensive could be launched, the increasing weight of French guns on the left bank was already taking a terrible toll. Many of these guns were old 155's, firing visually into the German ranks; and *VII Reserve Corps*, which had done so well initially on the Meuse flank, suffered prohibitive losses. The German wounded streaming back were 'like a vision in hell'. Franz Marc, the painter, wrote in a letter from the Verdun front on March 3: 'For days I have seen nothing but the most terrible things a human mind can depict.' He was killed next day by a French shell.

The French at this time were slowly wresting back air superiority over the battlefield. The dramatic step had been taken of assembling some 60 of the French air aces, including Brocard, Nungesser, Navarre and Guynemer, and banding them together into the famous *Groupe des Cigognes* (the Storks). Altogether the French brought in 120 aeroplanes over

Carrying their few bits and pieces, French prisoners move away from the Verdun front

Bibliothek für Zeitgeshichte

Verdun, to set against the Germans' 168 planes, 14 observation balloons and four Zeppelins. Poor German tactics and French verve tipped the scales, and in the next two months, while the German balloons were shot down in flames, French observation planes were to contribute greatly to the success of their artillery, who once more had eyes. The German aces, Boelcke and Immelmann, failed to upset the French supremacy, which the new Nieuport plane later helped to consolidate.

It was clear to the French that the Germans would attack on the left bank now; but the first days of March slipped by, and the French were given time to assemble four divisions on that side and one in reserve, under General Bazelaire, the French VII Corps commander. As the German attack still failed to materialise, Pétain was heard to remark, 'They don't know their business.' All the same, when it came, on March 6, starting with a bombardment comparable to that of February 21, the Germans at first had considerable success. They took the Meuse villages of Forges and Regnéville and advanced towards the ridge of the Mort Homme on its north-eastern slope; but their attack on the northern side was held by a wall of fire from the French guns.

The French 67th Division, defending the north-eastern approach to the Mort Homme had given ground too readily following the terror of the German bombardment, and over 3,000 of its men had surrendered by the end of the second day. It took a brilliant bayonet charge by a superbly disciplined regiment at dawn on March 8 to hold the Germans on this flank of the hill and force them to delay the final assault. In fact, the front now established on this north-eastern approach barely

shifted for the next month.

The Germans had planned a simultaneous attack on the right bank of the river, with Fort Vaux as their objective. The difficulties of bringing up ammunition to the guns over the bogged and cratered ground delayed them for two days. They then blundered in thinking the fort had been taken and marched up in column of fours straight into the French machine gun fire.

The 'mincing machine'

The terrible frontal assault on the Mort Homme from the north began on March 14. With German reserves flowing in more freely, there seemed no limit to the men and shells they were willing to expend to gain possession of this desolate hill. Attack and defence were evenly matched and the fighting swayed forward and back, until a deadly pattern of combat was established. It continued in this sector for the next two months. There would first be several hours of German bombardment; then their assault troops would surge forward to carry what remained of the French front line, which was no longer trenches but clusters of shell-holes. When the German attack had exhausted itself, ground down by the barrages from the French guns on the Bois Bourrus ridge to the south-east, the French would counterattack within 24 hours and drive the surviving Germans back again. But each flow and ebb of the tide brought the German high water mark a little further forward. The cost was terrible: by the end of March, 81,607 Germans had been killed and 89,000 French; and in this compressed and shelterless battle area a high proportion of the casualties were senior commanding officers.

Here, on the bare slopes of the Mort Homme, there were no woods or ravines to

favour infiltration, so that the Germans had lost the advantage they had gained elsewhere by their expert use of these tactics. Their flame-throwers had become suicide weapons, for the pioneers were easy targets in open country; and the French flanking fire was even more crippling on these naked hill-sides.

The Germans, in attacking the Mort Homme with the object of eliminating the threat to their exposed Meuse flank, now found their new right flank on the Mort Homme itself a target for crippling French shell-fire—this time from the neighbouring hill on the west side, Côte 304. So this too would have to be attacked and taken. The German army, even more reluctantly than the French, was being sucked into this ever expanding battle in the Verdun salient. The German Command was still not facing the full implications of its extended attack. Replacements were not being provided: the divisions were being kept too long in the line, and the gaps in the ranks merely filled with unseasoned youths. The French, in contrast, accepting the German challenge in its entirety, were committed to a scheme of rapid rotation of units at Verdun. Pétain, facing this same problem of exhaustion, was insisting, through his *Noria* system of reliefs, that no division remained more than a few days under fire. As a result, two-thirds of the French army was to be fed through the 'mincing machine' of Verdun, and its reserves drained. This was what Falkenhayn had hoped and expected—but he had not anticipated that the German army would also be 'bled white'.

A key position at the base of Côte 304 fell to the Germans on March 20, after French deserters had given them information on the defences. But this success brought no further advantage, only staggering losses under the French machine gun fire. Signs of exhaustion and unwillingness to attack were increasing in the German ranks.

The German *Fifth Army* command structure was now simplified in preparation for a final assault. Mudra, a corps commander from the Argonne front, was put in command of the right bank, and Gallwitz, a gunner, fresh from successes in the Balkans, in command of the left bank. On April 9 a massive simultaneous attack was mounted on Côte 304 and the Mort Homme. The Germans reached only a secondary crest of the Mort Homme; and on Côte 304 the French guns kept up their relentless, devastating fire on the exposed German flank. Pétain, recognising that an extremely dangerous attack had been held, issued next day his famous Order No 94, which he ended with the exhortation: *'Courage! On les aura!'* ('We'll get them!').

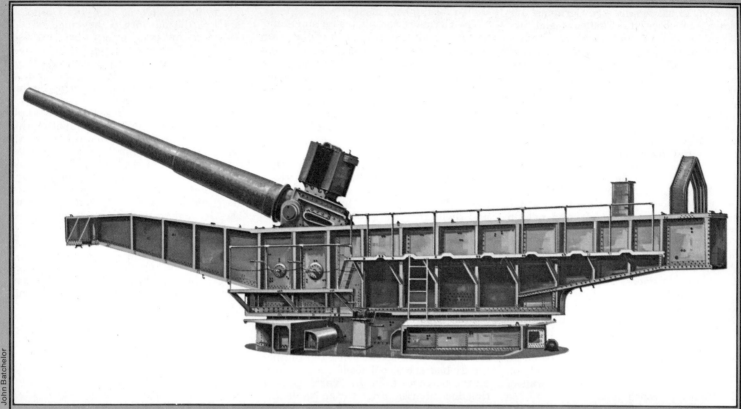

John Batchelor

The big guns—rivals at Verdun

Above: One of the two 38-cm long-range (29 miles) naval guns earmarked by the Germans to interrupt communications behind Verdun.
Below: A French 370-mm howitzer, part of France's inadequate defences

Imperial War Museum

THE SOMME BARRAGE

British and French tactics for the Somme offensive hinged upon a massive and prolonged artillery barrage. Despite recent examples to the contrary, it was hoped that an intense seven-day bombardment would destroy all German resistance and leave the way open for easy infantry penetration. Accordingly, the British and French marshalled over 1,300 heavy guns and more than 2,000,000 shells, and began to pound the German line. By the seventh day most of the shells had been fired with an effect so stunning that nearly all present were convinced that there could be nobody left alive in the German trenches. *Major-General Anthony Farrar-Hockley.*

The Somme barrage gets under way — British howitzers in action

So much tragedy, so much emotion now pervades our view of the First World War that we tend to overlook the perspective of the men involved, not least of those engaged in the direction of the war. For example, at the end of 1915 the British cabinet was appalled that they were not a whit nearer victory despite a year of hard fighting and a casualty list exceeding 500,000. But whenever they talked of an armistice, their debate was frustrated by an intolerable fact: Germany, the aggressor, the invader, occupied almost all Belgium and a rich tract of France. So, on December 28, the cabinet approved a paper of intention prepared by its war committee, which included this paragraph:

Every effort is to be made for carrying out the offensive operations next spring in the main theatre of war (France and Flanders) in close co-operation with the Allies, and in the greatest possible strength.

Yet within a week their apprehensions of mass slaughter persuaded them to hesitate. At the end of this paragraph, the following was added:

although it must not be assumed that such operations are finally decided on.

Weekly – sometimes daily – until April 1916, Mr Asquith and his cabinet colleagues, severally and collectively, discussed ways and means of defeating the Germans without heavy loss; all the while agreeing jointly with the French that 'we have to destroy the morale of the German Army and nation'. At length, with the crocuses bursting through the grass in St James's Park, with all prospect of a spring offensive past, the decision had to be taken. On April 7, the Chief of the Imperial General Staff, Sir William Robertson, was able to telegraph to General Haig in France that the government approved a combined offensive with the French in the summer.

The planning staffs of the Allied commanders-in-chief in France had not of course been waiting on this authority. Since February, Joffre and Haig had struck a form of bargain: they would

△ A British shell explodes in the German
trenches prior to the attack on La Boiselle,
half way along Fourth Army's front
◁ The lanyard is pulled, and one of
the 1,732,873 shells expended by the Allies
in the preliminary bombardment is fired from
a British 8-inch Mk V howitzer

jointly undertake a major assault on the German defences on either side of the River Somme. They do not appear to have been attracted to this sector by any special strategic prize. It was more a matter of convenience: along the river the British and French armies joined together. When the bargain was made, Joffre believed that a common action round this junction point would make it easier for him to retain control of the timing and direction of the offensive. Haig's agreement was qualified by private reservations. While General Rawlinson and the headquarters of the newly created Fourth Army studied the Somme lines, preparations for offensive operations in other British sectors continued. Haig's preference was for a summer campaign in Flanders; but he was not anxious to be drawn into the battles of attrition which Joffre believed to be necessary. But while the planners of both nations were still busy at their maps and sums, their chiefs were reminded that the initiative did not lie with them alone. At 0430 hours in the morning of February 21, 1916, the Germans fired the first shot of their opening barrage at Verdun. After drenching the

French defences with high explosive for eight hours, the German and French infantry locked in a dreadful struggle.

The events at Verdun soon put the British and French at variance in the matter of timing for their summer offensive. As more and more French divisions were drawn in to bolster the crumbling French line across the Meuse – from which they emerged exhausted and depleted after a period of days – it was only reasonable from the French point of view that the British should be asked to hasten the date of the Allied offensive. A little after 11 o'clock on the morning of Friday, May 26, Joffre came to Haig's headquarters to ask directly for an opening date. 'The moment I mentioned August 15,' Haig wrote in his diary, 'Joffre at once got very excited and shouted that "The French Army would cease to exist if we did nothing till then". The rest of us looked on at this outburst of excitement, and then I pointed out that, in spite of August 15 being the most favourable date for the British Army to take action, yet in view of what he had said regarding the unfortunate condition of the French Army, I was prepared to com-

Imperial War Museum

mence operations on July 1 or thereabouts. This calmed the old man, but I saw he had come to the meeting prepared to combat a refusal on my part, and was prepared to be very nasty.'

In the previous winter, the French offensive had been conceived as a massive venture employing 40 divisions from the Somme southward to Lassigny. It was hoped that the British would employ 25 divisions to the north of the river. Now on the last day of May, the French President, Prime Minister and Minister for War came from Paris to see Haig with the brief that 'one can and one must foresee the eventuality of the British army conducting the offensive'. The change of view had already been recognised by Haig. Recounting the events of the meeting in his diary, he remarked: 'The slow output of French heavy guns was pointed out and the need for supplying Verdun with everything necessary was recognised, and I said that *in view of the possibility of the British having to attack alone,* it was most desirable to bring to France the Divisions which the Allies held at Salonika. Beat the Germans here and we can make what terms we like!'

Haig was now in difficulty. Having agreed to attack on the Somme so as to combine directly with the French it was now clear that France might not participate. There was a British plan for an independent attack in the sector but it showed little promise. Flanders beckoned Haig strongly still.

'D' day agreed

Joffre was aware from the reports of his liaison staff at the British headquarters that Haig was not enthusiastic for the Somme project. He feared that, lacking a French presence, the British might withdraw altogether from offensive action in 1916. The Russians had achieved nothing in their assaults round Lake Narotch and were asking urgently for heavy ordnance to help them in a new operation. The Italians were calling similarly for help after their defeat in the Trentino. The needs of these allies might offer a ready excuse for British inactivity in France. But Joffre misconstrued Haig's reluctance; the question in the latter's mind was not 'whether?' but 'where?' When Joffre wrote blandly on June 3 to ask what notice Haig required to open the offensive on July 1, he obtained a straight answer—'12 days' notice for an attack on July 1 or later,' and a direct question, 'how many French divisions will be taking part?' Since Joffre's staff had been quite unable to find 20 divisions, the reply was vague: General Fayolle's Sixth Army would attack astride the Somme 'to support the British'. Now they had it. Fayolle had but 12 divisions and his assault frontage was limited to about eight miles. This contribution, necessarily diminished by the demands of Verdun, was but a fragment of the original scheme; but it was enough to ensure that the British would fight also on the Somme that summer.

The trenches of the British Fourth Army ran for 20 miles from the village of Hébuterne south across the Ancre to the Willow Stream by Fricourt, then east to the Bois de Maricourt. The wood was held by the French, Maricourt village just behind was shared by British and French companies. Two miles to the south lay the left bank of the Somme. The country here, both north and south of the river, is downland, open

grazing above rolling chalk hills with a scattering of woods. It had been a quiet sector. The woods and villages were mostly unbroken, the trees in full leaf; patches of the grassland were bright with larkspur.

Opposite both Rawlinson's and Fayolle's armies lay the *Second Army* of General Fritz von Below, comprised of three corps and, depleted by the removal of forces for the attack on Verdun, lacking any appreciable army reserve. In May, Below and his Chief-of-Staff, Grünert, became convinced that they were about to be attacked. From about Hébuterne to the Ancre, Below's regiments were on ground generally overlooked by the British though often hidden by woods and re-entrants. The Serre knoll was an important exception, offering an excellent view north-west to Hébuterne, west to the copses amongst the British trenches and south along the road to Beaucourt. Below the Ancre, the Germans held all the high ground to the valley of the Somme and south again across the Flaucourt plateau. Most importantly, the first line of German trenches had been cunningly dug amongst the many outcrops of the main ridge line, so as to command every approach from the west.

Aware of their enemy's advantages in observation, the British and French took care during their preparations, confining much of their movement and construction work to the night. They used wire netting, hessian and paint in camouflage successfully by day but had not the means or as yet the skill to deny all their activity to the air observers in aircraft or balloons. The very fact that the allied flying corps concentrated progressively to deny access to German aircraft and intensified their attacks on balloons confirmed that something of importance was happening to the west. The sightings of new roads, railways and hutments in rear, the glimpses of new battery positions, were more than enough to suggest what all this presaged.

While preparing to open the onslaught at Verdun, Falkenhayn had warned all army commanders that they must expect Allied relief attacks elsewhere. Since he would be unable to send reinforcements to any other point along the German line, he advised—and they would have needed very good reasons for ignoring his 'advice'—that each regimental front should be held in strength as far forward as possible. They should not surrender a metre of ground. If a section of trench should be lost it must be retaken immediately. The officers and men forward must understand that the order was to defend their position to the last man, to the last round; orders which would surely persuade them to fight tenaciously from the outset since they would realise that there was no hope of being told to fall back if pressure became intense. The consequences of this policy were twofold. First, as the majority of each division was within the range of the Allied field artillery, it was necessary to provide them with adequate shelter. Deep shafts were sunk into the chalk below the trenches and galleries were run out on either side, providing safe living quarters in dormitories for the soldiers and separate rooms for the officers. Fresh air was pumped in; stocks of bottled water, biscuits and tinned food were cached against emergencies when fresh rations could not be brought up daily. Secondly, surface strong points were provided a little in rear of each front trench to house

medium machine guns and many of these positions were constructed in reinforced concrete by the end of June. In certain areas, the strong points were connected by tunnels to form enclosed fortresses. Throughout the second line of defence—on average 4,000 yards behind the first—there were deep dugouts and a number of deep shafts and galleries. When Below warned Falkenhayn of his expectations, he was sent an additional detachment of 8-inch howitzers, captured from the Russians, and a reinforcement of labour units to speed the development of the third line, five miles behind the first. Between mid-March and the beginning of June, the German wire was consistently thickened in front of the forward defences.

Despite the progress made in preparations against enemy attack, Below and Grünert remained anxious. In each divisional area, almost all the infantry were committed to the first line; even the corps reserves had been tapped to man trenches at critical points between divisions. On May 25, Grünert proposed to Supreme Headquarters that they should make a preventive attack on the British front. He asked tentatively for more infantry. With the battle at Verdun at its height, there were none, though Falkenhayn made no objection to a local venture within their own resources. As encouragement, he sent two more batteries of artillery. By June 6, Below was more sanguine about the British threat but still very anxious about the French below the Somme. He reported that: *The preparations of the British in the area Serre-Gommecourt, as well as the increase of 29 emplacements of artillery in the past few days, detected by air photographs, lead to the conclusion that the enemy thinks first and foremost of attacking the projecting angles of Fricourt and Gommecourt. In view of the ground and the lie of the trenches, it is quite conceivable that he will attempt only to pin the front of 26th Reserve Division by artillery fire (that is, the ground immediately north and south of the Ancre valley), but he will not make a serious attack. To oppose (the French south of the Somme), XVII Corps is too weak, both in infantry and guns. Even against an enemy attack on a narrow front made only as a diversion the Guard Corps (on the extreme left of his line) is also too thin; it is holding 36 kilometres with 12 regiments, and behind it there are no reserves of any kind.*

It was on this day, June 6, that Haig accepted that he was committed to the Somme offensive. Perhaps due to the uncertainty, he had not given Rawlinson precise orders during the long preliminaries of reconnaissance and preparation, an omission which had inhibited Fourth Army commander. Yet, whatever his task was to be, Rawlinson was very clear by June that the enemy defences were formidable; any attempt to breach the two main lines would be costly. Imaginative and shrewd, he was also an ambitious man. In April, it is probable that he hoped to cover himself against failure by suggesting that: *It does not appear to me that the gain of 2 or 3 more kilometres of ground is of much consequence, or that the existing situation is so urgent to demand that we should incur very heavy losses in order to draw a large number of German reserves against this portion of our front. Our object rather seems to be to kill as many Germans as possible with the least loss to ourselves, and the best way to do*

The date is agreed. Early on Sunday morning the clouds cleared. It was Z day, Zero hour

Top: The British 8-inch Mk VI howitzer, the power behind the Somme barrage. *Range:* 10,000 yards. *Weight in action:* 9 tons 9 cwt

Centre: The British 6-inch 26 cwt howitzer. Designed in January 1915 to replace the 6-inch 30 cwt howitzer, this piece was coming into use by the time of the Somme. *Range:* 10,000 yards. *Weight in action:* 4 tons 6 cwt

Below: The guns at work — a painting by Hamlyn Reid entitled *Barrage on the Somme*

John Batchelor

Imperial War Museum

this appears to me to be to seize points of tactical importance which will provide us with good observation and which we may feel certain the Germans will counter-attack under disadvantages likely to conduce heavy losses.

Haig rebuked him for this limited view but it was June 16 before the strategic concept was finally made clear, when the principal objectives were issued to Fourth Army and subsidiaries to First, Second and Third, acting in support. In brief, Rawlinson was to seize the main feature in front of him between the Ancre and Montauban. If all went well, the cavalry under Gough should exploit any full breach of the enemy defence lines by passing through into the open country in the enemy rear while the infantry attack swung north towards Arras up the line of the enemy trenches. Alternatively, Haig's orders continued, 'after gaining our first objective as described we may find that a further advance eastwards is not advisable. In that case, the most profitable course will probably be to transfer our main efforts rapidly to another portion of the British front but leaving a sufficient force on the Fourth Army front to secure the ground gained, to compel the enemy to use up all his reserves and to prevent him from withdrawing them elsewhere.'

Between January and June, 19 divisions were sent to France to join the British Expeditionary Force. Eleven had seen active service before—some were Gallipoli veterans—the remainder were from the New Armies formed by Kitchener. By May, Rawlinson had 16 divisions in his army, including a share of those recently arrived, and had allocated three to each of his five corps headquarters. His plan envisaged an assault by two divisions in each corps with the third ready to exploit or reinforce. In retrospect, there is an ominous similarity in the orders issued within each formation for the first assault, even though the ground and the enemy in front of each division was markedly different. But it is easy to criticise, to forget that an operation was developing on a scale never attempted before by the British army. Its professional commanders still had much to learn in the handling of vast numbers of men; their communications were inadequate for the type of operation envisaged; many of the subordinate commanders and staffs were amateurs, the majority of the regimental officers and men raw. None knew this better than Haig. It may sometimes be said of him that he was overconfident in himself but he was seriously concerned to keep matters simple for the assault forces, without inhibiting the enterprise of the junior commanders. Stage by stage he discussed the tactics to be employed with Rawlinson, and with Allenby whose right flank corps was to attack Gommecourt as Rawlinson's men went forward.

Along the whole front from mid-June onwards there were to be local attacks and raids: no hint was to be given that one sector was more important than another. A barrage of some days' duration was to be fired by the Royal Artillery with the aim of breaking the enemy wire and destroying the other defences; eroding the strength of the German troops in the first line; reducing to as great an extent as possible their guns, howitzers and mortars by a great weight of counter battery fire; and striking the enemy's routes forward into the battle area. Because telephone lines were so often

Loading one of the 305-mm guns

cut, all British artillery fire for the assault was to be controlled by time. At a given time, zero hour, the guns would lift to let the infantry cross No-Man's Land into the first of the enemy trenches. After a time judged to be adequate to complete the subjection of these, the fire would lift again to permit a second advance. Provided the Germans knew and obeyed the rules of this theory, the general plan must succeed.

In the French sector, General Fayolle had borne with good humour the many visits paid to him. The British called periodically and a valuable friendship developed with Rawlinson and his headquarters. Amongst the French, Foch appeared often. As Commander of the northern group of armies it was his task to distil practical directions from the high sounding but vague orders issued by Joffre's staff for the offensive. Foch wished to delay the attack of Fayolle's army for at least several days after the British attack, believing that the French divisions would gain the advantage of surprise. Understandably, Rawlinson did not agree; one of Fayolle's corps, XX, was directly on his right on the north bank of the river. Any delay by the French here would uncover his flank. Foch continued to argue doggedly for his proposal and won eventually the concession that the I Colonial and XXXV Corps south of the river might attack a little after zero but on the same morning. The objective of XX Corps was Hem, where they expected to breach the enemy first line. The other two were directed to pass through both the first and second lines to capture the Flaucourt plateau. To support this attack, the French artillery was to fire for eight days beforehand, 117 heavy batteries supplementing the many field and medium guns.

The British artillery support for the offensive was unique. The preliminary bombardment was to be fired over five days, U, V, W, X and Y. On U and V days the gunners were to register and cut enemy barbed wire; on W, X and Y the shells were to rain on enemy defences while wire-cutting continued. Counter-battery fire would feature on every day. There would be checks and pauses to deceive the enemy, to persuade them that an assault had begun so that they would hasten up the steep ladders from their shelters to man the fire

strips only to be raked with shellfire. When these stratagems no longer tricked the foe, gas would be fired to persuade them that an assault had begun at last. The programme for Z day, the day of assault, was so arranged that the usual 80-minute bombardment at that time should be fired for only 65 minutes. It was hoped that the enemy would be sufficiently used to the former to stay below while the British infantry crossed.

This mass of shells and mortar bombs, high explosive, shrapnel and phosphorus gas was to fall upon a frontage of attack of 25,000 yards. 1,010 British field guns and howitzers, 427 heavies and 100 French pieces fired for Rawlinson's army. 2,000,000 shells and bombs lay stacked ready for their use.

Vague orders
As all the preparations went forward, there were a series of final crises among the Allied chiefs as to Z day. On June 12 the Germans renewed the assault at Verdun and Clemenceau sought to bring down Briand's government in Paris. Joffre urged Haig to bring forward the opening date to June 25. Haig agreed. On the evening of the 16th came a telephone call to say that Verdun was secure and Briand's government had won a vote of confidence. June 29, even July 1, would now be preferable, Joffre suggested, for Z day. After some demur, the 29th was accepted at GHQ. At 0700 hours on June 24, therefore, the first of the heavy guns was loaded and the gun position officer cried 'Fire'. The bombardment had begun.

Sunday the 25th was a fine warm day in contrast to two preceding days of summer storms. The Royal Flying Corps scouted or spotted for the guns. They found 102 enemy batteries and began early the counter-bombardment. But this was the last day of good weather for some time. Morning mist, low cloud and bursts of heavy rainfall hampered observation on the ground and in the air. Numbers of dud shells or fuses delayed still more the gunners' work. The high rate of fire wore seriously the buffer springs and smoothed those gun barrels which had already once been relined. On Wednesday, it was decided to go on firing until the 1st and assault on that day. Forward, the infantry trenches were manned by skeleton forces while the bulk of each battalion rested in such billets as could be found immediately in rear. On the Friday night, June 30, commanding officers consulted with their brigade commanders and there were some disquieting reports that long stretches of enemy wire remained uncut. Still the boom and bang of the bombardment, faithfully maintained by the weary gunners after a week, raised many hopes.

Early on the Saturday morning, July 1, the clouds cleared. Breakfast was eaten in the dark, kit was packed and dumped. Company guides began to lead their fellows forward across the dark wet ground. The German lines were quiet as the soldiers marched heavily into the foremost trenches, weighed down with full marching order: rifle and ammunition, grenades or bombs, a digging tool and perhaps some other special load. Below ground, the miners waited, sweating in the heat of the narrow tunnels leading to the mine chambers they had cut under the enemy positions.

The warm sun rose dispelling the mist. It was Z day. Zero hour approached.

For a third week, the gunners laboured . . .

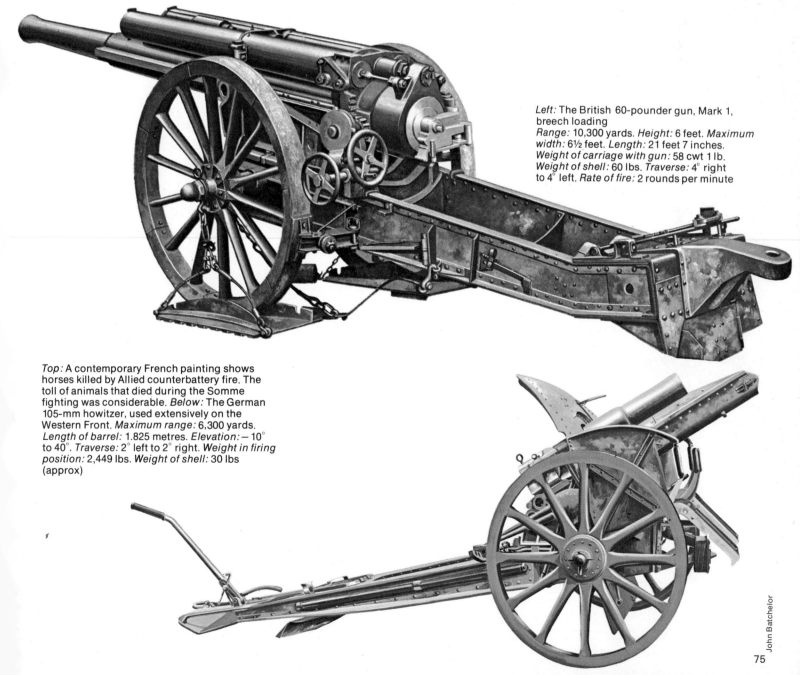

Left: The British 60-pounder gun, Mark 1, breech loading
Range: 10,300 yards. *Height:* 6 feet. *Maximum width:* 6½ feet. *Length:* 21 feet 7 inches. *Weight of carriage with gun:* 58 cwt 1 lb. *Weight of shell:* 60 lbs. *Traverse:* 4° right to 4° left. *Rate of fire:* 2 rounds per minute

Top: A contemporary French painting shows horses killed by Allied counterbattery fire. The toll of animals that died during the Somme fighting was considerable. *Below:* The German 105-mm howitzer, used extensively on the Western Front. *Maximum range:* 6,300 yards. *Length of barrel:* 1.825 metres. *Elevation:* — 10° to 40°. *Traverse:* 2° left to 2° right. *Weight in firing position:* 2,449 lbs. *Weight of shell:* 30 lbs (approx)

The weapons — large and small — of the attack 'organised for amateurs by amateurs'

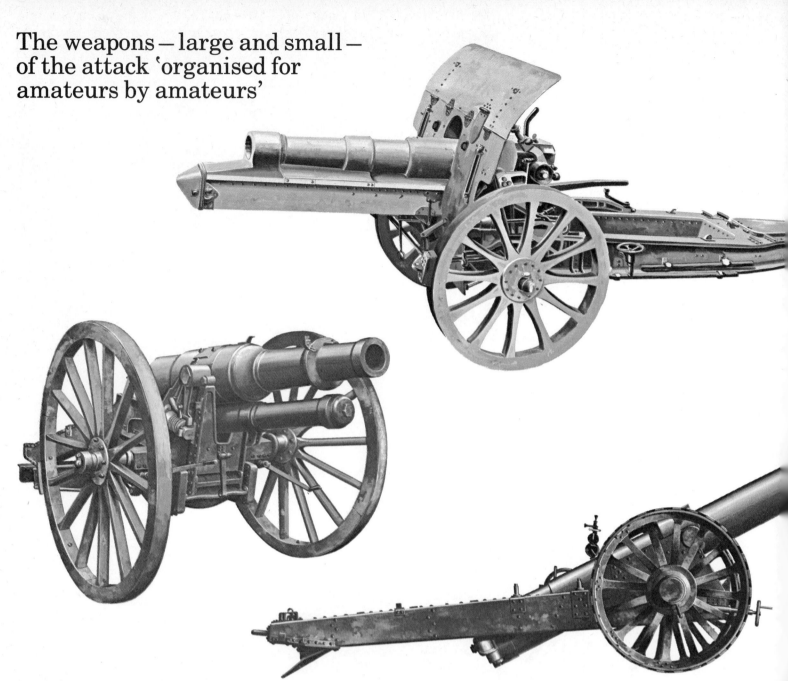

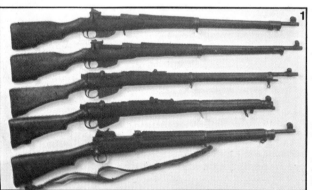

1 *Top to bottom:* Canadian Ross rifle Mark III .303 calibre; Canadian Ross rifle Mark II .303; .303 long Lee Enfield rifle Mark I; .303 rifle No 1 with short magazine; .303 pattern 14 (P14) rifle. The Ross rifles have been the cause of considerable controversy. Issued by the Canadian government to their troops, they often proved troublesome in muddy conditions and were frequently discarded in favour of the SMLE. However, this rifle had a large following and was used extensively for sniping purposes. 2 *Top:* .303 P14 rifle fitted with telescopic sight. Rifle .303−inch, No 1 Mark III, Short Magazine Lee Enfield with telescopic sight. 3 .303 Vickers machine gun Mark 1. Originally called the Vickers Maxim, this gun was adopted by the British in 1912 and has an extremely sound reputation.

Opposite page, top: The German 5·9-inch howitzer. *Elevation:* 0° to 45°. *Traverse:* 4° left to 4° right. *Maximum range:* 9,296 yards. *Weight:* 92 lbs. *Opposite page, left:* The French 120-mm short gun, made in about 1890 but nonetheless in service throughout the war. *Opposite* page, *bottom:* The British Mark 7 naval gun on an artificial mounting which adapted it for use on the battlefield. *Bottom:* The British Mark 1 9.2-inch howitzer. *Maximum range:* 10,000 yards. *Rate of fire:* 2 rounds per minute. *Below:* The latest toll of deaths on the Somme

British 158 786   103 000 German

French 49 859

July 2–14

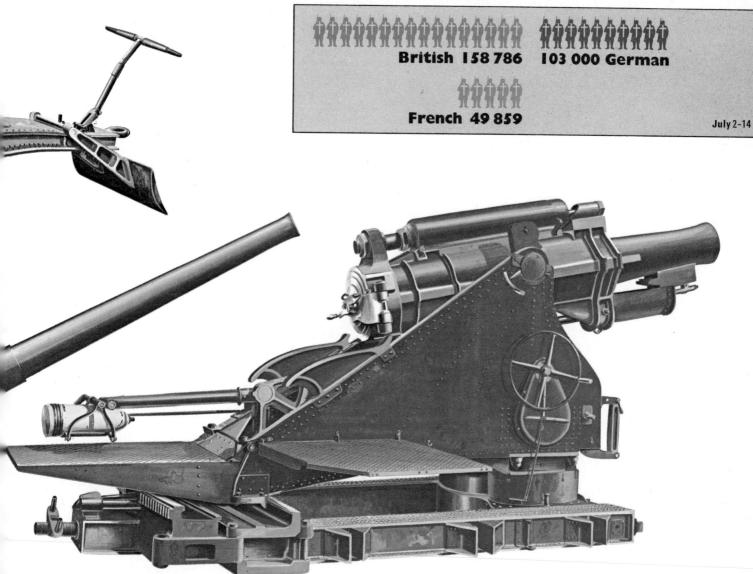

John Batchelor

4 *Top left:* Oval grenade – 'Egg'. *Top right:* Hand grenade, the Mills Bomb. *Centre:* An unidentified percussion grenade. *Bottom:* Grenade No 2, or Tonite Grenade. **5** *Top:* Two flare pistols by Scott and Co of Birmingham. *Left:* Wire cutters to be attached to a rifle muzzle. *Bottom:* One of the more bloodthirsty 'specials' improvised at trench level. **6** *Top to bottom:* Newton rifle grenade; a cheap form of Hales rifle grenade; Hales grenade; hand grenade. **7** The Webley .455 Automatic pistol Mark 1 No 2, with detachable stock and holster – again made by Scott and Co. **8** *Top left:* Webley .455 pistol Mark V. *Centre left:* Webley-Fosbery automatic revolver .455. *Bottom left:* Webley automatic .455 Mark 1 No 2. *Centre:* Webley-Green .455 revolver. *Right:* Webley .455 pistol Mark VI No 1

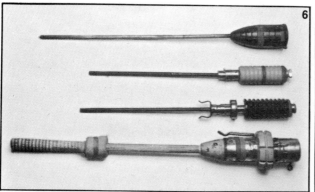

Jack Pia

Amid the reek of putrefaction, the sickly sweet smell of gas

Above: By this time a familiar sight on the Western Front—the explosion of the gas bomb. *Near right:* British machine gunners wearing their gas masks. They worked with the masks pinned to their shirts, ready to be pulled on when the gas bells clanged. *Centre:* Though only partially effective, the gas mask was an essential part of life at the front; here, men of the RGA operate a Fullerphone. *Far right:* A gas sentry ringing a make-shift alarm

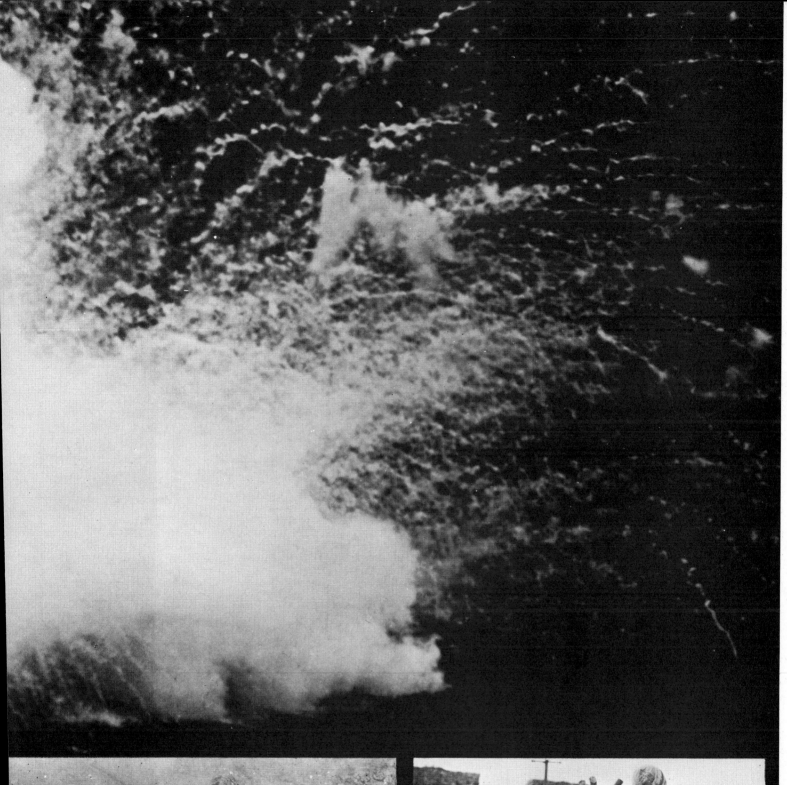

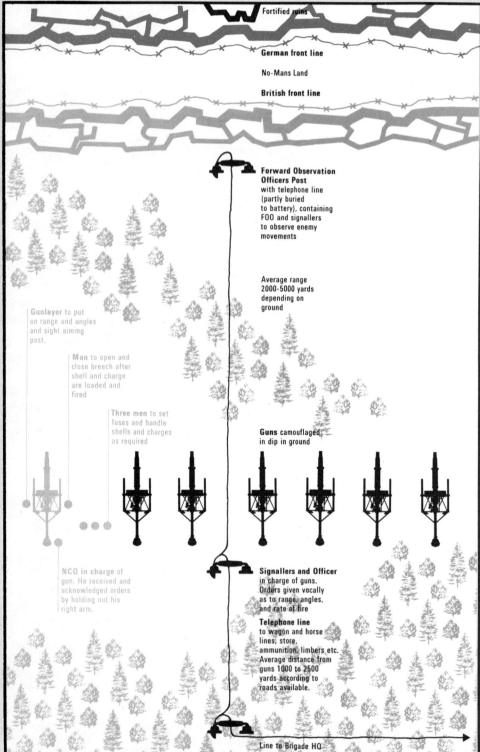

Fortified ruins

German front line

No-Mans Land

British front line

**Forward Observation
Officers Post**
with telephone line
(partly buried
to battery), containing
FOO and signallers
to observe enemy
movements

Average range
2000-5000 yards
depending on
ground

Gunlayer to put
on range and angles
and sight aiming
post.

Man to open and
close breech after
shell and charge
are loaded and
fired

Three men to set
fuses and handle
shells and charges
as required

Guns camouflaged
in dip in ground

NCO in charge of
gun. He received and
acknowledged orders
by holding out his
right arm.

Signallers and Officer
in charge of guns.
Orders given vocally
as to range, angles,
and rate of fire

Telephone line
to wagon and horse
lines, store,
ammunition, limbers etc.
Average distance from
guns 1000 to 2500
yards according to
roads available.

Line to Brigade HQ

The front now had the aura of a decaying suburb . . .

Top: Resourcefulness in adversity. Royal Engineers build a makeshift bridge over the Ancre swamps near Aveluy. Rain had been falling for several weeks, breaking up the ground, and the infantrymen were often wet to the skin even before they went into battle. *Left:* The chart shows the layout and establishment of an 18-pounder battery. *Above:* Exhaustion sets in: British troops sleep in their trenches near Thiepval Ridge. This area was not as badly damaged as that south of the Somme, but rain and cold had seriously debilitated the troops

On July 29, Sir William Robertson, the Chief of the Imperial General Staff in London, wrote to Haig:

'The Powers that be are beginning to get a little uneasy in regard to the situation. The casualties are mounting up and they were wondering whether we are likely to get a proper return for them . . . In general, what is bothering them is the probability that we may soon have to face a bill of 2-300,000 casualties with no very great gains additional to the present. It is thought that the primary object – relief of pressure on Verdun – has to some extent been achieved.'

The relief of pressure on Verdun had never been the primary object of the summer campaign; it had simply been the reason for hastening its opening. On this document, Haig wrote in his own hand, 'Not exactly the letter of a CIGS.'

Next day, August 1, he replied to Robertson in London, setting out the strategic advantages won during the preceding month by the British Expeditionary Force. In his diary on August 3, he summarised these views and his own intentions:

Principle on which we should act. Maintain our offensive. . . . *It is my intention:*

(a) *To maintain a steady pressure on Somme battle.*

(b) *To push my attack strongly whenever and wherever the state of my preparations and the general situation make success sufficiently probable to justify me in doing so, but not otherwise.*

(c) *To secure against counterattack each advantage gained and prepare thoroughly for each advance.*

Proceeding thus, I expect to be able to maintain the offensive well into the autumn.

It would not be justifiable to calculate on the enemy's resistance being completely broken without another campaign next year.

Some of Haig's critics have seized on this

THE SOMME
Debut of the Tank

Flers-Courcelette was Haig's third effort to break through on the Somme. This time he had a new weapon – the tank. Though slow and unreliable, the few tanks available gave the infantry worthwhile support, but all to no avail – the Germans were able to plug the breaches in their line. *Anthony Farrar-Hockley. Above:* D17, which helped to take Flers, on September 21, 1916

last sentence as a true indication of his state of mind: a growing pessimism as a result of the inability of his armies to make a breach in the German line. There is little evidence to suggest that he had lost confidence in the campaign. Just over four weeks had passed since Z-day. He had planned and stocked for a seven-week battle. There was evidence that Falkenhayn was weakening other sectors of the Western Front to reinforce the Somme; if he could judge the time nicely, it might yet prove possible to break through north or Arras in conjunction with a blow above the Somme.

The plans drafted at this time, the attempts to sound out the line to the north – however abortive – do not reflect a picture of a desperate commander but of one pressing forward hopefully in a great endeavour, his chief anxiety being to discern and exploit the opportunities of the battlefield. Certainly he was misled by the erroneous reports of his Chief of Intelligence, Brigadier-General Charteris, and by such subordinate commanders as General Monro, who directed an Australian and a British division in a 'sounding out' attack opposite Lille in July with chilling ineptitude. But there were indisputably masses of prisoners and an abundance of German dead still lay above and below ground between the Ancre and the Somme. The prisoners told of being drawn in from north and south to the Somme sector. Their stories of terror and disaster were corroborated by the quantities of German weapons and equipment captured or seen to be smashed in the portions of the German line now in Allied hands or under view.

Such fears as Haig had were more precise than those of the Cabinet. He was worried – and with reason – by the low standard of staff work: the British army had expanded a hundredfold, and while it had taken in many able and highly intelligent civilians as staff officers, the majority lacked the basic and essential knowledge which could only be gained in a fighting unit. In the fighting units, many of the best and most experienced leaders had been lost by death or wounds during the terrible actions since 1914. How could he ensure that the staffs directed, and the regimental officers and men fought, the battle ahead to gain a positive and advantageous end? The Commander-in-Chief's remedy was a keener scrutiny of his army commanders' plans, a readiness now to intervene at once if they did not conform to his requirements and the continual reminder to every headquarters he visited that they must make 'thorough preparation' for every operation.

One aspect of this preparation was the capture of certain German positions which remained as outworks of defence along the ridge: Thiepval, Mouquet Farm and the high point between Pozières and Martinpuich; High Wood; the cramped front of Delville Wood; Ginchy; Guillemont; Falfemont Farm and Leuze Wood on the long spur above the Combles ravine. A series of local operations was begun to take these positions during August. These operations, limited in scope and success, rarely excite interest or comment now, yet were among the bitterest actions of the war. Deter-

mined not to repeat a frontal attack on Thiepval, Gough attempted a pincer movement by V and II Corps while the ANZAC inched forward from Pozières, clawing into the German defences. High Wood remained impenetrable. The gains in Delville Wood were partially lost to a counterattack and so it became necessary to begin a recovery operation just when it was hoped the men there might be rested. After breathless, bloody scrambles along 'the terrible road into Guillemont, straight, desolate, swept by fire', the ruined village fell at last to the devoted 20th (Light) Division. Ginchy succumbed to an ardent assault by the depleted battalions of the 16th (Irish) Division. Slowly the line swung forward to the edge of Leuze Wood.

Spectacular advance

These engagements occupied August and the first part of September. Meanwhile, the French Sixth Army had made a spectacular advance along the north bank of the Somme, following the capture of Falfemont Farm by the 1st Norfolks on September 5. No longer hampered by the enfilade fire poured out from the site of the farm and the posts dug into the edge of the Combles ravine, no longer under the eyes of the German artillery observers in the cornfield above, the French left flank

moved on with a rush to catch up the divisions advancing through Cléry and Ommiècourt. An attempt by the Colonial Corps to take the southern outskirts of Combles as it passed failed; here the Germans were protected by masonry and steel above ground and found absolute safety from artillery in the ancient caves and galleries below. It was agreed between Foch and Haig that they should envelop the town and cut it off. But as Fayolle began to swing his line of assault to the north-east the weather changed, a series of heavy showers turning the tracks once more into mud. The process of reliefs was delayed, the forward movement of the guns hampered. It was the 12th before the French began to attack again. For a few hours, the assault was indecisive, as scores of small groups of infantry disappeared into the German front trenches where a ferocious struggle seemed to be taking place. Then, before noon, the brigade on the right of the Colonial Corps was seen to be in pursuit south-east of Le Forest, pushing swiftly on to capture the spurs above the Bois Marrières. The local trench lines were now broken open: the remainder of the Colonial Corps advanced north-east round Combles to capture the road to Rancourt. For a little while Foch urged Fayolle to push on to Sailly-Saillisel but the infantry was exhausted by its struggles through the mud and the intense German shell fire. Fayolle had to tell Rawlinson that he would be unable to assault again with the British when the main offensive began on September 15, but promised, as before, the fullest support of his guns. Not that the French had need to make apologies; they had carried the whole Allied line round in a spectacular turning movement. Small wonder that Haig agreed with the ever optimistic Joffre that they might now be on the eve of a great triumph.

Haig's plan for the third major offensive above the Somme envisaged three stages. In the first, the infantry of the Reserve Army and Fourth Army should capture what remained of the original German line along the ridge, their movement forward being helped perhaps by the 'tanks', the new armoured caterpillars of the Heavy Section, Machine Gun Corps. In the second stage, the Reserve Army would hold the left flank of the captured ground, while Rawlinson would dispose a force to defend his own right. The remainder of Fourth Army would complete the breach of the German line somewhere between the Albert-Bapaume road and the village of Flers, and once this was completed a cavalry corps of five divisions would pass through into the country behind, the third stage.

Above: The ruins of the main street of Flers, reduced to rubble but not, as yet, obliterated by the British and German guns. *Below:* Over the top: an officer leads his party out of the comparative safety of a sap into the bullet-swept waste of No-Man's Land and the German lines

In concept, it was the strategic plan of July 1, though now on a narrower front of about 12,000 yards. Of the 14 fresh divisions, eight went to Rawlinson, and six to Gough, with a provisional allotment of 36-42 and 18-24 tanks respectively.

Gough continued his attempts to envelop the rubble of Thiepval through the second half of August into early September. He was very much aware of the formidable defences which he ordered II Corps and I ANZAC to overcome, but kept them at it so as to be able 'to shoulder the enemy off the western end of the ridge when the offensive begins', as he told Birdwood, commander of the ANZAC. As a secondary benefit of his preliminary operations, he hoped to gain space on either side of the Albert-Bapaume road. Shortly, the Canadian Corps would arrive to relieve the Australians and from their narrow territory, perhaps 2,500 yards across to the east of Pozières, he had to launch forces which would prove irresistible. Thus, to their very last day in the line, the Australians were kept at the deadly task of securing Mouquet Farm and the *Fabeck Graben*. The rain which checked Fayolle added similarly to their burdens and, in the end, denied them success. They were not sorry to see the Canadians coming in to take over.

Reliefs were taking place similarly along most of the Fourth Army front. High Wood remained in German hands. The south-east corner of Delville Wood contained three German machine gun posts, a tiny salient connected to the switch trench of Armin's defences by Hop and Ale Alleys, Pint Trench and Lager Lane. Elsewhere, the preliminary operations had cleared the start lines on which Haig had insisted. All formation plans were complete by September 11 and the guns began the bombardment at 0600 hours on the 12th.

A repeated error

Rawlinson's first plan, a deliberate and limited advance on each of three successive nights, had been rejected by Haig. He foresaw the same loss of initiative which, as he now knew, had held back Congreve on July 1. Rawlinson's attempts to justify his proposal on the grounds that night operations would preserve the secret of the tanks were not accepted. GHQ orders on August 31 reminded both Gough and Rawlinson that: *The exploitation of success to the full during the first few hours is essential to a decision and it must be impressed on all Corps and Divisional Commanders that the situation calls for great boldness and determination on their part. It lies with them to feel the pulse of the battle and to turn favourable opportunities at once to the fullest account ... The necessity for great vigour and determination in this attack, and the great results that may be achieved by it, must be impressed on all ranks as soon as considerations of secrecy will permit of their being informed of what is required of them ... The remarks ... apply more particularly to the Fourth Army as regards the earlier operations, during which the Reserve Army, acting as a pivot, will have a more limited scope ...*

Rawlinson's revised plan subscribed to these requirements. While assaulting along his entire front early on Z-day, aided by the tanks, his main effort would be directed towards the capture of Geudecourt, Lesboeufs and Morval before nightfall. These three villages lay beyond the German trench system—a few shallow unwired and unconnected defences apart—and their possession should certainly mean a breach had been opened. We do not know whether Rawlinson believed this operation to be practicable or whether he simply offered it in the belief that it would be more acceptable to his Commander-in-Chief. Some of Rawlinson's comments to friends and subordinates, some of his documents may make us suspect the latter. Why else should he be wondering whether '. . . we can maintain ourselves on this high(!) waterless(!) and roadless plateau after we have captured it'?

On August 27, Rumania entered the war on the Allied side and almost at once began operations against Austria-Hungary. The event persuaded the Kaiser finally to call Hindenburg and Ludendorff from the Eastern Front to give alternative advice on the future conduct of the war. In the aloof style of the third person which he used for his war memoirs, Falkenhayn tells us that he '. . . had to reply [to the Kaiser] that he could only regard this summoning of a subordinate commander, without previous reference to him, for a consultation on a question the solution of which lay in his province alone, as a breach of his authority that he could not accept and as a sign that he no longer possessed the absolute confidence of the Supreme War Lord which was necessary for the continuance of his duties. He, therefore, begged to be relieved of his appointment.' The Kaiser acceded—had perhaps counted on the request for relief—and by a happy arrangement sent his former Chief of the General Staff to command the forces of the Central Powers against Rumania.

The new brooms arrived to see the Kaiser at Pless on the morning of August 29, where General von Lyncker, senior imperial aide-de-campe, informed them that '. . . *Feldmarschall* von Hindenburg had been appointed Chief of the General Staff of the Field Army and that I [Ludendorff] was to be Second Chief. The title "First Quartermaster-General" seemed to me more appropriate ... but in any case, I had been expressly assured that I should have joint responsibility in all decisions and in all measures that might be taken.'

On September 5, the two victors from the Eastern Front visited the front against the West. Due to the continuing onslaught on the Somme, they made for Cambrai, where Crown Prince Rupprecht of Bavaria had recently established an army group headquarters to control all forces engaged in the battle. All army commanders and chiefs-of-staff in the West were present. From the Somme, *Colonel von Lossberg in his serious way, and Colonel Bronsart von Schellendorf with his usual vivacity, supplemented General von Kuhl's report of the battle ... The loss of ground up to date appeared to me of little importance in itself. We could stand that; but the question how this, and the progressive falling off of our fighting power of which it was symptomatic, was to be prevented, was of immense importance.*

I attached great significance to what I learned at Cambrai about our infantry, about its tactics and preparation. Without doubt it fought too doggedly, clinging too resolutely to the mere holding of ground, with the results that the losses were heavy. The deep dugouts and cellars often became fatal mantraps. The use of the rifle was being forgotten, hand-grenades had become the chief weapon, and the equipment of the infantry with machine guns and similar weapons had fallen far behind the enemy. The Feldmarschall and I could for the moment only ask that the front line should be held more lightly, the deep underground works be destroyed, and all trenches and posts be given up if the retention of them were unnecessary to the maintenance of the position as a whole, and likely to be the cause of heavy losses.

A policy reversal.

So the policy for defence was to be profoundly changed—almost reversed. It is arguable that Falkenhayn's rule of holding everything had sapped the strength from the assaults of the Allies; but if it had done that it had sapped even more the long-term power of the Germans to resist. In the counterattacks on the Australians at Pozières and during the defence of the line, for example, the *18th Reserve Division* had lost 178 officers and 8,110 soldiers between July 24 and August 9. In every purely defensive action from Mid-August to September 10 from Thiepval to Courcelette, the average loss in each battalion was just over 40% of its strength, this proportion being that of those men killed, mortally wounded or sufficiently seriously incapacitated to be evacuated behind the corps casualty stations. Within 72 hours of Hindenburg and Ludendorff leaving Cambrai, new orders had been received in all regiments along the Western Front concerning the policy for defence. Just prior to the third British offensive, therefore, a high measure of tactical initiative was restored to corps and divisional commanders in the German line.

Five German divisions now awaited the assault between the Ancre and Combles, the latter area being occupied by the *185th*, which had been in the line on July 1 and had now returned, reinforced, for a second tour. All formation commanders expected an offensive. Though the Allied air forces still held their superiority, a few German scouts had penetrated their patrols and the reports of pilots and observers confirmed those of the kite balloon crews that roads and dumps were everywhere expanding. The latter had spotted some 'large armoured cars' moving behind the British front—probably tanks—but the observation does not appear to have excited German Intelligence. British and Australian prisoners captured in the preceding fortnight had little to tell; and the French falling into German hands by Combles thought that their allies wanted only to capture the ground immediately on the left. But when the Allied bombardment began at 0600 hours on the 12th and when the extensive night-firing programme took up the work at 0630 hours that evening and the reports of wire-cutting came back from every sector, the expectation of assault became conviction.

A common mood appears to have entered the Canadian, New Zealand and British troops waiting for Z-hour; a mood which overtook them when it became confirmed that the 15th was to be Z-day. Between those small numbers who, for one reason or another, were absolutely fearless and those who were haunted sleeping and waking by terror, the majority seem to have been possessed by a form of exhilaration tempered by anxiety. The veterans admitted that

Above: Success! As the French attackers sweep into a trench, its late German defenders flee.
Below: Albert Cathedral, and its famous 'leaning Virgin'. Legend had it that the war would end shortly after she finally toppled. The prediction did come true—though the British helped . . .

they were impressed by the extent and weight of preparation. Novices were eager to be engaged in the fighting.

In the last hour of daylight on the 14th, the assault battalions finished their meals, put on their webbing, heavy with ammunition and marched quietly forward into their assembly trenches. A proportion of the support battalions who followed them later were tired after carrying loads of ammunition and supplies forward during the day. It was cool enough for those men at rest in the trenches to wear greatcoats that night. But while some slept, the roads and tracks around Pozières, Bazentin le Petit, Longueval, Ginchy and Guillemont were busy as guides, porters and messengers passed to and fro in the moonlight.

The roaring engines of the tanks were scarcely noticed against the rumbling night bombardment. Guides with torches led forward the ungainly caterpillars, but not all succeeded in reaching their starting points. The most successful were the six supporting the Canadians. As planned, three arrived at the rendezvous 500 yards north of Pozières and three by the mill site on the road to Bapaume. Other than these, two took up their station behind the 15th (Scottish) Division for the assault on Martinpuich, two came into position between the two assault brigades of the 50th (Northumbrian), below High Wood were three (instead of four), four reached the 2nd New Zealand Brigade, seven of ten took up station for the 41st Division between the northern end of Longueval and the track leading north from Delville Wood, one of three reached the north-east of Delville Wood, one the road running away from the wood to Ginchy and from Ginchy to Leuze Wood, only nine of the 16 allotted to XIV Corps arrived.

The single tank which had stationed itself correctly on the road leading from Delville Wood to Ginchy was under the command of Lieutenant H. W. Mortimore. It was numbered D1 — tank number 1 of D Company. At 0520 hours it lumbered into action, firing towards the Germans as it moved. Behind it were two rifle companies of 6th King's Own Yorkshire Light Infantry. The few Germans in Hop Alley were astounded by its appearance and quickly surrendered. Covered by the tank, the remainder of the KOYLI hurried on towards Ale Alley to complete the capture of this longer trench, prior to the main assault. Before they could reach it, they were caught in enfilade by machine gun posts in several nearby craters. All the officers were killed or wounded but the surviving sergeants took command, saw to the bombing of the crater posts, cleared Ale Alley and had their men ready for zero. Lieutenant Mortimore and his crew, the first ever to take a tank into action, were following them when, by chance, they received two direct hits by shells. Two of the men inside were killed but the remainder of the crew jumped clear.

The creeping barrage
The sun rose at 0540 hours on the morning of the 15th. Behind, the guns and howitzers and mortars boomed and banged as the British and French batteries main-tained their fire; overhead the shells and bombs swished: beyond, they burst in flashes of flame and splashes of smoke. There was no abatement as Z-hour approached at 0620 hours. To the minute, the infantry began to leave their trenches to draw up to the edge of their own field artillery fire which at zero plus six minutes was due to start creeping ahead of them at the rate of 50 yards a minute until halting just beyond the trenches of the first objective.

The six tanks allotted to the Canadian assault had been ordered to advance in two groups of three on either side of the road to Bapaume. When they began the final run up in darkness a number of the infantry ahead were already fighting — ejecting the last of a series of bombing raids into their trenches begun at 0300 hours that morning, apparently seeking Allied prisoners. At Z-hour, six assault battalions followed the creeping barrage north-east towards Courcelette, keeping pace with their barrage, searching out swiftly and vigorously each German trench until, by 0700 hours, they had captured the Sugar Factory 1,200 yards beyond their start line. Indeed, they had made such excellent progress that they had out-stripped the tanks, one of which had broken down behind the line, while three others had become stuck or bogged among the German trenches captured. The two survivors came on faithfully and caused a number of Germans in positions on either flank of the Sugar Factory to surrender as they came forward. Embarked on a tide of success, the Canadians pushed on steadily through the day to capture Courcelette, despite increasing shellfire from the German medium guns and howitzers. North-west of Pozières they secured two extensive stretches of *Fabeck Graben* by dusk and brought their line within grenade range of Mouquet Farm.

Next to the Canadians, the 15th (Scottish) Division hastened forward at zero hour, passing into its own creeping barrage at several points in its eagerness. Behind it came one tank — the other had been hit by shellfire just after zero — which gave much help in the clearing out of captured trenches. At 1500 hours that afternoon, the 6th Cameron Highlanders advanced to clear the last of the Germans from Martinpuich. Beside a huge bag of prisoners, the division had taken during the day a field battery and a howitzer in an action lasting 11 hours.

To the right of the Scottish division was the 50th, and right again the 47th. Before the 47th stood the splintered trunks of High Wood. General Pulteney, commander of III Corps, had ruled, against the advice of the divisional commander, that the tanks should enter the wood to assist in its capture, a decision that is only comprehensible when it is known that he had never come sufficiently far forward to take a close look at this dreadful obstacle dominating his front. Of the four tanks to hand, one ditched at the outset, the leader of the others reached the nearest German trench, which it raked with its machine guns, before a shell set the hull on fire. The other two were unable to find a way through the broken trees and sought a route towards the east. In the confusion, they broke back into their own lines to fire on men of the waiting London Regiment.

The consequence of this setback was that the flank and rear defences of High Wood were not assailed. As soon as the British barrage had passed, the machine guns on the reverse slopes opened a rapid fire on the 50th Division advancing briskly from the west. Caught in enfilade, the 50th was driven back. There it remained while the assault battalions of the 47th struggled vainly at the edge of the trees. About 1100 hours, the brigadier of the 140th Brigade — on the right of the 47th Divisional front — brought his trench mortars into position to open rapid fire. In seven minutes they fired 750 bombs and while the last of these were still in the air the Londoners advanced again on the wood with grenades and rifles in every expectation of a grim fight.

But suddenly the garrison had had enough. Exhausted, some demoralised, stained and filthy, the Bavarian soldiers stood up by sections, platoons and finally companies to surrender. Two months and one day had passed since High Wood lay open to General Horne for the taking. Now at last it passed into British hands.

General Horne's corps was now next in line to the east. As a consequence of Rawlinson's change of plan, he had the

The central sector of the front, with the skeleton of High Wood in the background, after it had been cleared by units of 47th Division

Imperial War Museum

crucial task of capturing the Gird Trench and Gird Support covering Guedecourt village, 4,000 yards north-north-east of his trenches. East again was Cavan's XIV Corps, directed to open the right hand side of the corridor as the Canadians and III Corps were to open the left.

Cavan's corps had three divisions in the line: left to right, the Guards, the 6th and the 56th. Two tanks were available to go forward with the 56th on the extreme right flank to assist in the capture of Bouleaux Wood. The assault battalions were not confident as zero hour approached; many knew that the guns had not cut the wire lying in hollows in front of the German positions. Helped by the tanks they scrambled into the nearest German trenches but were caught on the wire whenever assaults were launched across the Ginchy-Combles road. One of the tanks was hit by a shell while firing into The Loop, 300 yards from the edge of Combles. The other caught fire after it was bombed by a party of the *28th Reserve Regiment*.

Pulverising bombardment
The 6th Division faced the notorious Quadrilateral, held by the *21st Bavarians*. Several efforts had been made to capture this system of fortified posts following the fall of Guillemont and Ginchy but all had failed. Fresh troops, the pulverising bombardment arranged to open the offensive—more powerful than the Royal Artillery had ever fired in its long history—and the presence of the tanks raised hopes that it would fall on September 15. The fresh troops—the 6th Division—did not fail to leave their trenches promptly. The artillery had pounded the target area, but, as on every previous occasion, the majority of the shells had exploded before or beyond the cunningly sited galleried machine gun emplacements. One tank alone reached the rendezvous, from which it blundered into the 9th Royal Norfolks, waiting quietly for zero hour, opening fire on them by mistake. A company commander bravely approached the flustered crew to point out the error. Without waiting for zero hour the tank advanced along the railway line, fired into Straight Trench above the Quadrilateral and then retired, damaged. At zero, the infantry of 6th Division advanced, the creeping barrage of doubtful value as it was split by lanes left open for the absent tanks. Four battalions assaulted and lost heavily; two followed and their casualties joined those already thickly scattered along the machine gun arcs of the Quadrilateral. Having broken the 6th Division's assault, the Bavarians fired south to deny movement by the 56th Division into Bouleaux Wood, and northward into the flank of the Guards.

The eyes of the Guards' commanding officers were on the village of Lesboeufs and the triangle of roads leading north-west from it to Guedecourt. This was their objective, 3,600 yards from their trenches. Such tanks as had reached Ginchy either broke down, were quickly damaged, or disappeared in an irrelevant direction. Like the 6th, the Guards' infantry advanced alone at zero. But the creeping barrage was unbroken and perfectly timed. Thirty yards behind their own shells, four assault battalions advanced steadily over the cratered ground. As soon as they drew level with the Quadrilateral, however, the right flank came under heavy fire; the left was

meanwhile being caught in enfilade from Pint Trench and Lager Lane. Next, Straight Trench opened fire. The leading companies began to double forward, charging into the German trenches. Behind came elements of the support battalions. Dug-outs and shelters were searched and then the survivors took stock. Three-quarters of the officers had been killed, two thirds of the soldiers. The senior commanding officer, Lieutenant-Colonel John Campbell of the 3rd Coldstreams, held a council to reorganise the force and to decide how far they had advanced. Mistakenly they believed, and signalled to the rear, that they were on the final objective.

Some time before 1100 hours Campbell realised his mistake and led a second advance. They had lost their own barrage and at once attracted German shells. The ground was torn beyond recognition: no stone of Lesboeuf's stood sufficiently in silhouette to guide them. When at last they stopped, they were a little over half way to the village; they had advanced just far enough to open the right flank for the breaching operations of XV Corps.

Horne had three divisions of XV Corps in the line: the New Zealanders, the 41st and the 14th—the latter two in or just forward of Delville Wood. Helped by their tanks, ardent in advance, the infantry of these formations stepped quickly into the first German position, the more readily as the foremost platoons were actually inside their own barrage. The machine guns in High Wood hit 80 to 90 New Zealanders during the advance but by 0700 hours the first objective was completely captured. German shelling added to the confusion in the captured positions.

When the advance was resumed, the New Zealanders were hampered by uncut wire. Machine guns from the direction of Flers began to engage all groups moving in the open. Yet despite continuing losses and the progressive break-up of units, the infantry swept on so that, a little before 0820 hours, two small bands of the 12th East Surreys reached the wire covering Flers village. Of the two officers present, both subalterns, one was killed while crossing; the other scrambled on to search out the remaining Germans.

About this time a German field gun, concealed nearby, opened fire. Detachments of the West Kents, Surreys and Hampshires still behind the Flers wire drew together while one or two young second-lieutenants discussed what they should do. Then a tank appeared, D17 under command of Lieutenant Stuart Hastie. Slewing round on its caterpillar tracks, the tank advanced, 'fire spitting from its guns'. Three other tanks were seen to be raking the eastern defences of the village. In a spontaneous movement, the infantry parties ran forward to follow D17 into the ruins.

Overhead, aircraft of 1st Brigade, Royal Flying Corps, saw a triumph and reported it to Horne's headquarters at 1100 hours. The aim of the offensive had been achieved in a morning: a complete breach had been opened in the German line.

Good liaison
During the morning, the corps' commanders were in their headquarters, from which they could speak by telephone to Rawlinson at Fourth Army or forward to their divisional commanders. Thanks to the daring determination of the Royal Flying

Corps' pilots and observers, the corps and army commanders had more timely and accurate information of events in the line of battle than the divisional generals. Thus Cavan knew quickly that Colonel John Campbell and the foremost Guards were not at Lesbouefs. He knew that the 6th and 56th Divisions were checked before the Quadrilateral and at Bouleaux Wood, and made several attempts to help them on with the corps' artillery. Similarly, Horne at XV Corps headquarters took action to exploit his success. He spoke to General Lawford of the 41st Division at 1120 hours and told him to push on at all costs. The corps' reserve, 21st Division, was told to begin moving forward its leading brigade. Before noon, then, action was being taken to exploit success.

Unfortunately for the British, the situation in the breach was deteriorating. While the German *First Army* headquarters ordered every reserve unit forward of Bapaume to be committed to seal the breach, corps and army artillery switched the majority of their guns on to the area of Flers and the approaches to it through Longueval and Delville Wood. In the ensuing shellfire, four of the six commanding officers surviving were killed or wounded and with them many of their seconds-in-command, adjutants and company commanders. Early in the afternoon, the brigade-major of the 122nd Brigade, Gwyn Thomas, came forward to Flers to reorganise the fragments of units.

He arrived just in time. The lack of any communication in the little garrison—perhaps 80 strong—had induced a sense of uncertainty and isolation. Major Thomas took a company of engineers into infantry service, drew in two companies of the Middlesex and part of the brigade machine guns to establish a defence along the line of the third objective. The posts on the left were united with the New Zealanders. To the right of the village, Brigadier-General Clemson came up with about 200 men and returned for the remainder of the Middlesex. He and Thomas were anticipating their army commander's wishes for, already, Rawlinson had issued orders to reorganise and consolidate ready for the attack next day.

The sky darkened with rain clouds; rain fell intermittently through the night. Next day, there was insufficient infantry to mount a full assault and time was spent in exploiting whatever local opportunities offered. The gun and howitzer batteries moved forward: the bulk of the cavalry moved back. All day the German guns maintained heavy fire on the sector of attack. The clouds hampered the counter-battery work of the aerial observers. The Germans moved in five relief divisions.

The rain continued, British reliefs were delayed and the date for a renewed assault was successively postponed, eventually cancelled. A heavy counterattack fell on the French which, though repelled, employed the last of their immediate reserves.

The battle of Flers-Courcelette was thus a one-day affair in so far as it was a battle of promise. It was a novel operation because tanks were used for the first time. It was a remarkable action for the valour and endurance of those engaged.

THE TANK STORY

Foreseen before the war by H. G. Wells and other fiction writers, it is possible that the advent of armour was delayed by the mistrust and apathy of the military authorities. Here *Kenneth Macksey* discusses the delays that held up the introduction of the revolutionary weapon, the tank

On June 1, 1915, Lieutenant-Colonel Ernest Swinton heard that his appointment as 'Eyewitness' to the British army in France was to be terminated. That same day he submitted a paper to GHQ espousing the cause of 'Armoured Machine-Gun Destroyers'. Pointing out that it might be possible to blast a way with artillery through the German trenches he wrote: 'this is not at present within our power' – but he was adamant that the employment of 'petrol tractors on the caterpillar principle, . . . armoured with hardened steel plates proof against the German *steel-cored, armour piercing and reversed bullets,* and armed with – say – two Maxims and a Maxim 2-pounder gun' would enable machine guns to be engaged on advantageous terms.

Swinton was following up the idea he had proposed in 1914 and which, by various and devious channels through GHQ, the War Office, the Admiralty and industry were beginning to take shape in the concept of a new type of fighting vehicle. Not that fighting vehicles in themselves were an innovation: from the third millenium BC there had been chariots and a succession of vehicles armoured by hides, wood or metal, and for the past 20 years or more most of the ingredients from which a heavy motorised fighting vehicle could be assembled had been in existence either on the drawing board or as actual hardware in the hands of practical operators. The original spark ignition engine, devised in 1885 by Gottlieb Daimler, had been developed to the stage at which over 100 horse-power could be produced by a relatively compact power plant; various types of caterpillar tracks or footed wheels were in commercial use – particularly in the United States, where machines built by the Holt Company were in use helping to open up undeveloped territory. The British firm of Ruston Hornsby had also made a notable contribution but had then sold its patents to Holts in 1912. Armoured cars were in action on all fronts, manufactured by Austrians, Germans, French, Italians, British and Russians for use on open ground, though not where the trench systems made movement by wheeled vehicles almost impossible. Holt tractors were being employed as heavy gun tractors by the British in France.

One of the first tanks to go into action, a Mark I of C Company, ditched near Bouleaux Wood. Note the shell hole in its side, and the two wheels at its rear. The latter were an aid to steering

If armoured fighting vehicles were to play a part in trench warfare they had first to be defined, then designed and constructed, and finally put in the hands of leaders and men who understood their characteristics and who had the strength of purpose to persuade less enlightened individuals to make use of the new weapon. To the engineers, the feasibility of building a caterpillar fighting vehicle had never been in doubt, but already there had been frequent misunderstandings on the part of those conventionally minded General Staff officers who were responsible for devising means to break the trench deadlock. How could it be otherwise? In their semi-ignorance of science, the General Staffs in most armies regarded the technologists with ill-concealed suspicion: even the development of modern artillery and machine guns had taken several decades to push through in the teeth of opposition from those who worshipped the creed of physical combat. It could not be expected that professional General Staff officers would submerge their differences with the technologists overnight and allow them to produce a completely untested weapon system on the say-so of rabid enthusiasts who had nothing to offer by way of a practical demonstration of the feasibility of their ideas. A convincing working model was required, but first an agreed specification was fundamental to design.

The 'Trojan Horse'

By June 1915 the Landships Committee, set up by Winston Churchill in February under Mr Eustace Tennyson d'Eyncourt, the Director of Naval Construction, had gone beyond the original concept of a kind of 'Trojan Horse' transporting a trench storming party of 50 men into the German lines. With a better understanding of conditions at the front, the committee now realised that a machine big enough for this purpose would be highly vulnerable to artillery fire and extremely difficult to move to the front through narrow village streets and across weak bridges. At this same point Swinton's dialogue with GHQ began to make progress, for whereas he had at first been rebuffed by Sir John French's chief adviser on military engineering 'with a few sarcastic words

sufficient to strangle at birth any invention', he now got in his word with the newly-formed Inventions Committee of whose President, and others, he was a personal friend. A series of papers evolved, each demanding more than its predecessor as Swinton's thoughts were stimulated in discussion with experienced front line officers. By June 9 a specification had appeared which, not illogically, tied in closely with those being separately thought out by the Landships Committee. They were for a machine with:

- A top speed of not less than 4 mph on flat ground;
- The capability of a sharp turn at top speed;
- A reversing capability;
- The ability to climb a 5-foot earth parapet with a 1-in-1 slope;
- A gap crossing ability of 8 feet;
- A radius of action of 20 miles; and
- A crew of ten with two machine guns and one light QF gun.

Meanwhile the committee game was being played with verve and enthusiasm in England, where the Admiralty and War Office tossed the Landships Committee about, from one to the other. As the former began to retract from Churchill's early commitments and the latter got more interested, the Minister of Munitions, Lloyd George, became involved when questions of industrial commitment arose. Yet another shuttlecock was the RNAS Armoured Car Division which had pioneered mechanised warfare in Flanders and elsewhere, and had then been drawn into staking its survival on Churchill's sponsorship of landships. The men of No 20 Squadron had been the trials team for the early experiments and soon moves were afoot to transfer them to the army.

It was Swinton's specification, backed by the approval of GHQ in France, which gave real impetus to the technical enthusiasts in England. One moment the latter were dreading a direct order to stop all activities: the next, at the end of June, the Landships Committee had been converted to a joint Service venture with the War Office Director of Fortifications in the chair. Churchill, at this time, was under immense political pressure as a result of the Dardanelles fiasco, but continued to lend support at critical moments. In any case the drive was now coming from lower orders

—above all from the Secretary of the Landships Committee, Lieutenant Albert Stern, an RNVR officer whose civil occupation had been that of a city banker but whose wartime experience had been with the RNAS armoured cars. Stern, with a boiling force of personality and the requisite ruthlessness for cutting red tape, along with a useful technical insight, was the essential middle-man to control the inhibitions of politicians and bureaucrats, and to protect and encourage the designers and engineers while they created the new weapons.

The designers too were men of remarkable creative talent. Apart from D'Eyncourt and the members of his Committee, there was Major Thomas Hetherington, who had been Transport Officer of the Naval Armoured Car Division and who had run the naval experimental section established in the *Daily Mail* Airship Shed at Wormwood Scrubs. Then there was William Tritton, Managing Director of the firm of Foster's which had done original work on howitzer tractors and which, in July 1915, contracted for the development of a tracked machine. Responsible for the technical development was Lieutenant W. G. Wilson—an officer of somewhat unstable temperament, but, with Tritton, a first-class creative, practical engineer. A host of existing projects and prospective tracked models along with power plants and armament had to be investigated, rejected or accepted for further evaluation. Then they had to be tried out before a final design could be settled to satisfy Swinton's specifications which, for want of anything else, now represented the official requirement.

A complex armed vehicle

Holt tractors seemed to offer the best possible foundation for design after a succession of other contenders from the Big Wheel solution to Pedrail tracks and a Killen-Strait tractor had failed to make the grade. Even so, variations on the Holt theme also came to grief either because they were too weak or seemed unduly complicated and time consuming just when the need for rapid development became more urgent after each successive failure on the Western Front. When Swinton was writing his specifications, the French offensives in Artois were grinding to an awful halt and the Russians were being routed in Gorlice. But the General Staff specification in itself made complications inevitable, not only because it had shelved the concept of using a 'Trojan Horse' and, instead, demanded a complex, armed vehicle which would be a fighting platform, but also because the requirements for obstacle crossing were well beyond the capability inherent in Holt tractors. These, in the first instance, had been designed for working only on relatively unchurned ground and certainly not the cratered moonscape then being stirred up along the length of the Western Front.

When a development contract was placed with Foster's on July 24, it was fairly clear what existing components might be available for a production machine. There would be a Daimler 105-hp engine, hardly powerful enough in the event but which was already in quantity production: there was plenty of armoured plate and no great shortage of machine guns, even if the actual types were not the most suitable. Swinton's idea of a 2-pounder pom-pom had to be dropped as it was not available. Luckily, however, the navy offered sufficient 57-mm (6-pounder) guns, along with ammunition, as a substitute. The missing link—quite literally—was a suitable track plate and suspension system. The Holt track system was neither robust enough nor did it provide sufficient ground clearance to meet the specifications. Claims that another American type of track—the Bullock—might serve had to be deferred until a set could be shipped across the Atlantic. Meanwhile, Tritton and Wilson put their heads together and, on August 11—only three weeks after receiving the order—a prototype, to be known as the Tritton, began construction. It was modified, after persistent thought and discussion, by a succession of improvements which eventually turned into a tracked metal box called 'Little Willie'. 'Little Willie' was intended to have a rotating turret which would mount the pom-pom: its track was to be the Bullock, when it arrived, carried on a low slung suspension, but from the start it was clear that this 10-foot high vehicle would be top heavy, have very little ground clearance and only a slim chance of crossing an 8-foot gap. As a test bed it would be invaluable—as a fighting vehicle it would not do, and already Wilson was sketching the design of a machine with radically different lines. This was to be a rhomboidal shaped machine with the tracks carried right round the hull, but it would be so high that all hope of a rotating turret had to be dropped in favour of mounting the guns in 'sponsons' blistered onto the sides of the hull. In August it was renamed 'Big Willie'.

In July Swinton had come back to London to become acting Secretary of the Dardanelles Committee of the Cabinet. The Dardanelles operation was about to have its last fling at Suvla and would soon be over; thus those who wanted to win the war in the West were recovering the initiative and so trench fighting machines were also on the upward grade. There is little doubt that Swinton gave them as much attention as he did his proper job, using the powerful contacts inherent in his job to further their progress. Indeed, Hankey, for whom Swinton was acting, had brought him in specifically to speed things up—and this in itself was enough to give the members of the Landship Committee an attack of jitters in case the new broom would sweep their work away. As a matter of fact they were unaware that the specification to which they worked had been Swinton's handiwork. Fortunately the first meetings between Swinton and D'Eyncourt and then with Stern and the other members of the team were wholly amicable. Each came to appreciate the others' virtues, and with Swinton at the hub of decision and well-versed in the travails of the War Office, a vulnerable chink in the designer's bureaucratic armour was filled. From now on Swinton could manipulate the government committees while D'Eyncourt and Stern handled industry and Wilson.

The great day was approaching when the Tritton would take its first hesitant steps. On September 8 it crawled on the factory floor at Lincoln. On the 10th it crawled again—right off its Bullock tracks—and again there was track failure on the 19th when a trial before Swinton, the Landship Committee and the head of the Inventions Department of the Ministry of Munitions were there to watch. But these were practical men who understood that no brand new design is likely to work first time and the sight of a wooden mock-up of Wilson's 'Big Willie' gave each one of them hope of better things in the offing.

Birth of the 'Tank'

Thus far, the trials had proved what all had guessed—the incipient weakness of the track. But such was the genius, enthusiasm and drive of Tritton and Wilson that they had solved even that problem in three days by building and satisfactorily testing on the bench an entirely new, light-weight pressed-steel plate. Here was the missing link and this, as Stern was to write, was 'the birth of the tank'.

Everything now had to be concentrated upon making a viable fighting vehicle—the choice falling upon 'Big Willie', which had still to be built. On September 29, GHQ was represented at a meeting in London at which detailed specifications were settled—10-mm of frontal armour with 8-mm on the sides, a crew of eight men (of whom no less than four would be needed to co-operate in steering and gear-changing), a speed of 4 mph and acceptance of the Admiralty offer of 57-mm guns. That same day Stern asked Tritton to press ahead with the construction of 'Big Willie'—and there may have been some in the GHQ party who ought to have been thankful, for already it was clear from all their offensives culminating with Loos that artillery was not the answer to the trench deadlock.

'Little Willie' continued to operate as a test bed, all hope of it being a combat vehicle having been abandoned. However, on December 3 it proved, in a running trial, how much more reliable the new track was—but not until January 16, when 'Big Willie' ran for the first time, was there real reason to celebrate.

Thus far only those intimately associated with the production of the new vehicles had seen them. The strictest secrecy shrouded every operation and even prevented Foster's workmen from being awarded War Badges to prove they were engaged in work of national importance. Bereft of status and, in cases, the recipients of 'white feathers', the men began to leave, though Foster's were not at first allowed to give proof of the nature of their work in order to arrange for an issue of badges. Essential though secrecy was to prevent the Germans getting to hear, it was also a threat to the future introduction of the weapon. Until those in authority and those who would have to use the machines had seen them, they could neither be expected to understand their characteristics nor to give backing for priority in finance, production and manning.

The time for a demonstration to the representatives of both the government and the army had arrived. Already, however, GHQ was better educated about the possibilities of fighting machines, for in France, as a mere regimental officer after resigning the appointment of First Lord of the Admiralty, was Churchill, and Churchill had taken it upon himself to write, for the Commander-in-Chief, a long paper entitled 'Variants of the Offensive', which, among other things, drew a rather inaccurate picture of the state of tank development. Haig had read it on Christmas Day and sent a Major Hugh Elles to England to find out more about tanks. Elles duly reported favourably, though at first GHQ thought only in terms of an order for 40.

But GHQ were not represented at the key demonstrations which were held under conditions of great secrecy at Hatfield Park on January 29 and February 2, 1916. At the first, which was by way of being a rehearsal for the second, only those who had been closely associated with the initial experiments were present: for them it was the justification of their efforts. To the second came the great—Kitchener, the Secretary of State for War, Lloyd George, the Minister for Munitions, McKenna, the Chancellor of the Exchequer and many others in power—those personalities, in fact, who could most help or hinder the decision to enter full-scale production and service.

'Pretty mechanical toys'

The demonstrations put both 'Little' and 'Big Willie' through their paces over a stiff obstacle course which, apart from German fire and a bottomless bog, fairly represented the worst terrain that the Western Front could produce. Already at these demonstrations the vehicles were being called 'tanks'—the name which Swinton had given them for security cover, and by which they were to be known evermore. From the first they made a deep impression, though much play has been made of Kitchener's remarks to the designers that the tanks were 'pretty mechanical toys' and that 'the war would never be won by such machines'. Those close to Kitchener said that he did this as a cover so that the trials would not get talked about, but it could also be that, by a provocative remark, he could delve deeper into the minds of men who were, after all, trying a 'hard sell'. The fact remains that immediately afterwards, Kitchener asked Stern to go to the War Office as Head of the Department which was to see to the production of tanks (a logical request since, in December, the Ministry of Munitions had declined to make them)—and it was only later that Lloyd George stepped in to take Stern into the Ministry of Munitions, whose proper responsibility it was to build weapons. No matter: by February 12, Lloyd George had signed a charter of Stern's dictation in which it was settled that an initial order for 100 'Big Willie's' should be placed.

An impression has sometimes been given that the authorities, in the shape of Kitchener, Haig and others, were generally opposed to the introduction of tanks. Yet whenever this was the case it was rarely without a reason—either because the tank enthusiasts had failed to explain some vital point adequately, or because those same enthusiasts were expecting miraculous results from a totally untried device. When moving into the unknown some sense of caution is entirely desirable. In fact the tanks, by the standards of progress registered by many another technical innovation in the past, positively galloped ahead. In February Swinton was writing the foundations of a tactical doctrine: The simplest and surest way of destroying entrenched machine guns was 'by rolling over the emplacements and crushing them'. If that did not suffice they could be crushed by moving close to 'pour in shell at point blank range'. For communications to the rear, 'one tank in ten should be equipped with small wireless telegraphy sets', while others might lay telephone cable as they advanced and yet others depend upon visual signalling. The tanks would be very vulnerable to artillery fire and therefore must be supported by counterbattery fire.

Thereafter Swinton described in detail the way tanks should be prepared for action, transported to the front and integrated with the infantry so that each would combine with the other in making best use of the special attributes—but at this stage Swinton saw no further than the original concept that tanks were there to help the infantry make a hole in the German lines. 'It seems, as the Tanks are an auxiliary to the infantry, that they must be counted as infantry and in an operation be under the same command.' Although Swinton envisaged the German lines being penetrated as deep as the artillery zone in one day, his vision stopped short of outright exploitation which still, in the minds of the majority of soldiers, remained the prerogative of cavalry.

At this point, just when the British were crossing the dividing line between dreams and reality, the French, quite unknown to the British and in total ignorance themselves of British endeavours, were placing an order for 400 *chars d'assaut* of their own. Here, the efforts of their 'Swinton', an artillery colonel called Jean Estienne, to experiment with a tracked armoured vehicle had been twice foiled in 1915. Only in December 1915, after the total failure of the French offensives, was he given permission to seek a solution in conjunction with the Schneider Creusot factory. But the French were nothing like as thorough as their allies. Picking on a Holt tractor suspension as the best caterpillar system available, they immediately ordered an armed and armoured box to be perched on top without so much as em-

Above: The Daimler-Foster artillery tractor. This was the basis of the original 'big wheel' armoured vehicle idea. *Right:* The Killen-Strait armoured tractor, which failed to win approval as it could carry little armament and could not cross trenches. *Below left and right:* Prophetic—the de Mole 1912 landship

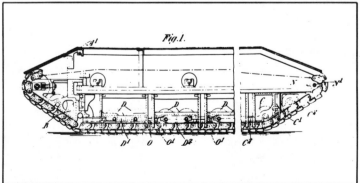

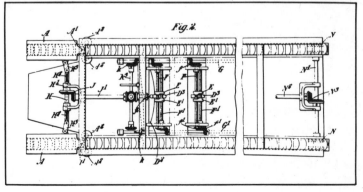

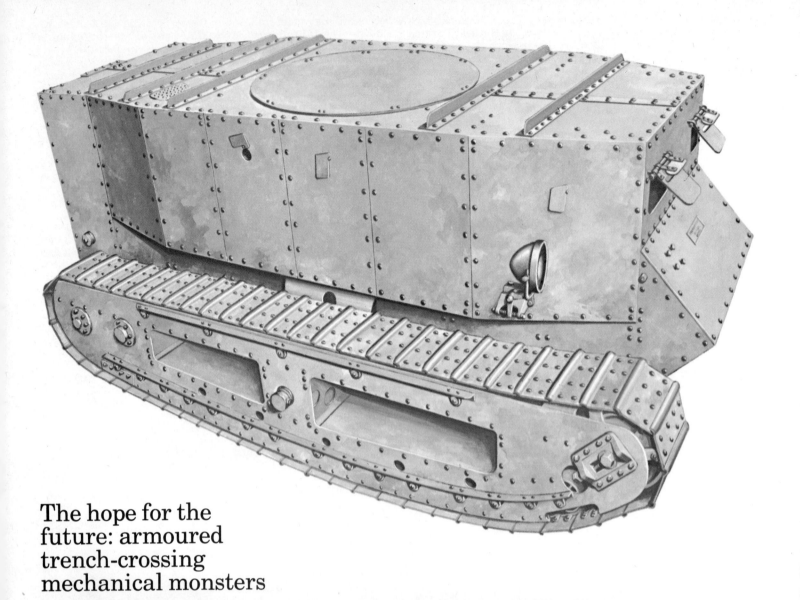

The hope for the future: armoured trench-crossing mechanical monsters

Above left: Almost but not quite there—'Little Willie', built by
the Foster works. It was unarmed, but embraced all the principles of
later tanks, such as tracks running the whole length of the fully-
armoured body. The type was originally envisaged with a revolving gun
turret on top, but as the silhouette of the vehicle would then have
been too high, the idea was dropped. 'Little Willie' had track frames
12 feet long, weighed 14 tons, had a crew of five and a top speed
of three and a half miles per hour (one and three-quarters across
country). *Above:* The Holt caterpillar tractor, which first suggested
the idea of the tank to Colonel Ernest Swinton. The design was American,
and Swinton saw the type in France, where it had been adopted by the
Royal Artillery as a heavy gun tractor soon after the beginning of the war.

Left: The old and the new: cavalry and the vehicle destined to replace
them, the example here being C19 of the Heavy Section, the Machine Gun
Corps on September 15, 1916. *Below:* The first real tank, 'Mother'.
Experiments had shown that earlier tank designs had too high a centre
of gravity and too short a track length to cross broad trenches, so
Tritton and his collaborator Wilson reverted to the 'Big Wheel' idea,
but stretched out to a rhomboidal shape to keep the centre of gravity
low and to give the tracks as much length on the ground, and so grip,
as possible. After the abandonment of the turret idea, sponsons were
fitted to the sides to accommodate the two naval 6-pounder guns carried
as main armament. Steering was by differential braking of the tracks,
and the vehicle could cross a 9-foot trench with a 4½-foot parapet

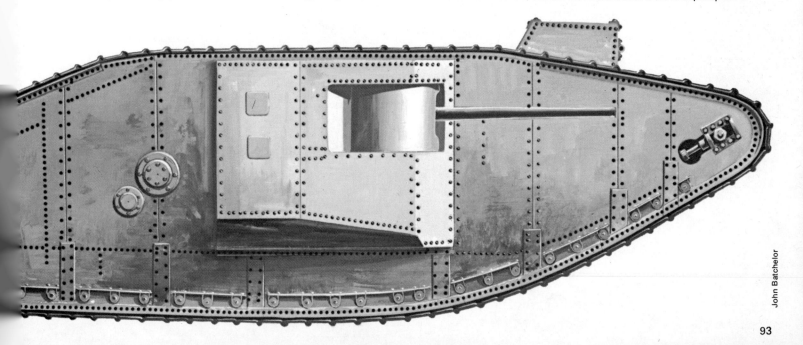

John Batchelor

barking on a trial to see if this would result in a combat-worthy machine. Thus they engineered a vehicle which was little better than the original Tritton—and which, incidentally, looked very like the Tritton in some respects. But even though they had taken a hazardous short-cut, the French were still a good six months behind the British with no hope of entering action until 1917. The British, meanwhile, were digesting Swinton's tactical doctrine and thinking in terms of tank action in the summer of 1916.

When Haig spoke to Swinton in London on April 14, he not only expressed general agreement with the proposed tactical doctrine but was anxious to obtain as many of the machines as possible by June 1—in time for the opening of the Somme offensive. This, of course, was asking for the moon. Production had hardly begun and the men to crew the tanks had yet to begin training—particularly since training could not realistically start until the first machines had been delivered, which would be in June. The earliest any might be expected in France was August 1 —and that was the height of optimism.

Meanwhile the British crews were being recruited—a major problem in itself from a nation to whom mechanical vehicles were still rather in the nature of a possession of the rich and the sporting. Since the first home of the new force was to be sited where the Motor Machine-Gun Service had its depot and that Service was in process of reduction, it was inevitable that these experts found their way into the new establishment. It was also sensible that, from a security point of view, their rôle should be disguised under the title of Armoured Car Section of the Motor Machine-Gun Service instead of, as originally, the Tank Detachment. It was hoped that many of the RNAS, who had been involved with the experiments, would transfer to the army, but faced by loss of status, a reduction in pay and an imprecise description of their future work, they drew back. From the motor trade, however, and with the assistance of the editor of *The Motor Cycle*, Mr G. Smith, many well-qualified tradesmen were recruited—though their knowledge of soldiering was precisely nil.

Training in driving (using 'Little Willie' at first), gunnery and rudimentary tactics went on throughout the summer in which the artillery's palpable ineffectuality was demonstrated daily at Verdun, on the Somme and anywhere else where men dared to rise above trench level. Pressure from GHQ to send out the tanks before they were ready was increased, and training had to be curtailed in order to meet a deadline to be in action on September 15, 1916.

The first batch of 50 machines was in France by August 30 and from then on could only prepare for a battle whose nature was absolutely beyond the comprehension of their crews. Enthusiasts they might be, but few had seen action before or were trained soldiers. One tank commander wrote: *I and my crew did not have a tank of our own the whole time we were in England. Ours went wrong the day it arrived. We had no reconnaissance or map read-* *ing . . . no practices or lectures on the compass . . . we had no signalling . . . and no practice in considering orders. We had no knowledge of where to look for information that would be necessary for us as Tank Commanders, nor did we know what information we should be likely to require.*

Picture the difficulties facing these men in their unlit steel boxes, filled with fumes and exposed working parts, jerked to and fro as its unsprung suspension heaved them across country, deaf from the din of engines and armament and practically unable to communicate among themselves or with the infantry outside. The commander could do little by way of assistance to his gunners to find and destroy targets and had to concentrate on driving the tank by hand signals to the man who changed the gears and the two men, one to each track at the rear of the tank, who worked the levers which 'braked' the track to change direction. It was remarkable that they moved at all, let alone fought, and particularly so when it is remembered that almost constant maintenance and adjustment was required of the machinery.

And once in France and embarked on a slow journey to the front, the crews found themselves alternately the victims of outlandish security measures or objects of intense curiosity under constant demand to give demonstration of the tanks' ability, with a consequent wear and tear on machines which were going to be difficult enough to keep going for the few miles up to and over the start line. There was total distraction of crews in their efforts to prepare themselves and their charges for action. Many were exhausted before they fired a shot in anger.

Nevertheless, on the night of September 13, those tanks which were fit crept forward, guided by white tapes, to their battle assembly positions, creating astonishment in all those who heard them spluttering and groaning in the dark. Next night they moved again to their battle positions and now they entered, for the first time, a genuine area of war-torn ground the like of which the training area simulations in England had not quite reproduced. Here many of the 30-ton machines, grossly underpowered and unsophisticated as they were, broke down as drivers, tense with anxiety, made errors. Gradually, the initial force of 50 was whittled down to a mere 36. But at dawn on the 15th those few were ready and, in the half-light, a new era in warfare opened.

Further Reading
Edmonds, J. E., *Military Operations: France and Belgium 1916* (Macmillan)
Liddell Hart, Sir Basil, *The Tanks* (Cassell 1959)
Stern, A., *The Log of a Pioneer*
Swinton, E. D., *Eyewitness*
Williams-Ellis, C. and A., *The Tank Corps* (Country Life 1919)

Infantrymen and their new aid, the machine that could crush the almighty power of the emplaced machine gun, barbed wire and well-sited trench

THE FRENCH TANK FORCE

On April 16, 1917, the first day of the Nivelle offensive, the French first used tanks in battle — seven months after the British first fielded them on the Somme. *Richard M. Ogorkiewicz* discusses the origins and early development of French armour

The creation of a tank force in France came shortly after that in Britain. It developed independently of the technological advances being made in Britain, and in a less dramatic but more complex way.

As in Britain, the idea of the tank was not conceived in France by any one man but emerged more diffusely, in response to the contemporary military situation, out of the existing technological possibilities. In fact, there were at least three sets of ideas which contributed to the development of the first French tank and hence of the first French tank units.

The initial set of ideas arose out of the development of the armoured car, the first of which had been exhibited as early as 1902 at the *Salon de l'Automobile* in Paris and although progress up to 1914 was very slow, once war began more and more armoured cars were built. Thus, by August 1914 the French Ministry of War had already ordered its first 136 armoured cars and within a month they began to be attached to French cavalry formations. However, these early, improvised armoured cars could operate only on roads, and in consequence, when the opening, mobile stages of the war were succeeded by more static warfare, during which the roads were cut or blocked, their usefulness quickly came to an end.

The short period of operation of the first armoured cars was, however, sufficient to

Below: Chars Schneider move up to the front

95

arouse interest in the possibility of developing them further, so that they could be used beyond the confines of open roads. The most significant move in this direction was taken by the French armaments firm of Schneider, which took out a patent for an improved type of armoured vehicle in January 1915 and then began to consider the design of another armoured car, fitted with tracks instead of wheels.

The French technicians' interest in the use of tracks for armoured cars was prompted by a visit to England in January 1915 by Brillié and another Schneider engineer to examine the potential of the American-built Holt tractor for hauling heavy guns. As a result of this visit, the Schneider Company purchased two tractors from the United States which were tested at its works at Creusot in May 1915. From these tests came the idea of using the lighter

of the two tractors, the 45-horsepower 'Baby Holt', as the basis of a tracked armoured car capable of moving off the road and crossing obstacles. In July, Schneider engineers began to work on the design of such a vehicle and in August they produced a layout drawing of its chassis, which was essentially that of a lengthened and widened 'Baby Holt'. However, a month later further work on the design of the new type of fighting vehicle was interrupted by the temporary ascendancy of other ideas.

The new ideas were prompted by the immediate tactical problem of cutting through the belts of barbed wire with which the opposing armies protected their trenches. The wire obviously had to be cut before any infantry attack could succeed and this fact turned thought to various wire-cutting or wire-crushing devices. The leading proponent of this trend of thought was J. L.

Breton, a parliamentary deputy who, in November 1914, submitted designs for a mechanical wire-cutter mounted on a wheeled agricultural tractor. His designs were favourably received by the technical branch of the Engineer Corps and, after some experiments, the Ministry of War placed an order in August 1915 for ten of the Breton wire-cutters for further trials. But it soon became clear that the performance of the Bajac wheeled tractor on which the cutter was mounted was inadequate and in consequence Breton was put in touch with Brillié in order to explore the possibility of mounting his wire-cutter on the 'Baby Holt' currently being tested at Creusot. This was deemed to be possible and the Ministry of War went so far as to place an order with Schneiders for ten armoured tracked tractors fitted with wire-cutters. However, further trials conducted

Left: Schneider tank. The girder fitted to the front of the vehicle was designed to force barbed wire under the tracks to be crushed. *Below:* Schneiders in action

DAZY

Brown Brothers

with the 'Baby Holt' from December 1915 to February 1916 showed that the wire-cutters were superfluous, since the tractor could crash through barbed wire by itself. This realisation coincided with the emergence of yet another set of ideas and the development of tractor-mounted wire-cutters was seen to be a blind alley and came quickly to an end.

Sanction from Joffre

The third set of ideas for the tank's development came from Colonel Estienne, an artillery officer who had already made a major contribution to technology warfare. Estienne's idea for a tracked armoured fighting vehicle began to take form in the latter part of 1915. In a letter to Joffre on December 1, 1915, Estienne asked for an interview to explain his idea of a tracked armoured vehicle which would facilitate the infantry's advance. At the time, GQG were daily receiving suggestions on how to achieve a decisive victory, most of them fit only for the waste paper basket. But a proposal bearing the signature of a widely-respected officer deserved—and was accorded—some consideration. On December 12 Estienne was summoned to present his ideas to Janin, the general responsible for equipment at Joffre's headquarters. Eight days later he was on his way to Paris to explore, with Joffre's tacit approval, the possibility of constructing the proposed vehicle.

The favourable attitude of Joffre and other French generals ensured that the ideas would be quickly translated into practice. Thus, within four weeks of writing his letter, Estienne was able to get in touch with Brillié, whom he managed almost immediately to interest in his ideas. Brillié, in turn, having studied the earlier schemes for the tracked armoured car and for the wire-cutter vehicle, was quickly able to produce the layout of a vehicle corresponding to Estienne's ideas. In consequence, on December 28, 1915 Estienne was already able to present to Janin a more concrete scheme which planned for the construction of 300 to 400 vehicles. Again Joffre responded quickly and favourably, recommending to the government department concerned that trials be pursued without delay. Then, on January 18, 1916, he received Estienne's personal report which was then submitted to his staff who received it favourably. In consequence, on January 31, 1916, Joffre wrote to the Under-Secretary of State for War asking for 400 vehicles built along the lines proposed by Estienne.

The responsibility for the supply of the vehicles was vested in the army technical services. These operated slowly and were not too well disposed towards a project which did not originate from them. Nevertheless, on February 26, 1916, the Ministry of War placed a firm order with Schneiders for 400 vehicles. It is interesting that this order was placed only two weeks later than the original British order for tanks and involved four times as many tanks. But the production of the French tanks took longer and the first was completed much later than the first British tank.

The specifications of the vehicle which Schneiders eventually produced did not differ very much from those originally laid down by Estienne. For security reasons it was at first referred to as an artillery tractor and was aptly called the *tracteur Estienne*. However, after Estienne adopted *chars d'assaut* as a generic name for what in English began to be known as tanks, they were called *chars Schneider* or less frequently, *chars CA*. The main difference between the new vehicle and the original specification was that it had a short 75-mm gun instead of a 37-mm gun, and a crew of six instead of four. In principle, it amounted to an armoured box-hull on a modified 'Baby Holt' chassis. The hull had 11.5-mm plates at the sides and rear but at the front

Above: French tank soldier—collar patches of 81st Heavy Artillery Regiment on service dress tunic beneath the leather combat jacket

Below: **Schneider**—*Crew:* 6; *Armament:* one 75-mm gun (90 rounds), two Hotchkiss machine guns (3,840 rounds); *Armour:* 11.5 mm max (17 mm later); *Speed:* 3.7 mph; *Weight:* 13.5 tons; *Engine:* Schneider 70 hp; *Length:* 19 ft 8 in; *Width:* 6 ft 6½ in; *Height:* 7 ft 10 in; *Range:* 30 miles

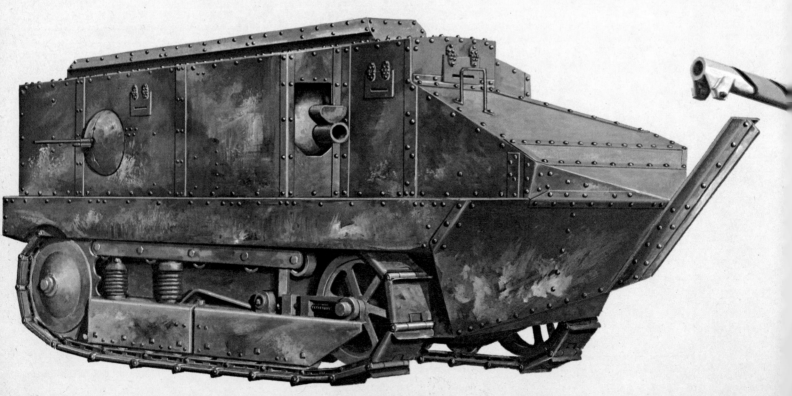

an extra plate was later added which brought the total thickness to 17-mm making it virtually immune to all rifle-calibre bullets. In the front right hand corner of the hull was mounted the short-barrelled 75-mm gun which could be fired over an arc of 20 degrees to the right of the axis of the tank. In addition, in each side plate there was a large spherical mounting for a Hotchkiss machine gun. Within the hull there was stowage space for 90 rounds of 75-mm gun ammunition and for 3,840 machine gun rounds.

With all its equipment the Schneider weighed 13.5 tons. It was powered by a four-cylinder water-cooled Schneider engine which developed 70 horsepower and gave it a maximum speed of six kilometres per hour. In this respect, the Schneider was comparable to the original British tanks. But it was greatly inferior to them in one other important respect, namely trench crossing performance, because its tracks were very much shorter than the overhead tracks of the rhomboidal British tanks. The limited trench-crossing capability of the Schneider tank was seen to be a particularly serious shortcoming after the British tanks made their début on September 15, 1916. Once the Germans knew of the existence of tanks they widened their trenches to make them more effective as tank obstacles and this, in turn, made the French revise their ideas on the employment of tanks.

Estienne's original idea was that tanks should be used in surprise mass assaults on enemy trenches. The tanks were to attack ahead of the infantry and after crossing the first line of trenches one half of them was to keep the enemy down by their fire so that the infantry could follow through the gaps which the tanks opened in the enemy defences and take the trenches. But the possibility of achieving complete surprise was lost as a result of the use of tanks by the British army—a

Below: **St Chamond**—*Crew:* 9; *Armament:* one 75-mm gun (106 rounds), four Hotchkiss machine guns (7,500 rounds); *Armour:* 17 mm max; *Speed:* 5.3 mph; *Weight:* 23 tons; *Engine:* Panhard 90 hp; *Length:* 25 ft 11 in; *Width:* 8 ft 9 in; *Height:* 7 ft 8 in; *Range:* 37 miles

Above: Engine of a St Chamond. With exposed moving parts and hot pipes, the interior of the early tanks was a hazardous place

use which the French regarded with bitterness as premature—and this coupled with the inability of the French tanks to cross wide trenches, led the French to envisage new tactics for tanks.

Diminished rôle

In general, in the period leading up to the first tank action, less came to be expected of tanks than Estienne had originally envisaged. Thus, the rôle assigned to them became the more modest one of complementing the artillery fire by attacking enemy positions that had escaped the artillery bombardment and which were holding up the advance of the infantry. Moreover, tanks were to be accompanied by units of specially trained infantry who would help them cross obstacles. Tanks were not, however, intended to be tied down to the pace of the infantry when they were able to move ahead on their own.

Before the proposed tactics could be put to the test tanks had to be produced in number and this proved to be more difficult than expected. The first Schneider tank was delivered to the French army on September 8, 1916, but by November 25, when all 400 were expected to have been delivered, only eight were in the hands of the troops. Moreover, they had been built using mild steel, because of delays in the production of armour plate, and were therefore suitable only for training. By mid-January 1917 there were still only 32 training tanks. But by the time the

Above: St Chamond tanks in a French wood. *Below:* Schneiders, knocked out by German artillery and stripped by the French for spares. The first two French tanks had poor cross-country capabilities. The Schneider and the St Chamond could only cross trenches 6 ft and 8 ft wide

tanks were introduced into action, in April 1917, the French army was better off than the British army was at the same stage as it had more than 200 Schneiders.

The delays in the delivery of armour plate and in the whole tank production programme were due not only to the usual manufacturing difficulties associated with a new product but also to the dispersion of the available effort in two different directions. The second type of tank originated with the technical services who, once shown the way, thought they could do much better than Estienne and Brillié. It became known as the St Chamond tank, named after the firm which designed it and which in April 1916 received an order for 400 vehicles — and all this without any reference to Estienne.

On paper, the St Chamond tank was superior to the Schneider. It had a normal, instead of a short-barrelled, 75-mm gun and two more machine guns; it also had a longer track and an electrical instead of a mechanical transmission which made it easier to drive. However, at 23 tons, it was considerably heavier than the Schneider and consequently performed poorly over soft ground. Moreover, the front of its box-hull projected well over the front of the tracks which greatly reduced its trench-crossing capability. It also experienced more breakdowns than the Schneider and its production was no faster. Admittedly, the first vehicle was delivered at about the same time as the first Schneider but on

the eve of the first French tank action in April 1917 there were only 16 St Chamond tanks fit for battle. As it happened, none of them was used, the only St Chamonds to accompany the Schneiders into their first action being four unarmed vehicles which carried supplies.

The production of tanks was accompanied by the formation of the first tank units. The original scheme for their organisation was due to Estienne who submitted it to Joffre in March 1916. It was approved but it was not until July that Estienne was relieved of his artillery command on the Verdun front and attached to Joffre's headquarters to take charge of the formation of tank units — the *artillerie d'assaut*. Following Estienne's original ideas, the basic unit became a *groupe* of 16 tanks: each was divided on traditional artillery lines into four batteries which could be temporarily combined under one command when required.

The organisation of the tank units started in August 1916, at Marly near Paris. Soon afterwards a second training centre was set up at Cercottes, near Orleans, and a tank camp was established near the front, at Champlieu, which became Estienne's headquarters. The first *groupe* was formed at Marly in October and by the end of March 1917 the camp at Champlieu contained 13 *groupes* of Schneiders and two of St Chamonds.

The officers of these units came from various services within the army and some

even from the navy; the men for most part came from cavalry regiments whose reserve squadrons were being disbanded for lack of employment. They had all undergone the contemporary type of conventional military training but otherwise almost all were completely ignorant of technical matters. They had, therefore, to learn from scratch how to operate tanks and it was typical of the kind of muddled thinking which surrounded new technological developments at this time that, after a mere two to three months' training at Champlieu, they were thought ready for battle.

Further Reading
Deygas, F. J., *Les Chars d'Assaut* (Charles-Lavauzelle, Paris 1937)
Crow, D. (ed.), *AFVs of World War One* (Profile Publications 1970)
Dutil, L., *Les Chars d'Assaut* (Berger-Levrault, Nancy, 1919)
Duvignac, A., *Histoire de l'Armée Motorisée* (Imprimerie Nationale, Paris 1948)
Macksey, K. and Batchelor, J., *Tank* (Macdonald 1970)
Perre, J., *Batailles et Combats des Chars Français* (Charles-Lavauzelle, Paris 1937)

RICHARD M. OGORKIEWICZ graduated in engineering at London University and after a period working in the motor industry returned to the Imperial College of Science to become a senior lecturer in mechanical engineering. He is the author of two books and numerous articles on the development of armoured fighting vehicles. He has also lectured on the subject in Britain, the United States, Israel and Sweden.

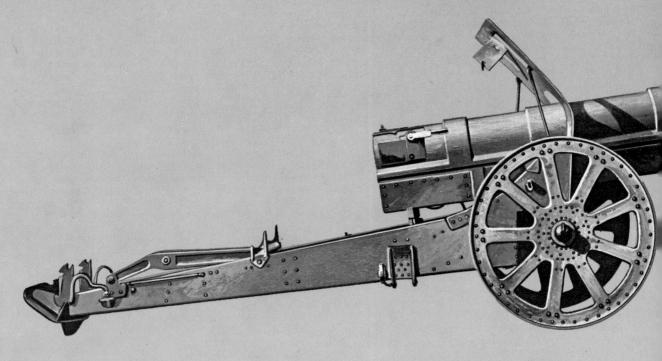

WESTERN FRONT WINTER 1916/17

In a winter which was first exceptionally wet and then unusually cold, both sides were content to rest and recover from the massive slaughters of the summer months. *John Keegan* describes the life in the trenches and the state of morale

OCTOBER NOVEMBER DECEMBER

Constant rain in the last half of the month turned the front line into a quagmire, hampering the supply of stores and ammunition and interfering in the work of the artillery. After the bloodbaths of Verdun and the Somme, the French, British and Germans welcomed the onset of winter as a time of recuperation. No large offensives were planned.

The bad weather continued. What had been an hour's walk in the summer had now become a gruelling ordeal—men carrying ammunition boxes slipped and fell in the mud; men carrying food supplies failed to reach the front line because they had fallen up to their necks in ice-cold water. Germany's manpower resources were becoming very stretched, and in November she was forced to call up the class of 1918— those youths who were born in 1898. By this means, and through the reduction of establishments, Germany was able to bring about a 20% increase in the number of its divisions by the end of the year.

Despite the continuing wet weather, the French General Nivelle had his 'last fling' at Verdun on December 15/16. But apart from this, the French tried to maintain a policy of inactivity along the whole front with the object of harbouring resources for a great offensive planned by Nivelle in the spring. Similarly, the Germans were relatively inactive as regards military operations, but otherwise they were extremely busy constructing the *Siegfried Stellung* (Hindenburg Line) in open country to the rear of the trenches.

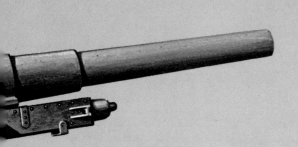

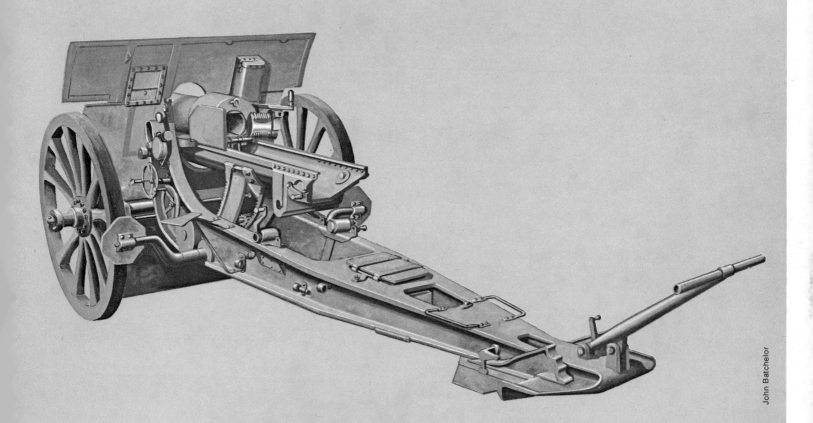

John Batchelor

JANUARY

On January 14 a frost at last hardened the ground sufficiently to restore comparative ease of movement over the battlefield, but it did not dry the ground so that the shortest spell of warmer weather brought chaos again. The frost, though preferable, brought its own discomforts and difficulties. It froze the water supply and forced the men to melt ice from shell holes. Most dangerously, the frost damaged the cordite propellant of shells and caused them to drop short. Before the frost froze the wet ground the British made three raids between January 10/12, two of which were successful.

FEBRUARY

The British continued small scale raiding into February. There was a large night operation on the 17/18th on the north bank of the Ancre, but the Germans were alerted and laid down heavy artillery fire. The British lost 2,700 officers and men and gained between 500 and 1,000 yards, while in simultaneous operations to the north of this Fourth Army won even less ground and lost 1,100 men.

MARCH

The Germans continued to increase the size of their army, raising the number of combat battalions on the Western Front from 1,289 in September 1916 to 1,422 in March 1917. Morale was still high, and the production of munitions had continued to rise. In the British sector the morale effects of the raids of the winter months were dubious, but the raids did reveal a new standard of skill in the divisions of the Kitchener armies. French morale was still holding up after a relatively quiet winter preparing for the Nivelles Offensive in the spring of 1917.

Above: The German 7.7-cm M 1916 field gun. This new model was longer than the M 1896 and could be fired at very much increased angles of elevation. *Range:* 9,405 yards. *Weight in action:* 2,750 lbs. *Muzzle velocity:* 1,571 feet-per-second. *Elevation:* + 38 degrees to −9 degrees 30 minutes. *Traverse:* 2 degrees right to 2 degrees left.

Left: The German 15-cm M 1916 (Krupp) field gun. *Range:* 23,500 yards. *Weight of gun:* 7 tons 3 cwt. *Weight of shell:* 28 lbs. *Maximum elevation:* 46 degrees

Below: The German 15-cm M 1913 long howitzer. *Range:* 9,296 yards. *Rate of fire:* 2 rounds-per-minute normal rate

John Batchelor

The guns of Passchendaele

Left: British 12-inch Mark 3 rail howitzer, an enlarged version of the 9.2-inch howitzer. *Range:* 14,500 yards (accurate to within 10 yards of target). *Maximum elevation:* 45°. *Traverse:* 120° right or left. *Weight of shell:* 750 lb. *Muzzle velocity:* 1,475 feet per second. Thirty-five Mark 3 howitzers were built and 138 12-inch howitzers of all marks. Total number of rounds fired by all marks—227,000

Right: Field artilleryman in shirt-sleeve order. He is wearing the regulation grey-blue shirt and riding breeches and carries one tool of his trade, a knife for picking stones out of horses' hooves. His boots and leggings, ideal for use in the Flanders mud, were issued to some field artillery units in 1917 but were withdrawn in 1918

Barry Evans

Below: British 12-inch Mark 9 railway gun. Only four of these guns were built. The first and second had a maximum elevation of 30° and a range of 28,000 yards. The third and fourth had a maximum elevation of 35½° and a range of 33,000 yards. The traverse of both types was 1°. *Weight of shell:* 850 lb. *Total weight of gun and carriage:* 152 tons. The Mark 9 was combined with one 9.2-inch gun to form a siege battery, manned by four officers and 67 men. The four Mark 9 guns fired a mere 5,000 rounds throughout their use in France

John Batchelor

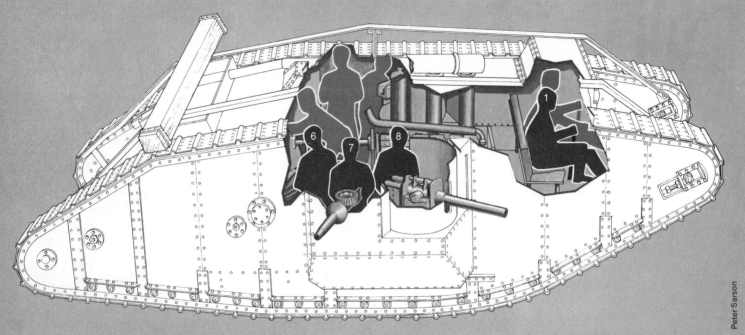

Peter Sarson

Above: The crew of a Mark IV tank: **1** Commander. **2** Gearsman. **3** 6-pounder gunner. **4** Lewis machine gunner (and loader for 6-pounder). **5** and **6** Brakesmen. **7** Lewis machine gunner (and loader for 6-pounder). **8** 6-pounder gunner. Moving over rough ground without suspension or any means of shock absorption, and at a maximum speed of 3.5 mph, tank crews were subjected to intense noise, vibration, smell and high tempera-tures (around 100 degrees Fahrenheit) which compounded with cramped space to produce an extremely exhausting fighting environment. *Below, left and right:* The cramped interior of the Mark V tank, looking forward on the right-hand side, and looking behind the engine from left to right. *Bottom:* An exterior view of the same Mark V, showing camouflage painting

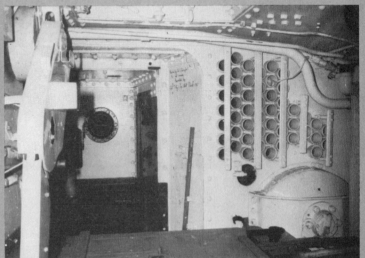

John Batchelor

Special Order No 6.

1. Tomorrow the Tank Corps will have the chance for which it has been waiting for many months, — to operate on good going in the van of the battle.

2. All that hard work & ingenuity can achieve has been done in the way of preparation

3. It remains for unit commanders and for tank crews to complete the work by judgment & pluck in the battle itself.

4. In the light of past experience I leave the good name of the Corps with great confidence in their hands

5. I propose leading the attack of the centre division

Hugh Elles.
B.G.

15th Nov. 1917.   Commanding Tank Corps.

Distribution to Tank Commanders.

At 0620 hours on November 20 1917, 381 tanks moved forward from their last assembly points, led by their commander General Elles flying his flag in one of the leading machines in the true style of a 'land admiral'. The wire was easily crushed, German troops in the front lines fled in terror, and a huge hole almost six miles wide and up to 4,000 yards deep was torn in the formidable Hindenburg Line. This was the dramatic birth of large scale tank warfare. Had Haig's armies at last discovered the real answer to barbed wire and machine gun war? *David Chandler. Above:* Brigadier-General Hugh Elles, Commander of the Tank Corps at Cambrai, painted by Sir William Orpen. *Above right:* Elles's Special Order, issued on the day before his Tank Corps went into action

CAMBRAI THE BRITISH ONSLAUGHT

Julian Allen

John Batchelor

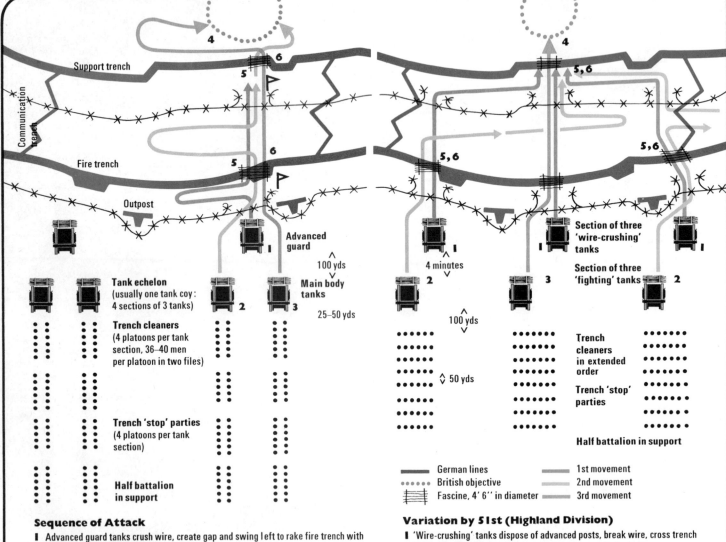

Support trench

Communication trench

Fire trench

Outpost

Advanced guard

Tank echelon
(usually one tank coy:
4 sections of 3 tanks)

Main body
tanks

100 yds

25–50 yds

Section of three
'wire-crushing'
tanks

4 minutes

Section of three
'fighting' tanks

Trench cleaners
(4 platoons per tank
section, 36–40 men
per platoon in two files)

100 yds

50 yds

Trench
cleaners
in extended
order

Trench 'stop'
parties

Trench 'stop' parties
(4 platoons per tank
section)

Half battalion
in support

Half battalion in support

▬▬▬ German lines	▬▬▬ 1st movement
•••••• British objective	▬▬▬ 2nd movement
▦ Fascine, 4' 6" in diameter	▬▬▬ 3rd movement

Sequence of Attack

1 Advanced guard tanks crush wire, create gap and swing left to rake fire trench with main and secondary armament.
2 These are followed by second tank in each section, which drops fascine, plants flag and crosses fire trench, then swings left to rake support trench.
3 The last tank in each section crushes second wire barrier, drops fascine, plants flag and crosses the support trench towards the objective.
4 All tanks then rally in the vicinity of the objective.
5 Infantry 'trench cleaners' then clear the German lines (2 platoons per trench) from right to left, aided by 2nd & 3rd tanks.
6 Infantry 'stop' platoons secure trenches, set up blocks and improve wire gaps, through which the reserves then pass.

Variation by 51st (Highland Division)

1 'Wire-crushing' tanks dispose of advanced posts, break wire, cross trench (with aid of fascine) and drive on to breach second wire belt.
2 Two flanking fighting tanks cross fascine bridges, then turn right to shoot up the fire trench.
3 The remaining fighting tank passes straight ahead to bridge support trench.
4 General tank rally.
5,6 Infantry, deployed in extended order, carry out normal 'cleaning' and 'stopping' rôle (from left to right).

The dramatic events in the vicinity of Cambrai in November 1917 occupy a special place in the history of the First World War in general and in that of the Royal Tank Regiment (formerly the Tank Corps) in particular. November 20 saw the birth of large scale tank warfare, and for a brief but glorious moment it seemed that a decisive breakthrough had at last been achieved, and that an element of strategic mobility was about to return to the Western Front after three years of trench warfare. In fact the success was to prove more illusory than real, and the ringing of church bells throughout Great Britain on November 23 (the sole time this was authorised before the Armistice) was decidedly premature, for the well-known conditions of stalemate and positional warfare were to reassert themselves within three days of the offensive's opening. Nevertheless, as John Terraine has written: 'After Cambrai there could be no further argument; the decisive weapon of land warfare in the mid-twentieth century had now definitely arrived. A new dimension of war was established. Haig's armies had at last discovered what the real answer to wire and machine guns was.'

Haig had been considering 'an attack of surprise in the centre with tanks, and without artillery bombardment' as early as February 1917. As the autumn of a year more dismal than any of its predecessors set in, the idea hardened into a firm resolve. Haig noted in his diary for September 16: 'I discussed with Byng some operations which he proposed. I told him I would give him all the help I could.' This laconic entry marks the practical genesis of the Cambrai offensive. General Sir Julian Byng, Commander of Third Army, was told that his command would be reinforced to a total strength of 19 divisions, and that the entire Tank Corps would be placed at his disposal. With these forces he was to launch a surprise attack on the German lines near Cambrai on a date – still to be confirmed – in November.

The choice of both place and time was carefully considered. The rolling chalk-based downlands of the Somme region promised far better going than the morass of the Flanders front. It had also been a relatively quiet sector of the front for some time, and the ground was comparatively free of shell holes. Furthermore, the main line of the Hindenburg defences (to which the Germans had retired earlier in 1917) swung away to the northwestward and then north in a vast sweeping curve, and the capture of the ridge near the village of Bourlon would provide a superb view over the Germans' rear areas almost as far as Valenciennes. A quick surprise attack here might yield useful results. The Cambrai plain had been suggested as a suitable site for a large scale tank raid by Lieutenant-Colonel J. F. C. Fuller, GSO 1 to Brigadier-General Hugh Elles, commanding the Tank Corps, in June 1917. Apart from being well drained, the area was bounded by two canals to west and east, the Canal du Nord, and the St Quentin respectively, and the Allied front included thick woods, more particularly near Havrincourt, and these would help conceal the final preparations. Intelligence reported only two German divisions in the line, with no more than five, and 150 guns, in support. A refined version of Fuller's plan earned Elles' full approval, and was forthwith forwarded to both GHQ and General Byng. Simultaneously, a second plan calling for a similar combined tank and infantry onslaught without a preliminary bombardment over the very same ground was being prepared by Brigadier-General H. H. Tudor, CRA to 9th (Scottish) Division. This scheme reached Third Army headquarters on August 23 via IV Corps. It was a plan based on these two schemes which won Haig's approval at a conference with Byng on September 16.

The timing of the offensive was of the greatest importance to the British Commander-in-Chief. At first it seems he envisaged Cambrai as a purely diversionary attack, intended to ease pressure from the embattled divisions still locked (in September) in the agony of Third Ypres. In fact, however, that terrible offensive was to come to an unmourned conclusion two weeks before the opening of Cambrai. Nevertheless, Haig well understood the need to strive for a success somewhere on the Western Front to set against the dismal tale of failures at Ypres, mutinies in the French army, the catastrophic situation developing on the Russian Front, and, more recently, the massive Italian disaster at Caporetto. There was no idea in Haig's mind of delivering a war-winning blow at Cambrai – he knew his resources would never permit a massive exploitation even if the attack proved a success – but he was aware of the vital need to give the Allied governments and peoples some evidence of at least one tangible success to compensate in some measure for the horrific casualty lists incurred in earlier failures. His object, then to quote the Cambrai Despatch, was 'a local success'.

Sir Julian Byng, not surprisingly, was more ambitious. Briefing his corps commanders at Albert on October 26, he painted a picture of a three-stage battle opening the way to greater things. In the first phase, some 409 tanks followed by infantry were to smash a five-mile breach through the Hindenburg Line in the sector bounded by the two canals; their initial advance would coincide with the laying down of an unregistered bombardment by 1,003 guns which would provide the Germans with no warning of what was afoot. Next, a 'cavalry gap' would be carved towards the River Sensée, involving the capture of the key communication centre of Cambrai, the domination of Bourlon ridge, and the establishment of bridgeheads over the Sensée itself. Thirdly, the victorious Third Army would sweep north and west up the German line towards Valenciennes.

Haig, as we have seen, was not so optimistic as his subordinate, and insisted on modifications to the plan. First, he laid down a 48-hour deadline for the initial breakthrough; if that was not achieved by the time limit, the offensive would be closed down. Secondly, he insisted that Cambrai should be masked, but not actually captured, in phase two; far more important, in his view, was the physical occupation of Bourlon ridge – and the wood and village of the same name – which would afford a superb view over the German rear areas for artillery observation posts at the earliest moment possible. Thirdly, Haig knew that the exploitation envisaged by Byng was beyond his means. The French, eager to share in such a promising operation, had placed five divisions at his disposal near Peronne, intending to roll up the German line southwards in the exploitation phase, but the British Staff were not too keen to accept this aid. As for Third Army, no less than 14 of its 19

Left: Trees cut down by the Germans across a road in the Flesquières sector—one reason for the check here which ultimately jeopardised the battle

Right: Lt-Gen Kavanagh. His cavalry were to follow through when the tanks broke the German line

Far right: Maj-Gen Harper, 51st Division. He substituted his own infantry/tank drill for Fuller's—another reason for the check before Flesquières

Below: 'Hyacinth', ditched on the first day in the German second line near Ribécourt

Imperial War Museum

Left: A British Mark IV, destroyed by German shell fire
Right: Mark IVs wait to move forward again across terrain (rolling chalk-based downland, comparatively free of shell holes) which promised good going for Fuller's standard tank drill (explained diagrammatically on page 110)
Far right: General Sir Julian Byng, Third Army. He was more ambitious than Haig in his estimate of the breakthrough possible at Cambrai. Haig had laid a 48-hour deadline for the initial onslaught — if the major objectives were not achieved the offensive was to end

divisions had been bled white at Third Ypres, and both men and resources were almost at exhaustion point. As Brigadier-General John Charteris, Haig's Chief of Intelligence, wrote: 'We shall have no reserves. We shall be alright at first; afterwards is in the lap of the God of battle.' At GHQ there was therefore little hope of a runaway success.

The Hindenburg Line

A word must here be devoted to the Hindenburg Line. The position comprised three lines of double trenches. The front line consisted of advanced posts, then a wide fire trench, with a support trench some 200 yards to the rear. A mile back was the official Hindenburg Support Line, once again two trenches deep, and two miles behind that was a third position, almost completed, known to the Germans as *Siegfried II.* Six belts of wire, the main one 100 yards deep, protected the main positions, and all support and rearward positions were similarly provided for. Holding the Cambrai sector of this strong position were the *20th Landwehr Division,* the *54th Division,* and part of the *9th Reserve Division,* which, together with the *183rd* further south, formed the *Gruppe Caudry* (or *XIII Korps*); this formation formed part of General von der Marwitz's *Second Army* (HQ at Le Cateau), which in turn was under Crown Prince Rupprecht of Bavaria's headquarters. The front line forces were not particularly strong, as Ludendorff subsequently admitted, for the German forces had also suffered considerably at Third Ypres. By a stroke of good fortune, however, the leading elements of the *107th Division,* being transferred from the Eastern Front, began to detrain at Cambrai on November 19, and were placed under temporary command of *54th Division* for use in any (as yet unforeseen) emergency.

General Byng entrusted the brunt of the initial phase of 'Operation G Y' to his III and IV Corps, with the Tank Corps to the fore. Lieutenant-General Sir William Pulteney (III Corps), had four divisions (6th, 12th, 20th and 29th), a cavalry detachment, and the 2nd and 3rd Tank Brigades at his disposal. His task was to engage three divisions in the assault phase, and with them secure the offensive's right flank (aided by VII Corps on

its right). First his divisions would be expected to breach the Hindenburg defences from Crèvecoeur to Bonavis, secure the line of the St Quentin Canal, and then force crossings at Marcoing and Masnières to form a bridgehead, through which the 2nd and 5th Cavalry Divisions of Lieutenant-General Sir C. T. McM. Kavanagh's Cavalry Corps would then advance to bypass Cambrai, sever its railway links, and pour on towards Cauroir, and Iway, to seize crossings over the Sensée. On Z+1 they would be followed by the 3rd and 4th Cavalry Divisions, whilst III Corps infantry formed a defensive flank from Gonnelieu to La Belle Etoile.

Meanwhile, Lieutenant-General Sir Charles Woollcombe (IV Corps), would initially have committed two divisions — the 62nd (West Riding) and the 51st (Highland) — to the battle, supported by the advance of a third — the 36th — on the left or northern flank of the offensive zone. They would be led into battle by the tanks of the 1st Tank Brigade, and it was ultimately decided that the corps would have 1st Cavalry Division under command. The IV Corps' task was to breach the German lines along its sector, capture Havrincourt, Flesquières, Graincourt and Cantaing in the process, and then pass through its reserve divisions and the cavalry for the all important attack on the Bourlon complex (ridge, wood and village) — Sir Douglas Haig's prime objective for the first stage of the battle. It was envisaged that the 1st Cavalry Division would aid in the isolation of Cambrai by establishing contact with the main Cavalry Corps to its right, before wheeling left to attack Bourlon village from the northeast.

Once these objectives had been achieved, Byng would commit his reserve (V Corps — the Guards, 40th and 59th Divisions) to secure the Sensée crossings, and create a bridgehead to the heights beyond. These troops would advance over IV Corps communications. All these associated operations would be supported by some 300 Allied aircraft, four squadrons being earmarked for close tactical support of the assault itself. Finally, on November 17, HQ Third Army notified all senior commanders that the offensive was to open at 0620 hours on Tuesday, November 20.

Meanwhile, feverish but well-concealed preparations were being undertaken at all

levels. One thousand guns were moved up and placed in camouflaged emplacements, and their crews instructed in the means of opening the bombardment without prior registration or ranging. Thirty-six special trains were steadily moving up the 476 tanks, first to the corps mounting area at Bray-sur-Somme, then towards their individual battalion 'lying-up' areas immediately behind the front. One by one, battalions were taken out of the line to train in tank/infantry tactics. The form of these followed the proposals of Fuller: tanks were to operate in groups of three, each vehicle carrying a heavy fascine of wood (weighing $1\frac{3}{4}$ tons), in association with four platoons of infantry, following in file at prescribed distances in two waves. The 'Tank Battle Drill' would be followed by every assault formation except the 51st (Highland) Division, whose commander, Major-General G. M. Harper, a soldier 'of extremely strong views' (according to Woollcombe) imposed a variation which was to have important repercussions on his sector.

Stage by stage the broad intention (but not the date) was revealed to lower formations. The final moves took place from November 17. The tanks (each battalion having a fighting strength of 36 machines supported by a further six) moved forward

Ullstein

Imperial War Museum

by night, crawling along in bottom gear (to reduce the volume of engine noise) following white tapes over the last 1,500 yards, while aircraft flew low overhead and machine guns maintained routine fire tasks further to smother the tell-tale rumble — with success. Camp fires were strictly forbidden by night and carefully controlled by day. On the 62nd Division's front, a false screen of trees and brushwood was erected surreptitiously along the forward edge of Havrincourt Wood, behind which the divisional artillery was sited within 2,000 yards of the German front line. However, the Germans were not so completely fooled as has sometimes been claimed. Although on November 16 von der Marwitz reported that he did not anticipate a major British attack, Crown Prince Rupprecht was not wholly satisfied. Early on the 18th a German raid captured an NCO and five men, who under interrogation revealed that an attack was being prepared against Havrincourt. Consequently, the German *54th Division* was placed on full alert from midday. On the 19th, German monitors picked up a snatch of a telephone conservation — 'Tuesday, Flanders . . .' As a result an infantry regiment and several reserve battalions were switched into the Cambrai area, while *Gruppe Arras* was alerted that an attack

on Havrincourt was likely the next morning, and that it might be accompanied by tanks — preceded, however, by a bombardment. From the early hours of the 20th, *Gruppe Caudry* was also stood-to. These moves, however, were mainly precautionary in nature and decidedly limited in scale. In sum, the Germans knew that something was brewing on the Cambrai sector, but failed to appreciate its scale or form. The German High Command was caught unprepared. As Ludendorff wrote: 'We were expecting a continuation of the attacks in Flanders and on the French front, when on November 20 we were surprised by a fresh blow at Cambrai.'

'Zero-hour will be 0620 hours'

By the hour before dawn on Tuesday, November 20, 216 tanks had been moved into carefully camouflaged positions along a six-mile front. A little way back waited a further 96 in support, and each of the nine tank battalions had six more in reserve besides three signal tanks per brigade. At appointed positions 32 special 'grapnel-tanks' waited (charged with tearing large gaps through the German wire), and two 'bridge-tanks' prepared to accompany the cavalry advance. Five divisions of infantry silently formed up behind their tanks, many officers donning privates'

uniforms and carrying rifles. Brigadier-General Elles took up his position in *'Hilda'* towards the centre of 6th Division's front, his tank flying the improvised green, red and brown battle-standard (signifying 'from mud through blood to the green fields beyond'). A final message went to all formations: 'Zero-hour will be 0620 hours.'

Precisely at the appointed minute the guns crashed out, and the first wave of tanks surged forwards into No-Man's Land, each 'main body' vehicle being followed at 25 to 50 yards by the infantry platoons. As the guns successively raised their sights to keep the shell-line 300 yards ahead of the advance, a two-mile smoke-screen soon blinded the defenders of the Flesquières sector, and a half-mile screen filled Nine Wood, supplementing the prevailing autumn mists. Good progress was soon being signalled on all sectors. 'The German outposts, dazed or annihilated by the sudden deluge of shells, were overrun in an instant', wrote D. G. Browne in *The Tank in Action*. 'The triple belts of wire were crossed as if they had been beds of nettles, and 350 pathways were sheared through them for the infantry.' Reaching the German fire-trench, the tanks carried through their drill, passing over fascines to attack the support trench in similar fashion while the infantry poured into the German positions to clear them with grenade, bullet and bayonet. The *Blue Line* was occupied by 1000 hours.

Progress was astoundingly rapid for officers and men used to measuring progress in yards. For the tank crews, the advance was exhilarating but far from comfortable. 'When it lurched it threw its crews about like so many peanuts,' recalled Captain D. E. Hickey, 'and they had to clutch on to whatever they could when we were going over uneven ground. The rattle of the tracked machinery produced the illusion of tremendous speed.' In fact, an average speed of 2 mph was deemed good going, but the lurches, restricted view, appalling din and fumes of the roaring engines placed a heavy strain on the tank crews, all wearing chain-mail masks as protection against metal splinters from inside of the hulls.

These apparitions crawling out of the mists of a November morning in such large numbers terrified the leading German

formations. Many turned to flee. 'Without exaggeration', wrote a German officer of the scenes behind the lines, 'some of the infantry seemed to be off their heads with fright.' It seemed as if the entire front was collapsing, as the tank second wave passed to the front through their rallying comrades, and headed for the *Brown Line* — the Hindenburg support positions. On many sectors the rate of advance was maintained unchecked. Away on the distant right, 12th Division had reached the *Blue Line* as early as 0830 hours, and within an hour all III Corps' initial tasks had been achieved. In front of Ribécourt, however, *'Hilda'* had become firmly 'ditched' and a frustrated Elles had to return to his headquarters, where he found time to send a telegram to Colonel Ernest Swinton (the creator of the Tank Corps) in London: 'All ranks thank you. Your show. Elles.'

General Harper's peculiar tactics

All this had been achieved with insignificant loss — except on the Flesquières sector of 51st Division's front, where, as we have already recorded, tank casualties had been heavy. Here officers were already complaining that an excessive interval had developed between infantry and tanks. Why was this? The arguments continue to this day, but it would seem that Major-General Harper must bear personal responsibility. This officer had substituted his own infantry/tank drill for Colonel Fuller's standard version. By Harper's method, the 'advanced guard tanks' were brigaded together, and ordered to push straight ahead over the fire-trench to achieve the maximum possible initial penetration, leaving the 'main body' tanks to support the infantry clearing operations. Even more important, Harper laid down an interval of 'not less than 100 yards' between tanks and following infantry, and insisted that the latter should advance, not by platoons in file, but by sections in extended order. The cumulative effect of these changes was that the lead tanks got far ahead of the infantry, and became easy targets, while far to their rear the Highlanders, in deployed sections, found far more difficulty in struggling through the wire than their comrades on other sectors, who passed through in file immediately behind their tanks, and thus kept in close touch with them.

Here was one reason for the check before Flesquières — which ultimately ruined the effectiveness of the battle. But there were other reasons too. There was no denying that Flesquières constituted the strongest sector of the German position opposite IV Corps. Placed as it was on a reverse slope, its protective ridge forming the head of a re-entrant, leading down to the Grand Ravine valley, which was in turn flanked by the villages of Havrincourt and Ribécourt, it can be understood why General Harper doubted the advisability of a straightforward frontal assault on such a position, with or without tank support. Other unfortunate coincidences created further difficulties. The master plan enjoined a pause to consolidate and regroup for the forces operating against the flanks of the Flesquières salient after gaining their *Blue Line* objectives, and this encouraged the development of a certain loss of momentum in the centre. Furthermore, for some reason the effectiveness of the bombardment by the Allied artillery and aircraft against the German artillery positions neighbouring the village left much to be desired, many of the German guns remaining in action against the exposed forward tanks.

The garrison of Flesquières sector, commanded by Major Krebs, consisted of three battalions, two machine gun companies, and five field gun batteries, supported by four batteries of howitzers and five of heavier guns. Parts of *213 Field Artillery Regiment* were also placed under the command of *54th Division* as they materialised from Cambrai with the leading troops of *107th Division*. General von der Watter, moreover, had specifically and effectively trained his gunners to engage moving targets over open sights. These guns, therefore, proceeded to mete out destruction to 'D' and 'E' Battalions of the Tank Corps. Some accounts claim that 16 tanks were knocked out *before* they even reached the Flesquières wire; the modern consensus would seem to suggest only seven. This notwithstanding, there is no doubt that by last light on November 20 no less than 39 tanks had been destroyed or disabled (two temporarily) on the sector — sufficient, as it proved, to ruin the success of the first day's fighting and even to blunt the effectiveness of Third Army's entire offensive by its side-effects.

Opposite page. Top left: A Mark IV ditched outside Bourlon, the northern-most limit of the British advance. Bourlon Wood was taken on the third day, but despite desperate attacks over the next four days the village itself never fell
Above: A Mark IV clears barbed wire, one of the most important functions of the new tank arm. Preliminary artillery barrages (there was not one at Cambrai) consumed up to 4,500,000 shells to achieve the same end
Left: Fontaine, near Bourlon. German troops march round an abandoned Mark IV on the third day

Haig and Byng called for new efforts, but tempers were becoming brittle. 'If you don't take Fontaine, General Harper', Sir Charles Woollcombe declared, 'God help you!'

The celebrated story of 'the gunner of Flesquières' to whom even Haig paid homage in the Cambrai Despatch, has become the subject of great contention. Some versions speak of a German Major who single-handed operated an abandoned field gun to devastating effect before being killed himself. 'The great bravery of this officer aroused the admiration of all ranks' (Haig). The Germans, however, have never officially recognised the existence of this heroic officer, but *Oberleutnant* Zindler's eye-witness account claims the honour for *Unteroffizier* Krüger of the *8th Battery*. This NCO was stationed 500 yards east of the village, and the historian Becke claims that it was his gun that knocked out seven tanks, and that his body was found beside his gun.

Thus there were many factors that contributed to the 51st Division's failure before Flesquières. However, it cannot be disputed that the division's reserve brigade was never sent into the battle on the first day. It was also unfortunate that an attempt to launch 18th Brigade of the 6th Division against the flank of the position from the neighbourhood of Premy Chapel also foundered. Both circumstances are capable of explanation; *Plan GY* laid it down that the Highlanders' reserve was to be used only against *Siegfried II* – which had not been reached on this sector; and 18th Brigade Headquarters, well forward in the battle zone, was found only after a long search, and daylight was fading before an attack could be extemporised. By that time the rain was pouring down, so it was decided to call the effort off.

One consequence of these setbacks was the failure to take Bourlon and its key ridge. An important rôle in this operation was entrusted to the 1st Cavalry Division (IV Corps). This force was under orders to pass up the Grand Ravin to Ribécourt, flank Nine Wood, press along the St Quentin Canal, and finally swing northwest to attack Bourlon. The cavalry commander, Major-General R. L. Mullens, on receipt of an erroneous message from the Cavalry Corps at 1130 hours to the effect that Ribécourt was still dominated by the Germans, adjusted his orders to incorporate an advance through Flesquières. This was obviously impossible, and by the time 2nd Cavalry Brigade, leading the division, had extricated itself from

the rear area of the blocked 51st Division and reassembled in the Grand Ravine, several hours had passed. Similar difficulties prevented any participation in the proposed 18th Brigade's attack, which was in any case hampered by command problems of its own. Thus 1st Cavalry Division spent a frustrating day; the only achievement was a useful raid by 'A' Squadron, 4th Dragoon Guards, which bypassed Cantaing and routed a German headquarters near Chateau-la-Folie. This was the most dramatic penetration of the 'Cavalry Gap' on the first day.

Meanwhile, on the left flank of IV Corps, the 62nd Division had made great progress up the Canal du Nord, capturing both Havrincourt and Graincourt, and ultimately 186th Brigade reached the Cambrai-Bapaume highroad. Parts of 36th Division made equal gains along their sector to the 62nd's left. As for the fortunes of III Corps later on November 20, these too, had continued to be blessed with success. The Hindenburg Support Line had been substantially breached by midday. On the extreme right, 12th Division commenced the formation of a defensive flank along Bonavis Ridge, whilst the 70 surviving 'runners' of the III Corps tank battalions led a further advance to the St Quentin Canal. Prominent in these operations were the men of 29th Division, who reached Nine Wood and Marcoing by midday. Further to the right, the canal bank was occupied, but not before the Germans had successfully demolished the main crossings. This caused some delay in establishing the 'cavalry gap' demanded by the plan; long lines of horsemen were already

Above: A British Mark IV brings in a captured German 5.9-inch naval gun east of Ribecourt. *Below:* British tanks, two of them carrying fascines, move on the German lines.

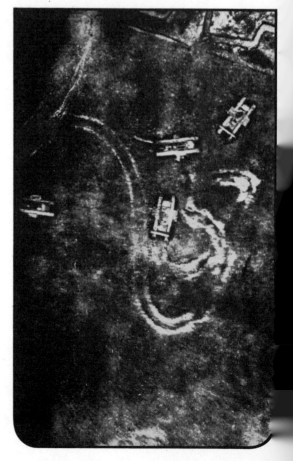

This page: 3.7-cm. German anti-tank gun *(bottom)*, manufactured by Rheinmetall. *Weight in firing position:* 385 lbs, including 24 rounds of ammunition. *Weight in travelling position:* 1,020 lbs. *Weight of shell:* 1 lb. *Barrel length:* 21.5 calibres. *Muzzle velocity:* 1,125 ft./sec.; later 1,420 ft./sec. *Range:* 2,866 yds. *Elevation:* −6° to +9°. *Traverse:* 21°. *Crew:* 3 men. *Centre picture:* German prisoners march past British anti-aircraft guns: the RAF lost air superiority at an early stage. *Below:* German anti-tank rifle, a model disliked by the troops because of its strong recoil and because the barrel quickly became unbearably hot. *Weight:* 35 lbs. *Barrel lenght:* 4 ft. 3 ins. *Muzzle velocity:* 2200 ft./sec. *Maximum range:* 70 yds. *Calibre:* 13-mm.

John Batchelor

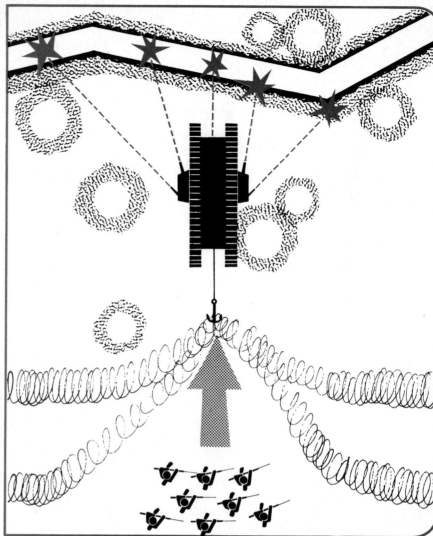

Imperial War Museum

moving up from Villers Plouich towards Marcoing, and from Gouzeaucourt towards Les Rues Vertes, but the canal line had still to be breached. This was, in fact, achieved about 1530 hours by 'B' Squadron of the Canadian Fort Garry Horse, which passed over an intact footbridge and routed a German battery, but in the end this intrepid formation was forced to stampede its horses and return to the bridge on foot under command of Lieutenant H. Strachan, who won the VC that day. Meanwhile, near Noyelles, the Reserve Squadron of the 7th Dragoon Guards also made a brief excursion to the north bank over three trestle bridges, but these relatively small scale successes, together with the exploit of the 4th Dragoon Guards already related, were the sum total of the cavalry's achievements on November 20.

The breakthrough is held
Apart from these setbacks, however, almost all objectives set for the 'break-in' phase of the battle had been achieved. A huge breach—almost six miles wide and up to 4,000 yards deep, had been torn into the Hindenburg Line; even *Siegfried II* had been severely breached on the Marcoing sector. This achievement had cost some 4,000 British casualties, but over 4,200 German prisoners had been taken, besides 100 guns. Much would clearly depend on the outcome of the next day's fighting. Overnight, however, the Germans were taking steps to check the rot. The intrepid survivors of the Flesquières garrison were withdrawn; the *107th Division* continued to come up from Cambrai, proving an in-

valuable source of men for patching the crucial sector between the Bapaume road and Masnières. They were still very anxious about the situation, being unaware of how many of the British reserves had already been committed to battle. 'Fortunately,' recorded the German High Command, 'the enemy was himself taken by surprise by the extent of his success.'

The renewed battle on November 21 from the British point of view is one of increasing exhaustion and a general reduction in forward momentum. In the early hours 51st Division had at last occupied a deserted Flesquières, and 154th Brigade passed through to attack Cantaing—captured, along with 300 prisoners, by 0030 hours. A number of tanks and part of the 5th Dragoon Guards shared in this success, which extended the breach driven into the *Siegfried II* position to a width of well over two miles. However, the desperate weariness of the infantry—and the tank crews—made a full exploitation impossible. Petrol, water and even ammunition were all in short supply. So the theme for November 21 was one of 'consolidation' rather than of dramatic gains. There were, indeed, a few setbacks. On left of IV Corps, a German counterattack retook Moeuvres from 36th Division. That evening, however, Sir Douglas Haig decided that the battle should continue, although his original time limit had been reached without the gaining of all key objectives—most especially the Bourlon position and the exploitation of the 'gap'—and despite the increasing evidence that the Germans were fast recovering from their initial surprise and be-

ginning to bring up reinforcements. However, the bolder course won. 'After weighing these various considerations, I decided to continue the operations to gain the Bourlon position.' Byng was accordingly ordered that IV Corps, supported by V Corps, was to continue its attack, whilst III Corps' offensive was to be 'shut-down' to a holding action on the right. The lure of Bourlon Ridge thus continued to exert its strong—and deadly—fascination over the Commander-in-Chief.

The next week's events were dominated by a herculean struggle for Bourlon. It was predominantly an infantry slogging match of the usual kind. For the British, the new phase opened badly with the loss of Fontaine on November 22 as German counterattacks slammed home against the tired 51st and 62nd Divisions. However, in rear the Guards Division and two fresh brigades of 36th Division were preparing to mount an attack to regain Moeuvres, while the artillery poured a relentless fire into Bourlon Wood and village.

Next morning (November 23), 100 tanks led 40th Division against Bourlon ridge, with 432 guns giving fire support. Bourlon, however, was now the responsibility of *Gruppe Arras* which had two divisions in the line and a further two (including the *3rd Guards*) moving up in support; there were also now 200 German guns to help contest the sector, while the arrival of Baron von Richthofen's famous 'circus' implied a serious challenge for control of the air over the battle zone. Meantime *Gruppe Caudry* had received two new divisions, and *Gruppe Busigny* had taken over the left of

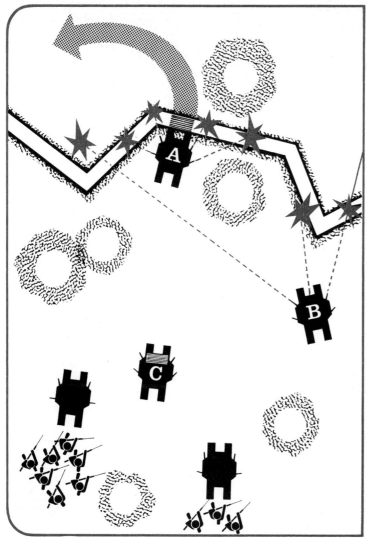

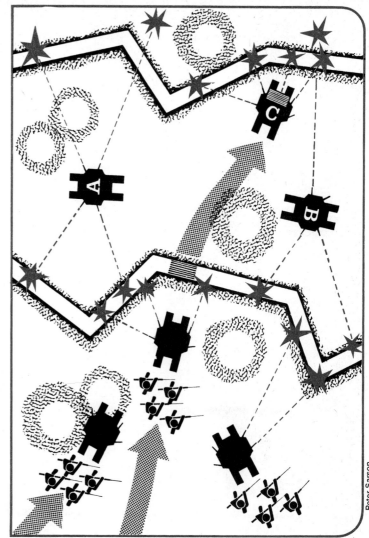

Peter Sarson

Opposite page. Top left: A fascine-laying Mark IV. The fascines, carried on the unditching rail and released from inside, were 4½ feet in diameter, 10 feet long and 1½ tons in weight. *Top right:* The rôle of the 32 special grapnel-tanks. *This page. Above and above right:* A closer view of Fuller's tanks battle drill (shown on page 110), showing in more detail the relationship between infantry and fascine-laying tanks

the German line to relieve its battered neighbour of all distractions.

The struggle for Bourlon reached its peak on November 23, when Major-General Sir John Ponsonby's 40th Division reached the crest of Bourlon Wood, but failed to penetrate the village beyond. In the space of 72 hours this single division was to lose 4,000 casualties. The Germans were suffering too, but Moser (commanding *Gruppe Arras*) noted that 'our artillery and aeroplanes are gaining the ascendancy more and more'.

Haig and Byng called for new efforts, but tempers were becoming brittle. 'If you don't take Fointaine, General Harper,' Sir Charles Woollcombe declared to the commander of 51st Division, charged with a new objective on the right of 40th Division, 'God help you!' Despite this admonition, little progress was ultimately made, and 40th Division's flank remained dangerously exposed. Shortages of manpower were being experienced, and Byng authorised IV Corps to use 1st Cavalry Division in infantry rôles. He also brought in the Guards Division to relieve the 51st.

A further major effort was ordered for midday on November 24; the gallant 40th Division and the dismounted 9th Cavalry Brigade were to be thrown against Bourlon at noon; the Guards were to attack the Fontaine sector, and the 36th and 56th Divisions were to clear German positions between Moeuvres and Inchy.

The battle was ferocious; at first it seemed that 40th Division might take the village, but by 1100 hours they had been repulsed save for one small party under Colonel Wade which clung to a segment of the village for two days. By this time the whole of the Cavalry Division was involved, but by last light the Germans had regained the northern edge of the Wood. Sir Douglas Haig might still publicly talk about a breakthrough by three cavalry divisions, and had managed to detain two divisions earmarked for transfer to Italy. Brigadier-General Charteris (his Intelligence chief) was aware of the true situation: 'Things have not gone well. Our troops are tired, and the Germans are getting up large reinforcements; we have none available.'

Byng continued to fling his weary and depleted divisions against Bourlon, but to no avail. He switched his divisions to and fro, set new deadlines, tried to inspire his commanders and men to new exertions, but it was by now beyond the power of Third Army. Colonel Wade was rescued on November 26, and heavy bombardments were hurled at the Germans, who by now had seven divisions in the line, four of them in the key sector, and their artillery strength was 500.

The last effort, as it proved, was made on November 27. The 62nd and Guards Divisions, accompanied by 30 tanks, advanced yet again against the Bourlon and Fontaine sectors. Casualties were heavy. Fontaine was partly overrun, but a German counterattack, ten battalions strong, proved too heavy, and the Guards Division was driven back to its start line. By this time the 2nd Guards Brigade had been reduced to 500 men. At the day's end, the British line in Bourlon Wood had been driven back to the crest amongst the trees.

After this failure the Commander-in-Chief was prepared to concede failure, and Third Army was ordered to close down its offensive. On November 28, IV Corps headquarters issued a significant order: 'Wire should be got up at all costs today, and wiring on the Bourlon perimeter systematically carried out by pioneers and R.E.' This signalled the end of the British adventure of Cambrai: the old conditions of stalemate had reasserted themselves.

But the history of the Battle of Cambrai still had another, final, act to run. German artillery were now hurling shells against the newly-extemporised British positions in Bourlon Wood. On other sectors, both November 28 and 29 were relatively quiet days—but to some this seemed more menacing than reassuring. These fears were proved correct when, at 0830 hours on November 30, an intense German bombardment rained down on the British lines. Soon alarm signals were being received from many sectors. As one sergeant of the 1st Royal Berkshires reported to his officer: 'SOS gone up in 27 different places, and the Bosche coming over the 'ill in thousands.' The German counteroffensive had begun.

121

CAMBRAI
THE GERMAN

Ten days after the tank surprise at Cambrai the Germans repaid the compliment with one which was similar in principle but different in method. Without a long artillery preparation, a short hurricane gas and smoke bombardment cleared the way for an infantry attack using new infiltration tactics. This was a foretaste of the offensives which were to smash through the British and French armies in spring 1918.
David Chandler.

COUNTERATTACK

Below: A German war artist's impression of the first day of the stunning counterattack

A battery of 77-mm's in action.
Shell fire was the greatest single cause of tank casualty at Cambrai,
accounting for over one third of all losses on the first day (65 out of 179 tanks)

Following the eventual stabilisation of the front after the exhaustion of General Sir Julian Byng's dramatic but brief offensive, it was not long before the German High Command was planning a counteroffensive. On November 27, while the battle of Bourlon Wood still raged, Ludendorff conferred with Crown Prince Rupprecht at Le Cateau. Meantime 20 divisions were being massed in the Cambrai area. Ludendorff was convinced that 'there has never been such an opportunity' for a successful blow, and insisted that all should be in readiness by November 30.

The plan envisaged a short, intense bombardment followed by a rapid attack by *Gruppe Caudry* and *Gruppe Busigny* in the general direction of Metz, the village immediately to the south of Havrincourt Wood. If successful, the German generals hoped that this would compromise the whole British salient and compel its evacuation. Meanwhile, *Gruppe Arras* (some seven divisions and 130,000 men strong) would fire heavy barrages and conduct demonstrations designed to distract the British IV Corps (Lieutenant-General Sir Charles Woollcombe), and once the eastern attacks had become well established, Lieutent-General Otto von Moser was to launch an attack with three divisions to the west of Bourlon Wood and drive southwards. Reserve divisions would be held in readiness to consolidate and exploit the gains on both sectors of the German attack. The minimum to be achieved was the recapture of the Hindenburg position, but some officers spoke hopefully of a major breakthrough.

November 28 saw the opening of a heavy bombardment against Bourlon Wood, 16,000 rounds of gas and high explosive shells being pumped into the ravaged area. The violence of this shelling served to convince the British High Command that the major German blow, if and when it came, would almost certainly be launched against the northern part of the salient. Not that warnings of other possibilities were lacking. Major-General J. S. Jeudwine, GOC 55th Division (VII Corps), whose troops were holding a six-mile front facing *Gruppe Busigny,* reported ominous German concentrations in the vicinity of Banteux and Twenty-Two Ravines. Indications of a coming attack included German artillery registration on hitherto unshelled areas, a marked increase in German air activity over the front line, and the measures taken to prevent Allied aircraft reconnoitring in the Banteux area. Yet there was little sense of urgency at either GHQ or Third Army Headquarters, and Byng authorised no preliminary movement by the slender reserves.

The British generals later denied vehemently that they were taken by surprise by both the scale and direction of the German onslaught unleashed on November 30, and attempted to blame junior commanders for lack of proper vigilance. Be that as it may (and subsequent research has not borne out the conclusions reached by the Cambrai Enquiry), in the words of the late Sir Basil Liddell Hart, the Germans 'repaid the tank surprise by one that was similar in principle if different in method'. When the German guns abruptly crashed

out shortly after dawn on November 29, there was no immediate indication of the line of the coming attack. All too soon, however, SOS signals were hurtling skywards and reports of attacks flooding into HQ VII Corps, as wave after wave of German troops (some 12 in all, the rearward lines being led by mounted officers) emerged one after another from the Banteux Ravine at the precise time (0700 hours) that Jeudwine had asked for the heavy bombardment. The focal point of the German attack was the boundary between III and VII Corps, and the brunt of the onslaught was borne by Major-General W. D. Smith's 20th Division and Major-General A. B. Scott's 12th Division—occupying the right of Pulteney's sector—and Jeudwine's over-extended 55th on the extreme left of VII Corps. The southernmost limit of *Gruppe Busigny's* attack was opposite Vendhuille, but the northern extension rapidly grew as, at about 0800 hours, the Germans committed *Gruppe Caudry* to the attack from Bantouzelle to Rumilly, and drove forwards towards Marcoing. Thus III Corps suddenly found three of its four front line divisions heavily, and then critically, involved.

The form of attack was novel on the Western Front. Behind the intense and short bombardment, the German infantry came forward in groups, employing infiltration tactics. Centres of strong opposition were deliberately bypassed by the leading formations, which pressed on towards the rear exploiting the line of least resistance. Low-flying aircraft, strafing and bombing, gave close support, and num-

bers of artillery batteries advanced immediately behind the infantry. Similar methods had been employed with success in both Russia (notably at Riga) and Italy. They worked well enough in France on this bleak late-November morning, and would be re-employed on a huge scale the following spring with even more telling effect. The speed of the German penetration took many formations completely by surprise. Major-General Sir B. de Lisle (GOC 29th Division) was almost captured at Quentin Mill. Headquarters of 35th Brigade (12th Division) was surrounded, but Brigadier-General B. Vincent rallied his staff and fought his way out back to Revelon Ridge, where he began to form 'Vincent Force' from men of many units in an effort to stem the tide of the German advance, which was overrunning the gun-line, capturing many pieces before they could be withdrawn.

The British line is torn open
Soon the rear area of III Corps was jammed with retreating transport and guns, and here and there the flood was joined by units of administrative troops. Numbers of front line battalions found themselves cut off, and communications were plunged into chaos. Generals desperately tried to find reserves to bolster the collapsing front. But the German advance swept remorselessly forward, and by 1030 hours an eight-mile sector of the British line from Les Rues Vertes to Vendhuille had been driven

in, the maximum penetration bringing the Germans to within two and a half miles of Metz after their capture, in succession, of Gauche Wood, Gonnelieu and Gouzeaucourt. The significance of Metz to the German plan lay in the fact that through it ran the most important route up the Grand Ravine to Flesquières and distant Bourlon Wood.

On the northern sector, meanwhile, another heavy engagement had flared up at about 0930 hours as Moser's *Gruppe Arras* began the secondary attack against IV and VI Corps. The German *49th Reserve, 214th* and *221st Divisions* advanced along a three-mile front running from Tadpole Copse in the west to the approaches to Bourlon Wood, supported by the fire of 500 guns. However, the Germans found themselves facing a maelstrom of artillery fire. In anticipation of just such an attack, Third Army had allocated IV Corps no less than eight division's worth of supporting artillery, together with corps and army heavy batteries, and these now made their weight of fire felt. Thus, when the German *3rd Guards Division* (east of Bourlon) attempted to probe towards Cantaing, the attack was disintegrated by gunfire, and by midday it had made no progress. On Moser's main sector of attack, however, the greatest pressure was felt by 56th Division (VI Corps), 2nd and part of 47th Divisions (both parts of IVth Corps). Despite heavy casualties, the German attacks kept coming forward. One group of eight British machine guns, part of 2nd Division, fired all of 70,000 rounds into the flank of successive attacks at close range

with dire effect. Nevertheless, so well pressed-home were the German attacks, that at 1030 hours IV Corps warned its artillery to be ready to fall back at short notice and took preliminary measures to garrison the Hindenburg Support Line in case it should prove necessary to abandon the front. For a time it seemed impossible that Major-General F. A. Davidson's 56th Division could remain in line, but it did hold despite relentless pressure and a growing shortage of ammunition. The Germans made some ground near Tadpole Copse, but nothing vital was lost. On the right of the 56th, the brigades of 2nd Division (IV Corps) also came in for a difficult time. For a period they found themselves out of contact with Major-General C. E. Pereira's headquarters, but 'A' Squadron of King Edward's Horse provided the GOC with 15 gallopers, who maintained a tenuous link with the embattled front line. Many units sacrificed themselves during these desperate hours, and many deeds of great gallantry were performed. The net result was that *Gruppe Arras's* attack proved abortive on November 30. The *Official History* justly describes this as 'an outstanding British achievement in the war on the Western Front'.

It must be recognised, however, that this local success was made possible only at the cost of absorbing a very high proportion of Third Army's resources in terms of both men and guns. Haig urged Byng 'to use his reserves energetically' as GHQ took steps to move four divisions from other sectors towards Bapaume and Peronne as a safeguard against a German breakthrough.

Below: Germans pose confidently on the wreckage of a British tank which a few days before had sent them running in terror. *Inset:* Another wrecked Mark IV, this one at Rumilly, captured by the Germans on the first day of their counterattack

But Byng had soon played almost all his cards; as we shall see, the whole of the Cavalry Corps was eventually committed in support of VII and III Corps, as well as 61st Division to man the Hindenburg Support Line. By 1100 hours it had also proved necessary to transfer the Guards Division from IV to III Corps command.

The major area of crisis remained on the eastern sector, where confusion was threatening to convert defeat into disaster. A series of circumstances staved off this danger. Brigadier-General Vincent's intrepid force on Revelon Ridge was first joined by 20th Hussars, who at once entered the fight in a dismounted rôle. Next to appear was 1st Guards Brigade from Metz, the first echelon of Major-General G. P. T. Fielding's division to be sent up from a rest area to patch the staggering line. The 1st Guards Brigade did more: extemporising a counterattack towards Gouzeaucourt, by 1130 hours it had succeeded in clearing the village, while to the south large numbers of British cavalry extended the *ad hoc* line. The recapture of Gouzeaucourt was consolidated by the arrival of 36 tanks of 'A' and 'B' Battalions, switched to the south after the abrupt cancellation of their movement order to the Somme. Then 5th Cavalry Division, after a considerable approach march from Peronne, passed through the Guards towards Villers Guislain, only to be halted by severe German fire. Snow was of the opinion that it would have been far better to unleash the cavalry against the overextended southern flank of the narrow and still disorganised German corridor extending from the east of Gouzeaucourt, but Byng insisted on committing the cavalry to frontal operations. This decision may have sacrificed a chance of a telling riposte to the German breakthrough, and the Germans were afforded an opportunity to consolidate their gains. Be that as it may, by dusk some sort of a British line had been cobbled together, linking the shattered right wing of III Corps with the left of VII. By last light, 3rd Guards Brigade had come up, with 2nd in support in Gouzeaucourt Wood. The 20th Division was rallying on Welsh Ridge, and on its left 29th Division was still in firm possession of Marcoing after a day of heavy fighting. Administrative confusion had still to be sorted out, and many men went hungry. Others were more fortunate. 'Our rations captured', noted Lieutenant-G. E. Chandler of 11th Essex (2nd Division), who had spent an exhausting day digging extemporised positions, 'but another division shared with us'.

On December 1 much of the drama of the day's fighting again centred around the events on the southern flank. With the intention of forestalling a renewed German attack, at 0630 hours 3rd Guards Brigade launched an attack towards Gonnelieu. It did not fare too successfully, but the appearance of a single tank at a critical moment near Green Switch Trench caused many Germans to surrender. Better fortune blessed 1st Guards Brigade which, aided by 16 tanks of 'H' Battalion, succeeded in retaking much of Gauche Wood and Quintin Mill. Mounted action by 4th and 5th Cavalry Divisions on the left of the Guards achieved little of positive value, but there were now 74 tanks available in support of III and VII Corps flank.

In the centre matters did not go well, both Masnières and Les Rues Vertes were lost, but the German impetus was by now nearly exhausted. The same was true of the northern side of the salient, where the immediate crisis was deemed to have passed by the end of December 1. Lieutenant-General Sir E. A. Fanshawe and HQ V Corps was able to relieve Lieutenant-General Woollcombe and HQ IV Corps—the original hand-over planned for the 30th having been postponed.

The Germans had not yet quite shot their bolt, however. After a reasonably quiet day on the 2nd, they advanced once again on December 3 and took possession of La Vacquerie from 61st Division, which had relieved parts of the exhausted 12th and 20th Divisions only the previous day. The same day, it was deemed necessary to withdraw from the British bridgehead on the east bank of the St Quentin Canal and reconcentrate 29th Division around Marcoing. Meanwhile, the German line assumed its final shape, running from south of Quentin Ridge to Welsh Ridge and thence to Marcoing.

Ludendorff stridently claimed 'an offensive victory on the Western Front' but admitted that matters had not developed 'quite as well as I had hoped'. High British staff circles were far from happy with the overall situation. On December 2, Haig visited the front and told Byng 'to select a good winter line' but not to issue any orders on the subject at that point. The problem was to find a truly secure position. Anxiety centred around the continued German possession of Bonavis Ridge; unless it could be recovered, Haig considered that Third Army's position would remain precariously unbalanced, and the whole Marcoing/Bourlon salient would be compromised. Yet there were no fresh troops whatsoever with which to attempt its recapture, and the thought of becoming involved in a battle of attrition through the winter was unattractive to say the least, and, as Haig pointed out to the CIGS, there would be no question of sending further aid to Italy under the prevailing circumstances. A hard decision had to be taken. On December 3 it was decided that the Bourlon salient must be evacuated; the orders were issued next day. On the night of the 4th/5th the evacuation began, the troops falling back to the 'Yellow Line', some 2,000 yards ahead of the chosen final position. Thus blood-stained Bourlon Wood, the Marcoing area, and long stretches of the Hindenburg Line, were all abandoned to the Germans. The operation was completed by the early hours of December 7. To the west of the Canal du Nord, the British line was almost back where it had been on November 19. In the centre, Havrincourt, Flesquières and Ribécourt remained in British hands, the new line here running along parts of the Hindenburg Support Line, but from Welsh Ridge it swung sharply away to the southwest, and the Germans were left with considerable areas of former British territory in their possession. Losses and gains more or less cancelled each other out.

Not unnaturally, the Germans were astounded by these unforeseen developments. 'We are jubilant,' wrote Moser; 'since 1914 the first withdrawal of the proud Briton'. Ludendorff later described it as 'a good ending to the extremely heavy fighting of 1917. Our action has given us valuable hints for an offensive battle in the west if we wished to undertake one in 1918.'

For the immediate future, however, the Germans needed rest and recuperation as much as their opponents. So ended the famous battle of Cambrai. For the British it had proved 'a sombre sunset after a brilliant sunrise' (Liddell Hart).

The cost of the 17 days of triumph and tragedy had not been light. From November 20 to December 8, the 19 divisions of Third Army had lost 44,207 men. Over 6,000 of these casualties, together with the loss of 103 field and 55 heavy guns, had been the cost of November 30. The German *Second Army* (ultimately 20 divisions strong) confessed to 41,000 casualties, and the loss of 145 guns, but probably excluded considerable numbers of lightly wounded. Of their total admitted loss, 14,000 were suffered during the counteroffensive. What had been gained by either side from these exertions? In terms of the Western Front, precious little ground or overall advantage. It is true that two German divisions from the Eastern Front, intended for Italy, had been diverted to Flanders, but the British programme for reinforcing their Italian allies had also been at least partially disrupted. On the other hand, both sides had learnt much of tactical value: on the British side the employment of tanks; on the German the value of infiltration tactics.

As might be expected, there was much criticism in London about the reasons why 'a resounding victory' had been allowed to degenerate into 'a disastrous rout' (Lloyd George). Characteristically, Haig did not flinch from accepting the ultimate responsibility. On December 3 he had written '. . . whatever happens, the responsibility is mine'. At the Cambrai Enquiry, several senior officers, including Byng, tried to blame negligence on the part of the fighting soldier. 'I attribute the reason for the local success on the part of the enemy to one cause and one alone', testified the commander of Third Army, 'namely—lack of training on the part of junior officers and NCOs and men.' Although General J. C. Smuts, consulted by the War Cabinet as an independent assessor, substantially agreed with this assertion, the verdict has been much modified with the passage of time. In the words of a recent historian of Cambrai, Robert Woollcombe, 'the High Command was responsible for the confusion, and the troops for the valour'. Perhaps the ultimate reason for the bitter disappointments of the period between November 30 and December 7 has been most aptly summarised by John Terraine: 'Two years of offensive strategy had diverted the British army's attention from defensive problems.' This was demonstrably true at Cambrai. This shortcoming, together with the weariness of the troops, the shortage of adequate reserves, and the failure of British military intelligence to estimate correctly the strength and intentions of the Germans, accounts in large measure for the final outcome of this celebrated engagement which brought the bitter year of 1917 to an unmourned conclusion.

Further Reading
Liddell Hart, Sir Basil, *The Tanks,* Vol 1 (Cassell 1959)
Terraine, J., *Douglas Haig* (Hutchinson 1963)
Woollcombe, R., *The First Tank Battle: Cambrai 1917* (Barker 1967)

The new tank arm had seized almost all of its 'break-in' objectives on the first day of Cambrai and had opened the way for a massive advance, but on the second day the old conditions of stalemate reasserted themselves and the advance degenerated into yet another slogging match. Why?
General Sir Michael Carver

FALSE DAWN
TANKS AT CAMBRAI

The arguments about the use of tanks at the Battle of Cambrai in November 1917 were to set the pattern for arguments about the use of tanks in almost every battle in which they were employed thereafter. Protagonists and antagonists were to adopt much the same positions and to remain as blind to and contemptuous of the views of their opponents as they were in the aftermath of the battle itself.

The criticisms levelled by the tank enthusiasts in general were, first, that a higher proportion of the total number of tanks available should have been kept in reserve, ready for use after the initial attack on the Hindenburg Line; secondly, that there were no reserves of infantry available for the same purpose, and finally that the cavalry were not only too far back to exploit the opportunity to break out when it was presented, but that, when they were brought up to the starting post, they were unwilling or unable to achieve anything. In sum, the magnificent opportunity created by the employment of tanks in conditions that, for the first time, exploited their potentiality, was thrown away because the higher command had insufficient faith in the tanks to back them up properly with adequate reserves and placed too much reliance on the cavalry.

A fair amount of this criticism is based on hindsight: some of it is distorted and self-contradictory; but in general it is sound. However, it is oversimplified and glosses over many of the real difficulties.

First of all it is important to clarify the aim of the battle—it changed significantly in the process of conception, and this had important effects on the soundness of its plan. Although Fuller's original idea, conceived as far back as June 1917, was for a joint Franco-British attack to capture St Quentin as a preparatory step for a further offensive in the spring of 1918, this was modified in discussion with Elles to a plan for a raid south of Cambrai 'to destroy the enemy's personnel and guns, to demoralise and disorganise him, and not to capture ground'. The area chosen was that on which the Battle of Cambrai was subsequently to be fought.

Fuller's proposal was for a force of three tank brigades of two battalions each, one, or preferably two, divisions of infantry or cavalry with additional artillery and two air squadrons. The raiding force was to form three groups, the principal one intended to 'scour the country' in the area between the Canal du Nord to the west and the Canal de l'Escaut (also known as the St Quentin canal) to the east, a frontage of about six miles, while smaller groups operated on each flank.

It was envisaged that the first line of tanks would go straight through to the Germans' gun positions and attack them: the second would crush the wire and work up and down the forward trench system in co-operation with the infantry: the third would make for points of special tactical importance and would help in the mopping up. Only three hours after the attack had been launched, the force would begin to withdraw, forming a rearguard in co-operation with aircraft to protect dismounted cavalry who, it was imagined, would be escorting large numbers of prisoners back to the British lines.

It is surprising that so eminent and perceptive a military critic as Liddell Hart should have looked favourably on this plan. To launch a major attack on the most suitable ground for tanks on the whole front: to give a dress rehearsal, as it were, of what one would hope to do in any subsequent major decisive action: to abandon it all after only a few hours, having inflicted damage which would certainly not be decisive and which would probably not take long to make good, inevitably leaving a number of one's own tanks on the abandoned battlefield: to commit all these elementary military errors would have been folly indeed.

The idea smoulders on

Fortunately this plan was never pursued, but with its acceptance of a limitation of resources it had a pernicious influence on the plans which developed from it. Brigadier-General Tudor, Commander Royal Artillery of 9th Division, was anxious to try an attack with no preliminary artillery bombardment or registration, relying on survey methods of gun-laying and on the tanks for crushing wire. He persuaded his superiors in IV Corps to look favourably on these ideas as the basis for an ambitious attack 'to advance to Flesquières ridge, swing left and roll up the German line to the Scarpe, with cavalry operating towards Cambrai'. This was whittled down at corps HQ to a more limited operation involving little more than the occupation of the front and support trench systems of the Hindenburg Line between the two canals, and the destruction of the forces, including their artillery, in that area. Three divisions, two cavalry brigades, eight field artillery brigades and Third Army's heavy artillery were considered necessary for this operation, and it was suggested that the Tank Corps should assess the number of tanks needed.

The corps commander, Lieutenant-General Woollcombe, sent this plan up to Third Army on August 23. Between then and September 6 Elles visited the area and found it very suitable for tanks. So much so that he thought that more could be achieved and that the tanks could clear the area up to the eastern canal. General Byng was fired with enthusiasm again

and had visions of his favourite cavalry at last coming into their own. He envisaged that the tank action would lead to a complete breakdown in the Germans' defence, as there were only two German divisions in the area and it was estimated that 48 hours must elapse before any reinforcements could arrive. By that time General Byng saw the whole Cavalry Corps deployed, galloping northwards to the River Sensée and breaking out northeastwards round Cambrai to cause havoc on the German lines of communication.

Cold water was poured on this glorious concept by GHQ, but General Byng was told to continue to plan the operation in detail in complete secrecy. As the Flanders offensive gradually bogged down in the mud, Haig began to look around for other means of keeping the offensive going in order to prevent the Germans from attacking the French, still reeling from their mutinies. His aim was not just altruistic. He was determined to avoid taking over any more front from the French or reinforcing them.

On October 13 the go-ahead for Cambrai was given. Byng was allowed to allot four fresh infantry divisions (of the five he said he wanted) to start training with the three tank brigades (less four battalions) that had been made available. A commanders' conference was held on October 26 at which the plan was divulged, the very day that Haig was told by the War Office to send two divisions immediately to stop the rot on the Italian Front as a result of the disaster of Caporetto.

The final plan for the battle as issued on November 13 was modified by Byng after comments by Haig. The aim of the operation was now to clear the whole area between the two canals right up to the River Sensée, which would involve an advance of about 15 miles as the crow flies. On the right III Corps was to secure and extend the right flank on the high ground west of the Escaut Canal. On the left IV Corps was to seize Bourlon Wood and Marquion, three miles further northwest on the Canal du Nord. The Cavalry Corps was to cross the Escaut Canal, surround and isolate Cambrai and cover the advance north and northeast of V Corps, which was to exploit the success as far as the River Sensée, sending troops to occupy the hills beyond it. Haig would then decide whether to advance towards Douai and Valenciennes, six and 14 miles respectively beyond the river. Haig insisted that Bourlon Wood should be captured on the first day and said that he himself would call the whole offensive to a halt after the first 48 hours, or even earlier, if the results gained or the general situation did not appear to justify its continuance.

This grandiose and ambitious plan was a far cry from the Tank Corps' proposal in June of a three-hour raid, penetrating only a few miles. In the light of it, it is difficult to accuse Haig of having had no faith in the tanks, although he can certainly be accused of expecting too much of his own arm, the cavalry. He might even be charged with having had too much faith in the tanks and of not allotting sufficient resources to see that the initial success expected of them could be exploited.

In the light of the general concept of the battle, as it had finally developed, we must now consider the criticisms made of the part the tanks had to play. The grand total

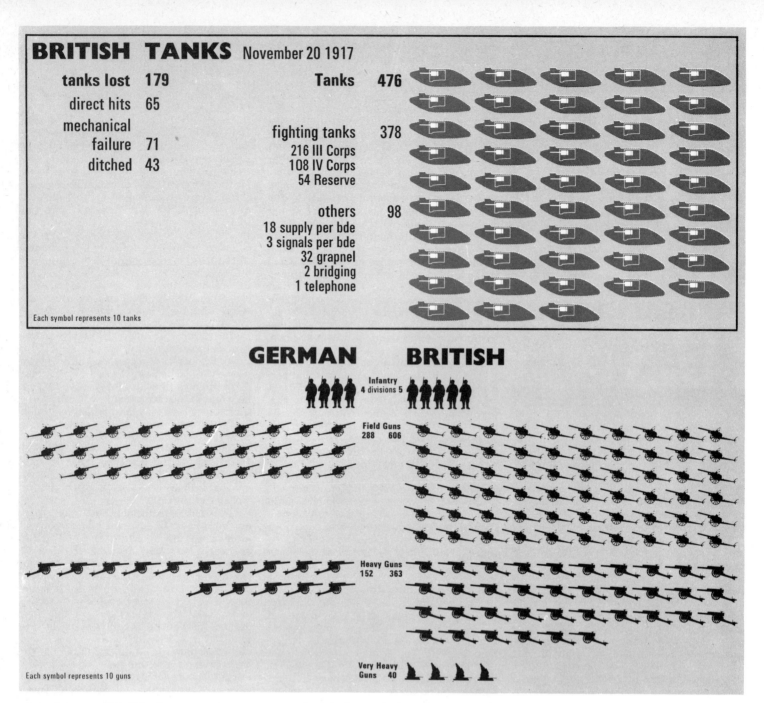

BRITISH TANKS November 20 1917

tanks lost	179
direct hits	65
mechanical failure	71
ditched	43

Tanks	476
fighting tanks	378
216 III Corps	
108 IV Corps	
54 Reserve	
others	98
18 supply per bde	
3 signals per bde	
32 grapnel	
2 bridging	
1 telephone	

Each symbol represents 10 tanks

GERMAN   BRITISH

GERMAN		BRITISH
	Infantry 4 divisions 5	
288	Field Guns	606
152	Heavy Guns	363
	Very Heavy Guns 40	

Each symbol represents 10 guns

---

of tanks was 476. Whether or not more could have been made available is not apparent and nowhere discussed. One must assume that HQ Tank Corps made available every tank that it was practicable to produce, man and maintain on the battlefield. From this total one should subtract 54 supply tanks; 32 fitted with grapnels for pulling away wire on the cavalry's path of advance; nine acting as wireless stations, one carrying telephone cable for HQ Third Army and two bridging tanks. This left 378 tanks, 324 in the frontline and 54 in reserve to replace breakdowns. Thus 324 was the figure available for specific tasks.

If, as has been suggested, as many tanks should have been kept in reserve as were allotted to the initial assault, the number of tanks available for the latter would have to have been reduced to about half of those actually employed. This would have been to put at risk the success of the attack on a 10,000 yard front of the formidable Hindenburg Line. On the right III Corps attacked with three divisions each of which had two battalions of tanks, one company of A Battalion with the 20th

being detached to 29th Division in reserve —the only reserve of tanks. The corps therefore had 216 tanks. IV Corps divided its three battalions (108 tanks) between its two assaulting divisions, 51st Division on its right, facing Flesquières, having one company more than 62nd Division on the left. In both corps approximately two-thirds of the tanks were to be used against the first objective, the first trench system of the Hindenburg Line, and one third, with survivors from the first attack, on the second objective which was generally speaking the support trench system, 3,000 yards further on. Some tanks, especially those directed on to the crossings over the Escaut Canal which the cavalry were to use, were given specific subsequent objectives.

The remarkable success of the initial assault was proof of the soundness and efficiency of the tactical plan and of the organisation of this phase. To have halved or seriously reduced the proportion of tanks to infantry or to yard of front would have prejudiced the chances of success almost as much as it would have done to have reduced the frontage on which the attack was launched.

The operations on the first day, November 20, proved that on firm ground the combination of close co-operation between tanks and infantry with an artillery fire-plan without preliminary bombardment or registration could overcome a defence which was taken by surprise both by the timing of the attack and by the scale of employment of tanks. How much of the success was due to the morale effect of the tanks and how much to the favourable conditions of poor visibility it is not possible to assess. But to suggest that significantly fewer tanks could have been used in the initial assault in order to retain more reserve for subsequent phases is to be wise after the event. In no previous action on the Western Front had attacks penetrated so far so quickly and achieved such decisive results. Too great an optimism as to the effect likely to be produced could have led to failure which would have been to the certain detriment not only of the acceptance and development of the tank, but also to the success of the following year's offensive.

As far as the tanks themselves were concerned, nobody could have expected more of them on the first day. The failures lay elsewhere. If 51st Division had not had peculiar ideas of its own which separated its infantry from the tanks, Flesquières might not have proved the stumbling block it did. The inflexibility of the command arrangements, and the primitive communications available, meant that this stumbling block was not removed, as it could easily have been, by a thrust from the flank or the rear. This prejudiced severely the possibility of achieving the Commander-in-Chief's aim of capturing Bourlon ridge on the first day, an ambitious aim in any case.

The failure of the cavalry

The other major disappointment was the failure to secure, develop and exploit a crossing over the Escaut Canal between Masnières and Marcoing in spite of the existence of a bridge capable of taking cavalry. Had the cavalry been brought up closer behind the attacking troops; had everybody made fewer mistakes and errors of judgement and omission; had the cavalry themselves acted boldly, perhaps the result might have been different. But the fact remains that, even after the barriers of wire had been left behind, the cavalry was both too vulnerable and too weak in fire and shock power on the battlefield to achieve anything when machine guns were present, unless the cohesion of the Germans' defence had totally broken down. The presence of the German *107th Division* carrying out a relief in the Cambrai area was an unexpected asset to the defence, and it is most unlikely that, had the cavalry succeeded in crossing the canal in any strength on November 20, they could have developed operations of any significance beyond it on subsequent days or even perhaps been able to maintain their position there. Nevertheless, in view of the Commander-in-Chief's intentions, and the optimistic assumptions on which the plan was based, the initial positioning and subsequent deployment of the cavalry is certainly open to criticism.

Given the basic failure to secure on the first day the two vital objectives on which the whole subsequent plan depended, what is to be said about the operations on November 21, and the following days? The available strength of tanks is relevant. Of the 378 tanks available on November 20, 179 were out of action by the end of the day, 65 by direct hits, 71 by mechanical troubles and 43 by ditching or other causes. Two hundred were theoretically available on November 21 — perhaps more, both because some of the mechanical troubles were cured, and because the grapnel-carrying tanks had completed their task. It must also be remembered that the decision as to what to do on the 21st had to be taken at an hour on the 20th early enough to enable orders, the distribution of which in 1917 was a slow business, to be issued all down the line, based on the scanty and inaccurate information available when dusk fell. In fact Third Army's orders were issued at 2000 hours, having been previously approved by Haig. In essence they were to complete the tasks which had not been achieved on the first day. Few other commanders would have made a different decision.

With all the advantages of hindsight it is now possible to judge that the decision of Haig and Byng was almost certainly wrong.

Many of their basic assumptions were false, especially as to what cavalry could achieve. They were over-optimistic, not only on that score, but also as to the effect on the Germans of even the unprecedented degree of success achieved by the tanks on the first day. If they had appreciated that they could count on success only where and when they could repeat the sort of operations that day had seen, they might have revised their original unrealistic plan on more realistic lines. The tanks could not cross the Escaut Canal. It was therefore hopeless to expect operations beyond it to be developed successfully. If the plan had been limited to making certain of securing a line west of the canal, including Bourlon ridge, as a jumping off place for a subsequent offensive, it might have succeeded and the ignominious withdrawal caused by the counterattacks at the end of the month been avoided. Even though the initial plan was hopelessly unrealistic, there was still a chance of achieving this more limited aim, if a change had been decided on at the close of the fighting on November 20. It would have meant a pause on the 21st while tanks and infantry were reorganised, and prepared for a second major thrust for Bourlon ridge, while the right flank was firmly held. Even if this had failed on November 21, it should have been possible to develop yet another attack within the next few days before all chance of success disappeared. Admittedly the line reached would have been awkward to hold, overlooked as it would have been between Fontaine and Flesquières by the high ground east of the canal south of Cambrai; but it would have been much easier and less vulnerable to hold than the deep and narrow salient that would have resulted if the original plan had been successful.

However unrealistic the plan may have been and whatever the possibilities of better results than those achieved, the repercussions of the Battle of Cambrai were far more significant than the immediate effects, local in place and time. The remarkable success of the tanks established them as a decisive weapon of war. It is astonishing that it was to be another 25 years, and half way through another world war, before it was finally recognised that horsed cavalry no longer had a part to play on the modern battlefield and that the tank had taken its place.

Further Reading

Dugdale, G., *Langemarck & Cambrai* (Wilding 1932)
Everest, J. H., *The First Battle of Tanks*
Liddell Hart, Sir Basil, *The Tanks*, Vol I (Cassell 1959)
Military Operations: France & Belgium 1917, Vol III (HMSO 1948)
Pree, Maj-Gen. H. D. de, *The Battle of Cambrai* (RA Institution 1928)

GENERAL SIR MICHAEL CARVER, GCB, CBE, DSO, MC, was born in 1915 and was educated at Winchester and Sandhurst. He was commissioned into the Royal Tank Corps in 1935 and during the Second World War he commanded the 1st Royal Tank Regiment in North Africa, Italy and France. Since the war he has held a number of important appointments, including C-in-C Far East, and has published three books — *Second to None, The History of the Royal Scots Greys; El Alamein,* and *Tobruk.* He is married and has four children.

FULLER Prophet of Armoured War

Lieutenant-Colonel (later Major-General) John Frederick Charles Fuller, DSO (later CB, CBE), who is remembered today as one of the most original and influential military thinkers of this century, was born at Chichester in 1878. Educated abroad and at Malvern, he was commissioned from Sandhurst in 1898 into the Oxfordshire and Buckinghamshire Light Infantry and almost at once ordered with it to South Africa. Appendicitis kept him out of the major battles of the Boer War and he spent the later guerrilla phase commanding block houses and reading widely. After serving in India in 1903/6, where he acquired what was to be a lifelong interest in its religions, he was appointed adjutant of a Volunteer battalion, the 10th Middlesex. This adjutancy gave him the chance to put his reading into practice, using real soldiers in tactical experiments, and also to begin writing. By the outbreak of war in 1914, besides having attended the Staff College, he had published articles on mobilisation, infantry tactics, weapon training and entrainment. The latter secured him the post, in August 1914, of railway transport officer at Southampton, from which he was later transferred to Second Army Headquarters at Tunbridge Wells. By the time he arrived in France in July 1915 the worst of the early fighting was over and he had established for himself a reputation as a Staff Officer of more than ordinary powers. Thus it was to turn out, though by no means at his own wish, that he was not to play any combatant rôle throughout the war. His creative and executive reputation grew during 1916, both through continued publication, notably of an article in which he satisfactorily defined for the first time the Principles of War, and through his creation of an Officers' School for the Third Army and then of a Senior Officers' Course for the BEF. He was therefore an obvious choice to fill the new appointment of Chief General Staff Officer (GSO 1) at Tank Corps Headquarters when it was created in December 1916. He came to the post, however, with no preconceived enthusiasm for armour. Indeed, his attitude has been described as 'caustically sceptical' towards tanks, which he regarded apparently as 'an adjunct' to the infantry and not as an arm of their own. His views were quickly to change and he was to infuse the Tank Corps staff with all the confidence he came to hold in the decisiveness of the tanks' future rôle, while himself working out from first principles both tactics and major operational plans. Of these the most important was that for Cambrai. [*John Keegan.*]

Imperial War Museum

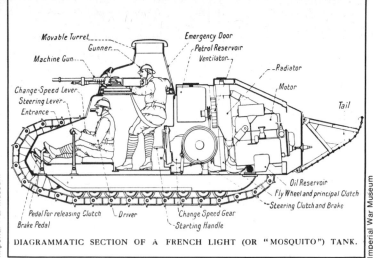

DIAGRAMMATIC SECTION OF A FRENCH LIGHT (OR "MOSQUITO") TANK.

Imperial War Museum

When they had recovered from their initial shock, the Germans dismissed the tank as a mere 'terror' weapon and devoted little time or energy to its development. Not so the British and French, some of whose officers clearly saw the possibilities of the new arm and were gripped by a sense of urgency. *Kenneth Macksey*

Top left: The British Gun Carrier Mark I. More a self-propelled gun than a tank, the first Gun Carrier was in service by January, 1917. The model pictured here carries a 60-pounder gun, although other Gun Carriers were fitted with a 6-inch howitzer. In the static war of the Western Front, the appearance of mobile guns caused considerable confusion to the opposing artillery spotters. *Above:* The French Renault 'Mosquito' tank. Equivalent to the British Whippet tank it had inferior cross-country characteristics but a good turn of speed. *Below:* The German A 7 V. A large armoured box on a Holt-type chassis weighing 30 tons, the A 7 V reflected the lack of importance attached to tank development by the German command. *Opposite page. Above:* A British Mark V, male. Successor to the Mark IV, the tank of Cambrai, the Mark V was the first heavy tank that could be driven by one man. It first appeared in 1918, and saw action in July of that year. *Below:* A British Mark IV with a 'tadpole' tail. The tail extension was added to increase the trench-crossing ability to 14 feet, but it proved lacking in rigidity and strength

TANK DEVELOPMENTS

Imperial War Museum

When the surviving handful of tanks were extracted from the Somme mud in November 1916 a rough poll of opinion taken from men at the front might have indicated that there was little future for the new weapon, so slight had been its effect. The Germans, recovering from their initial surprise, were almost unanimous in regarding tanks as a 'terror' weapon designed, more than anything else, to attack their soldiers' morale. Confident in their men's ability they were content to leave the solution of the problem to the artillery and infantry. One wrote that the troops 'soon learnt to know its vulnerable parts and attack it accordingly'. Another saw the tank as 'a comparatively easy prey for the artillery who detailed special guns to deal with it'. But at the British GHQ a more enlightened view prevailed. Here the tanks' initial mechanical failures and the inexperience of the crews were recognised as the inevitable results of untested equipment and organisation. Haig's approval of the new idea was reflected in his demand to the War Office for 1,000 new tanks and the setting up, on October 8, 1916, of a special central organisation in France to develop the fighting capacity of the tanks. Thus, long before the Battle of the Somme had ended, he brought into being the HQ of the Tank Corps and guaranteed the new weapons' future.

To head the new corps came Elles—the same sapper officer who earlier that year had reported favourably to Haig on the experiments in Britain. A few weeks later he was joined by his new Chief Staff Officer, Major J. F. C. Fuller, an intellectual and sceptical infantryman. Once convinced that the tank was a feasible weapon, it was Fuller, by his own example, who urged the other members of the staff not only to collect, synthesise and disseminate every scrap of information about tanks, but also to devise original and improved ways of employing them. Together this team was to turn an amorphous collection of men and machines into a coherent fighting force capable of working closely among themselves and in reliable fashion with the infantry who, incredulously, were beginning to regard tanks more as a help than a hindrance. Throughout the winter of 1917 the new organisation grew, publishing a flow of tactical notes, and issuing a flurry of technical directions to the crews and to the newly founded workshop organisation. The men of the Tank Corps were enthusiastic and quick to learn, but the indoctrination of senior officers and those who would enter battle with the tanks was more difficult. Progress with original ideas was bound to be slow in an army which still had much to learn about the older established weapons. The unimaginative employment of tanks at the Battles of Arras, Bullecourt, Messines and Passchendaele stemmed more from ignorance than bigotry. Even so, every obstacle seems to have been thrown in the tanks' path despite the appeals of the Tank Corps officers. Only lip service was paid to the principle that tanks, like any other weapon, did best when used by surprise and in large numbers rather than in pairs and even singly. More serious from the tanks' point of view was the sight of strengthened German defences, based on deep ditches and defended by field artillery, growing up to make the future task of Allied tank forces more difficult. They were not to know that, by November 1917, the Germans had virtually dismissed tanks as a serious threat. Indeed, the French tanks had also done badly on the Chemin des Dames

Imperial War Museum

and the newly arrived Americans were openly saying that there were no military prospects for a weapon of which the Germans had taken full measure.

But those who decried the deficiencies of the original machines overlooked the rapid changes that technology could effect. Even as the first thin-skinned British Mark I's were being delivered in June 1916, it was decided to attempt the construction of a shell-proof tank—an order which eventually resulted in the production of a machine with two to three inches of armour, armed with a 57-mm gun and five machine guns, weighing 100 tons and sardonically nicknamed 'Flying Elephant'. In fact, before trials commenced on this monster, it was rejected in January 1917 as being too costly. By then, however, it was realised that, although shell fire was the tanks' worst enemy, the chances of Mark I's survival were better than Kitchener and many others had first imagined. In fact the most dangerous threat to the Mark I, apart from its incipient unreliability, turned out to be the existing German armour-piercing bullet which easily penetrated the low quality British 10-mm armour plate. Based on battle experience, production of the earlier Mark I, Mark II and Mark III tanks was restricted to only 200 and the main effort concentrated upon building a more reliable and battle-worthy machine. Mark IV, when she began to arrive in France in April 1917, looked, at first glance, very much like her predecessors except that the tail wheels had been removed. She was still underpowered by the 105-hp Daimler engine and still required four men to drive and steer her, but in other ways she was much more formidable. The 57-mm guns were now short barrelled and better suited to use on rugged land than the original, longer naval types. Unfortunately the substitution of the Lewis machine gun in place of the original Hotchkiss and Vickers guns was a retrograde step, though not crippling. Range of action had been increased by carrying additional petrol, but the most important modification was an increase in the maximum thickness of the frontal armour to 12-mm— sufficient to defeat the Germans' armour-piercing bullet and thus to render their infantry wholly dependent on artillery for protection.

The sense of urgency

As the British were improving their original rhomboidal tanks, both they and the French were searching for means to increase both the number of tanks available and their speed across country. William Tritton made his proposal for a 'Chaser' tank in December 1916, seeing the need for a faster tank to exploit a breach, once it had been made in the German lines by the heavier tank. In effect he envisaged armoured vehicles in lieu of the horsed cavalry which had already shown its incapacity to operate in the open. From the Chaser was to come the development, throughout 1917, of the first Medium A's, later known as Whippets. These 14-ton tanks, with a speed of 8 mph (twice that of the Mark IV) and armed with four machine guns, were entering production just as the Tank Corps was being ground into the mud of Passchendaele, but were not to be ready for action until 1918.

The French became active in the construction of light tanks after the comparative failure of their early Schneiders and St Chamonds. On the suggestion of Estienne, in July 1916 an order had been placed with the firm of Renault for a simple, armoured machine gun carrier weighing six tons, capable of accompanying the infantry in the assault. By December the first model—called Renault FT—was ready and in March 1917 an initial order for 150 was placed. Unlike the British Whippet, which was given a good trench crossing capability of seven feet, the FT was struggling when crossing a six-foot gap and at first it was not received with acclaim by those in the French army who believed that an armament of a solitary machine gun was inadequate. Nevertheless, after successful trials in April 1917, coincident with the failure at the Chemin des Dames, the initial order was expanded to 1,000 and large scale production begun of the simplest and cheapest tank yet designed.

Nothing like the sense of urgency which gripped the Allied tank pioneers ever caught the Germans. Principal Engineer Vollmer of the Department A 7 V was instructed to design and produce a tank in November 1916, but because the General Staff remained so confident in the ability of the troops at the front to defeat tanks on their own, little enthusiasm was imparted to the work. Nevertheless, in January 1917 a wooden mock-up of the tank that was to be known as A 7 V had been produced and four days later an order for 100 was placed. But German industry was strictly orientated, as in everything else in 1917, to the demands of the basic military strategy. In the east, where they were on the offensive and the front was fluid, there was no overriding need for a 'storm tank'. Nor was there an immediate need for one in the west since

there the army stood on the defensive, although the possibility that, one day, they would have to attack and break the trenches, and might need tanks, seems to have been overlooked. Partly out of haste, but also because he lacked strong backing, Vollmer could do no better at first than mount a large armoured box on top of a Holt type chassis, based upon the one which had been demonstrated by Mr Steiner of Holts before the war and which, since then, had been lying, disused, in Austria. The 30-ton A 7 V, which appeared in prototype in October 1917, was therefore no more promising than the original French tanks on their Holt chassis. None would be ready for operations in 1917 and only 20 were ever made since an over-strained German industry was already fully committed to the manufacture of conventional weapons. Under these circumstances a project for a 150-ton heavy tank, launched in March 1917, can only be viewed as a distraction instead of part of a well thought out scheme.

In spite of the losses incurred by Allied tanks throughout the attritional artillery battles of the summer and early autumn of 1917, both the British and French tank forces improved in quality and quantity. The British took delivery of nearly 1,000 Mark IV's before November 20 and had about 450 ready for action that day. By then the French possessed some 500 Schneiders and St Chamonds although, in its phase of defensive recuperation, the French army found little use for them. The waste caused by mechanical unreliability overlaid almost every other consideration—even though most Allied machines were remarkably efficient considering the haste in which they had evolved. Nevertheless, any one tank which ran more than 20 miles without breakdown signified the performance of a notable feat even by a well trained crew. All crews had been hurriedly trained and few reached a high standard in 1917. Morale, however, was high. The Tank Corps had been fortunate in its leader in France and in the staff who served him. Not a brilliant man, Brigadier Hugh Elles nevertheless possessed the freshness of youth combined with a calm competence which gave men confidence. By gently fighting their battles at GHQ, he was able eventually to win the Tank Corps a square deal. If, in his GSO I, Fuller, he possessed a man with a caustic wit who rubbed up senior officers the wrong way, he also owned a genius —an officer endowed with genuinely original ideas who visualised a rôle for the tank far beyond its original concept as a mere instrument of assault. While guiding the tactical thoughts, training and operations of the corps through a trying year of innovation, Fuller was always searching for new ways to employ the tanks as the dominant technical weapon in a decisive battle—in a rôle subservient neither to infantry nor artillery. But within the framework of his most sweeping concepts, he made detailed plans like any good staff officer, demanding that tanks should never be launched into battle without thorough and minute preparation.

Thus when it was decided to give the tanks their first big chance in a massed attack, Fuller and his staff were ready with settled and well understood methods for moving tanks secretly and swiftly from one front to another. They had also collected detailed information about every piece of German-held terrain where tanks might have to be used and, in so doing, had created a reconnaissance organisation which always provided the crews with accurate information in plenty of time before action. Meticulous but simple drills had been devised for co-operation between tanks, infantry, artillery and aircraft in the assault—drills which, at Cambrai, were to revolutionise the whole scheme of assault. Where it was known that the tanks would have to cross the immensely wide anti-tank ditch running like a spine the length of the Hindenburg Line, great rolls of brushwood—called fascines—were made ready to be carried forward by the leading tanks to make the crossing possible.

This time when the tanks went into action each crew commander —unlike his predecessors on September 15, 1916—knew his task, understood the problems to be overcome and knew that artillery, infantry and aircraft were briefed to give specific aid to the tanks. And each crew member was trained to a higher standard than ever before, while the machines were vulnerable only to German artillery and stood a fair chance of reaching their limited objectives without breaking down.

Further Reading

Fuller, J. F. C., *Memoirs of an Unconventional Soldier* (Nicholson & Watson 1936)
Liddell Hart, Sir Basil, *The Tanks,* Vol I (Cassell 1959)
Macksey, K., & Batchelor, J., *Tank* (Macdonald 1970)
Nehring, W., *Die Geschichte der deutschen Panzerwaffe, 1916 bis 1945* (Propyläen 1969)
Ogorkiewicz, R. M., *Armour* (Stevens 1960)

Villers-Bretonneux

As the failure of the *Georgette* operation in the north became more and more obvious, Ludendorff resolved to make one last attempt to gain control of the strategically-important town of Amiens in the south, by capturing the vantage point of Villers-Bretonneux, about ten miles to the east. An element of surprise was vital to the success of this last phase of the *Michael* offensive, and was provided by a detachment of 13 tanks: machines which the Germans had been unusually slow to manufacture since their appearance on the Somme in 1916. The result was the first tank 'duel' of the war. *John Vader*. *Above:* German infantry suffered heavily in this new form of warfare. Crude anti-tank defences pitched against the manoeuvrable British Whippets were useless: two German battalions were nearly wiped out.

THE FIRST TANK-TO-TANK ENCOUNTER

By the end of March 1918 Ludendorff knew that his offensive across the old Somme battlefield had failed. He had run out of steam: the Germans had outrun their communications and their fire power was insufficient to maintain the thrust. Amiens, however, remained as a valuable objective. The Germans' drive had brought them to within a few miles of the plateau south of the Somme, and from the edge of the plateau near Villers-Bretonneux they would be able to overlook Avre, Noye and the vital road and rail junctions of Amiens.

'*Fini retreat — beaucoup australiens ici*', a Digger was reported to have informed a peasant woman during the retreat of Fifth Army, when the AIF 3rd Division was moving forward. 'You Australians think you can do anything,' said an artillery brigadier to Lieutenant-Colonel Lavarack, chief of

the divisional commander's staff, 'but you haven't a chance of holding them.' The morale of the Australians was high, the opposite to what the Germans expected when *Eighteenth Army* issued the order: 'The enemy for the moment has only inferior or beaten troops opposite us.' The Germans too were dispirited, as Captain G. Goes later reported:

The fighting during the closing days of March had left no doubt that the great offensive battle was threatening to become a battle of attrition on the largest scale. Everywhere along the front of nearly 120 miles the infantry had been compelled to have recourse to the spade, and saw the spectre of hated position warfare rising before it. The situation was very much more unfavourable than before March 21: no trenches, no dugouts, streaming rain for days, which had

turned the devastated area behind the troops into a morass, and caused the roads and tracks laboriously rendered serviceable to break up again. More and more reports trickled in of fresh enemy divisions springing up round the fragile salient which had been won; more and more the artillery fire swelled in volume; higher and higher rose the losses of the Germans in dead, wounded and sick, and more and more did the spirit of their attack evaporate.

Preparations were indeed in progress at another place, over in Flanders, for another surprise attack; nevertheless, the Supreme Command decided to try its luck with Amiens once more. Perhaps now, after a breathing space, after the arrival of fresh divisions and the replenishment of the ammunition dumps, the thing could be done.

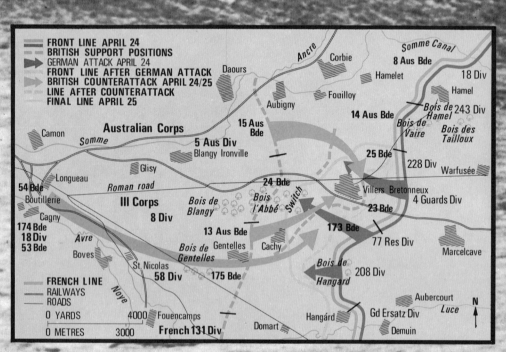

Map legend:
- FRONT LINE APRIL 24
- BRITISH SUPPORT POSITIONS
- GERMAN ATTACK APRIL 24
- FRONT LINE AFTER GERMAN ATTACK
- BRITISH COUNTERATTACK APRIL 24/25
- LINE AFTER COUNTERATTACK
- FINAL LINE APRIL 25

- FRENCH LINE
- RAILWAYS
- ROADS

0 YARDS 4000
0 METRES 3000

Somme Canal
Ancre
Corbie
8 Aus Bde
18 Div
Daours
Hamelet
Hamel
Fouilloy
Bois de Hamel
243 Div
Aubigny
14 Aus Bde
Bois de Vaire
Bois des Tailloux
Australian Corps
15 Aus Bde
Camon
Somme
5 Aus Div
Blangy Ironville
25 Bde
228 Div
Warfusée
Glisy
Roman road
24 Bde
Villers Bretonneux
Longueau
4 Guards Div
54 Bde
Boutillerie
III Corps
8 Div
Bois de Blangy
Bois l'Abbé
Switch
23 Bde
Cagny
173 Bde
77 Res Div
174 Bde
18 Div
53 Bde
Avre
13 Aus Bde
Gentelles
Cachy
Marcelcave
Boves
Bois de Gentelles
St. Nicolas
58 Div
175 Bde
Bois de Hangard
208 Div
Noye
Aubercourt
Luce
N
Fouencamps
Hangárd
Gd Ersatz Div
Domart
French 131 Div
Demuin

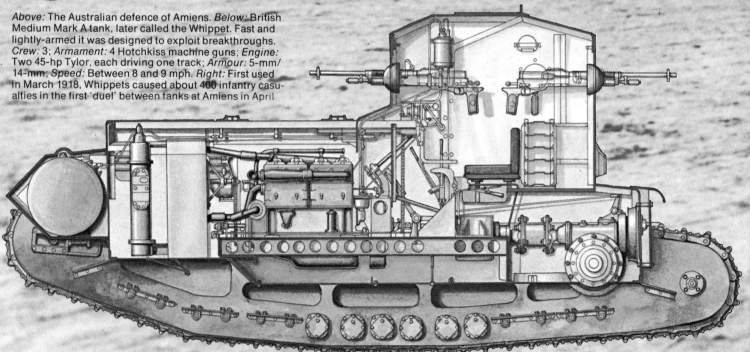

Above: The Australian defence of Amiens. *Below:* British Medium Mark A tank, later called the Whippet. Fast and lightly-armed it was designed to exploit breakthroughs. *Crew:* 3; *Armament:* 4 Hotchkiss machine guns; *Engine:* Two 45-hp Tylor, each driving one track; *Armour:* 5-mm/14-mm; *Speed:* Between 8 and 9 mph. *Right:* First used in March 1918, Whippets caused about 400 infantry casualties in the first 'duel' between tanks at Amiens in April

They benefited by fine weather during the first two days of April, but it began to rain on the 3rd, and the suspended attacks were renewed in the final offensive which began the following day. But the Allies had been forewarned by captured prisoners. Just before dawn, shrapnel and gas shells were bombarded across the front line and the artillery and back areas, followed at first light by swarms of infantry from *Eighteenth, Second* and *Seventeenth Armies.*

The British army too had had time to re-adjust its front. The sector attacked was held by two understrength divisions, the 14th and 18th, the latter without its artillery. The Australian 9th Brigade was also in the area, and the centre ahead of Villers-Bretonneux was held by its widely extended 35th Battalion, with the other three battalions behind the town in close reserve. They were to face the strongest attack yet made against Australians: five German divisions attacked under good cover across the valleys running parallel to the front.

The fresh *9th Bavarian Reserve Division* drove hard at the Australian battalion which stopped them with a heavy fire from machine guns and rifles, with good support from the guns. The 14th Division, to the north, suffered a particularly heavy bombardment and was driven back, and the Germans were able to take Hamel and to swing south, threatening to envelop the Australians from that flank; they withdrew to a support position a mile east of Villers-Bretonneux. The AIF 33rd Battalion then moved up to extend the left of the 35th, and 6th Cavalry Brigade was used to fill gaps in the line. The AIF 15th Brigade, guarding bridges, sent some of its men to help support the 18th Division, which managed to stand fast until a second attack, made in mid-afternoon, forced the line to retire. When the Germans reached the 1870 war monument on the outskirts of the town it looked as though the way to Amiens was about to be opened to them.

The Germans' first tank attack

The Germans' hopes were dashed by a remarkable charge by the reserve AIF 36th Battalion, joined in its attack by several British infantrymen on its right, part of 35th Battalion on its left, and later by some Londoners and the cavalry. Attacking on the run with bayonets flashing, the unexpected reinforcements first stopped the Germans advancing from Monument Wood then drove them back to old trenches more than a mile from the town.

The 15th Brigade was brought across the Somme to hold the vital heights north of the town, and the 33rd Battalion again went forward, protected on a flank by the mounted 17th Lancers and aided by a Canadian motor machine gun battery which dealt with German machine guns. By dusk the line east of Villers-Bretonneux was secure, and during the night the 9th Brigade made advances to gain ground which would improve its field of fire, a job which entangled the 34th in some stiff fighting before establishing itself in the old line that had been held by the 35th Battalion. South of Villers-Bretonneux the Germans drove the French back two miles beyond the Avre and were in sight of Amiens. The River Somme ran some five miles north of Villers-Bretonneux, and the Germans, expecting that the Allied line in

Imperial War Museum

front of the town would crumble, launched two divisions against the New Zealanders and Australians positioned north of the river. The Germans were thoroughly beaten here also. Further north, the Australian 4th Division fought a vicious battle on April 5, this resulting in little change in the line at a cost of some 1,100 casualties among its two brigades. The AIF 9th Brigade's casualties at Villers-Bretonneux were 660.

In a message to General Foch, Sir Henry Wilson predicted further German moves against Amiens: 'In my opinion, the proper course for the enemy to pursue now is as follows: place Amiens town and railway and junctions under his guns so as to deny all serious traffic, then mass an attack of 40 to 50 divisions against the British between Albert and the La Bassée canal.' During the first half of April, fighting raged all the way to the coast, both sides suffering catastrophic losses in several battles, before the Germans again became active at Villers-Bretonneux — disabling some 1,000 reserves behind the town with mustard gas shells. Before the Australians were relieved by two battleworn British divisions they witnessed the shooting down overhead of the German ace Richthofen.

The change over by the divisions was barely completed when the Germans made their first big attack with tanks — 13 of them — which broke through the British lines manned, in many cases, by 'boy recruits'. The attack began at dawn on April 24 in the manner to be exemplified in the Second World War — infantry following the tanks and rolling up the line right and left. The tactic was used on either side of the town, across good tank terrain, and the attack was pressed on, regardless of an enormous barrage laid down by Allied guns. Villers-Bretonneux, Abbey Wood and Hangard Wood were lost as the Germans moved forward as far as the junction of the Avre and Luce. One factor in the German success was the presence of fog, which they thickened with smoke; another was that there were no tank defences so that when a tank straddled a trench it could fire on the defenders.

The presence of the German tanks drew out British tanks, of which there were 20 in the area — 13 Mk IV's and seven Whippets. The 1st Section, 1st Battalion, went into action with one male, mounting two 6-pounder guns, and two females, mounting machine guns, and although the male was disabled by a shell the crew brought a gun into action and knocked out one of the German tanks. Another surrendered when the concentrated fire from the 58th Machine Gun Battalion put it out of action by 'the splash of the bullets'. More tanks of the 1st and 3rd Tank Battalions assisted the 1st Sherwood Foresters in their counterattack, and a charge by light Whippet tanks

Above: The shattered bodies of German soldiers killed during the hour-long artillery bombardment accompanying the Australian recapture of Villers-Bretonneux. Below: Scene of the first German tank attack. German A7V's in the field before Amiens, throwing up clouds of chalk dust

scattered the German infantry across the plateau, causing at least 400 casualties as they made two runs through battalions of *77th Reserve Division*. One Whippet was knocked out, three were disabled and three got back from this successful mission which originated when a scout plane dropped a message that the German battalions were resting in a hollow in front of Cachy. It was to have been a tank-infantry assault but, when the 23rd Brigade was ordered to counterattack with armour it had no men available, and when the 8th Division cancelled the order the tanks were not informed.

Lieutenant-General Butler of III Corps considered that the 8th Division was capable of dealing with the situation and decided not to accept the offer of the AIF 15th Brigade's commander, 'Pompey' Elliott, who was fuming to be allowed to strike. The brigade had been stationed about a mile north-west of the town, kept in reserve and ready to counterattack if the town were lost. While the formal messages were being passed, time was lost and the Germans were allowed to dig themselves into Abbey Wood and the heights that gave such an advantage over Amiens, good reasons for the 'tempestuous' Elliott to be 'fuming and petitioning' for action, as the AIF's historian observed.

It was predictable, nevertheless, that the brigadier would get the job, along with another Australian brigade commander, Brigadier-General Glasgow of the 13th, which was given the task of forming the southern pincer; the 15th formed the northern. When the brigade commanders met Butler they were told that he wished the attack to be made by daylight from the south northwards, and Glasgow said, 'If God Almighty gave the order we couldn't do it by daylight.' It was agreed, fortunately, that an attack across an open plateau, in front of machine gun fire, could barely succeed, and that the Australians' plan to attack at night without previous bombardment spoiling the surprise would be a better plan. Rawlinson and Butler decided that the attack should begin at 2200 hours when the moon would be giving some light to the field of action.

The 15th Brigade was to advance southeast across the northern side of the town; the 13th Brigade was to move across the southern side of the town; and the 54th Brigade was to move eastwards on the southern side of the 13th. The two Australian brigades were given the 2nd Northamptonshire and 22nd Durham battalions and three tanks to mop up when the town was recaptured. The artillery was to maintain a standing barrage on well defined targets until an hour after the initial attack began, then the heavy batteries were to lift 500 yards beyond the town while the field artillery lifted 300 yards.

One factor contributing to the early German successes near Amiens was the fog, thickened by smoke bombs. But the attacking tanks soon drew the British Mark IV's and Whippets into action, and a 'duel', the first of the war, began. *Above:* A captured Mk IV advances past a flamethrower

Before clouds covered the full moon there was too much light: the 13th Brigade was seen assembling south of the woods, and the Germans opened fire with numerous machine guns as the Australian advance moved forward. The flank was stopped until Sergeant 'Charlie' Stokes urged his platoon commander to attack the Germans in the woods – which was not part of the plan. The commander, Lieutenant Sadlier, and Stokes led their men in a bombing attack which quickly destroyed six German posts, allowing the southern pincer movement to go forward, and earning for the platoon commander a Victoria Cross, for the sergeant a DCM. As the brigade moved on it was joined by the Bedfordshire battalion of the 54th Brigade; its other two battalions – 9th London and 7th Royal West Kent – had lost touch, one forced to return to its start line, the other stopped by machine gun fire from shell holes and was finally dug in about halfway to the objective.

The 13th Brigade did not quite reach its objective, but it was in a position to ensure the success of the pincer movement if the 15th did its job. After the holdup at the wood, the 13th's 51st Battalion suffered heavy casualties going through the wire at Cachy Switch, where it was enfiladed by German machine guns. After one gun was captured, Vickers were set up to keep the others occupied and the Australians moved on, occasioning 'a panic which could not be overcome in time, as the reserves failed to come up quickly enough', as the Germans later stated. There were many leaderless Germans wandering aimlessly about who were killed or captured, but the opposition became very strong; an attack from the town endangered the 51st's flank until the

50th came up to drive them back. When these two battalions aligned themselves they were near a railway cutting south of the town, an exit which was eventually to allow some Germans to escape.

The 15th Brigade was late getting away from its start line and there were the usual confusions associated with night operations. One company could not even find the assembly point. The actual start was almost two hours late, so the Germans described this move as a second attack. There was also some confusion about exactly where lay the objectives – the Hamelet road and Hamel road. The first was reached after an uneventful mile tramp, then the Germans, alerted by scouting activity, began to let fly their flares and small arms. The 59th was the leading battalion and when its CO yelled 'Charge!' the command was rapidly passed down the line so that a great yelling, cheering rush was made on the Germans, now visible under the lights of their flares and the glare of burning houses. In their panic, the Germans forgot to lower their sights as the Australians with fixed bayonets rushed the enemy trenches and machine gun nests. At 0130 hours they were at the Hamel road having lost fewer than 150 men.

Two companies pushed on to make contact with the 13th Brigade which they expected to find near the old British front, but the 13th was not within sight or hearing and the 57th's companies returned to the road, where positions were dug facing Villers-Bretonneux. The mopping up battalions found that the town was still too strongly defended and their work was delayed. Not far away the French had prudently accumulated reserves in case the British could not handle the Villers-

Bretonneux situation, yet when General Debeney was informed of the plan to counterattack he promised to co-operate if it were postponed until the following day. When Rawlinson informed him that the action should take place that night he merely offered to send the Moroccan Division in to occupy the reserve line, freeing some of 58th Division's reserve.

Before the artillery was directed on to the railway cutting escape route that lay in the 1,500 yards between the two pincer ends, the Australian brigades began to close in on the town, assisted by four British infantry battalions and three tanks. By noon the town was occupied and the mopping up was left to the Durhams and Northamptonshires. Because of considerable confusion in the German lines, a counterattack by the German divisions ordered for 0700 hours was postponed twice, and then finally abandoned through lack of fresh troops. Just before 1000 hours Glasgow sent in his reserve battalion to close the gap, an operation which met with such fierce resistance that it was postponed until that night, when it was carried out with few casualties. The original British front had not been recovered, but Amiens was secure.

Further Reading
Bean, C. E. W., *Official History of Australia in the War* (Canberra)
Bean, C. E. W., *The AIF in France, 1918* (Angus & Robertson)
Cutlack, F. M., *The Australians: Final Campaign, 1918* (Low, Martson 1918)
Monash, Gen. Sir John, *The Australian Victories in France, 1918*

Imperial War Museum

Right: A German flamethrower crew moves up in support of one of Germany's few tanks

Inset: Elfriede, the first German tank to be engaged and captured by British tank forces. Found overturned and abandoned in a sandpit she was minutely examined by Allied Intelligence

'BIG BERTHA' BOMBARDS PARIS

Designed to coincide with Ludendorff's spring offensive on the Somme, Germany's engineers fulfilled a remarkable briefing: the creation of a colossal gun with a range of nearly 75 miles – aimed on Paris. *Jean Hallade. Below:* 'Big Bertha' under construction in the Krupp artillery department

In spring 1916, while the battle of Verdun was reaching its culmination, an important meeting was held in Berlin in the presence of General Ludendorff, Head of the Supreme War Council. The naval officers commanding the heavy-calibre guns on the Western Front asked him if he would approve the costs and éxpenditure of material and personnel necessary for the construction of a gun with a range of 62.1 miles (100 kms).

Ludendorff has often been depicted as a reserved and cold personality, with no great liking for advanced technical ideas. And yet the general reacted to this idea immediately. At that time, the front passed within 55.9 miles (90 kms) of Paris, and Ludendorff must have grasped the importance of this project. To be able to damage the morale of the people in the rear areas with a giant cannon, co-ordinated with a large scale offensive, would give striking proof to the world both of German technical capabilities, and of her ability to continue the war.

The project passed rapidly to a practical stage and the navy obtained the utmost co-operation from the Krupp Works' Artillery Department management. At the beginning of the war, Krupps had already astonished the world when the 420-mm mortars built in its foundries demolished the concrete masonry of the forts of Liège and Maubeuge. But between a 420-mm mortar, firing at a maximum range of under 9.3 miles (15 kms), and a piece whose calibre was still to be determined, which could fire a projectile a distance of 62.1 miles, many problems still remained to be solved. Gradually, in silence and secrecy, the navy's Artillery Direction, in collaboration with Krupp, pushed the project to a very advanced stage of studies allowing the construction of the supercannon.

Nine months had elapsed since the Berlin meeting in which the study of this fantastic cannon had been authorised, when a brief telegram arrived from Ludendorff: 'In subsequent work on very long range pieces, please take as your basis a range of 74.5 miles (120 kms) instead of 62.1 miles.' This astounded all those engaged on the project. There was colossal work involved in resuming the study taking into account these extra 12.4 miles (20 kms). Everything already calculated had to be recalculated from scratch. What was the reason for this modification of the range demanded in Ludendorff's telegram? It was due to the fact that a withdrawal to the Hindenburg

Line had been decided by the German High Command, increasing the distance from Paris to the front by 12.4 miles. These new technical problems provoked intense activity at Krupps and soon a model designed by Professor Rausenberger, Director of the Artillery Department, fulfilled all requirements. There now remained only to build the gun.

Immediately the project was given a name. It was called *Wilhelm Geschutz,* 'William's Gun', in honour of the Kaiser. Sometimes it was called 'Long Max', but

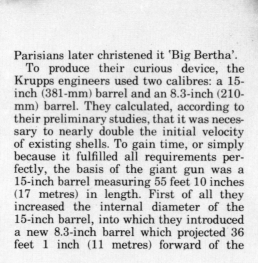

Parisians later christened it 'Big Bertha'.

To produce their curious device, the Krupps engineers used two calibres: a 15-inch (381-mm) barrel and an 8.3-inch (210-mm) barrel. They calculated, according to their preliminary studies, that it was necessary to nearly double the initial velocity of existing shells. To gain time, or simply because it fulfilled all requirements perfectly, the basis of the giant gun was a 15-inch barrel measuring 55 feet 10 inches (17 metres) in length. First of all they increased the internal diameter of the 15-inch barrel, into which they introduced a new 8.3-inch barrel which projected 36 feet 1 inch (11 metres) forward of the

muzzle of the original gun. The forward part of the gun was then reinforced with a barrel-sleeve. Finally—and this was a great innovation—they screwed on a chase (smooth tube) 19 feet 8 inches (six metres) long at the end of the gun. This chase was not rifled and its perfectly smooth internal bore prolonged exactly the bottom surface of the rifling in the main tube. The gun had a total length of 112 feet (34 metres) and an overall weight of 138 tons.

Once the gun had been made, they had to devise a special powder for it. It had to

Opposite: left to right: The 8.26-inch calibre shell; *Zusatzkartusche* (auxiliary) and *Vorkartusche* (main) cordite charges, both wrapped in silk; and *Hulsenkartusche,* encased in brass

weight. During its railway journey, the tube had a length of only 91 feet 7½ inches (28 metres), the smooth extension being mounted later in the firing position. An enormous truck-mounting comprising 18 axles carried the rifled and nondismantable section whose overall weight was 256 tons. The assembly of the 19-foot 8-inch (six metres) smooth tube took place near the spot which the gun was to occupy. This operation was complicated and delicate. The smooth tube was screwed on to the rifled tube with the help of a dismantable gantry moving on two parallel railway lines to the right and left of the main railway line on which the truck-mounting lay. A platform was necessary to receive the truck-mounting. It consisted of a ring-shaped casing, which could be dismantled into six sections for purposes of transportation, on which was laid a turntable mounted on ball bearings. The lot rested on a solid concrete foundation.

'Tomorrow, open fire on Paris'

During the long truce of winter 1917/1918, and despite tentative peace proposals, the Germans did not remain inactive, and they continued with their great project for an offensive along the entire front, supported by long range artillery fire aimed at demoralising the French. The place chosen to install the gun in battery was the forest of Crépy-en-Laonnois near Laon. The railway station of Crépy-en-Laonnois lies on the railway between Laon and La Fère. That is where they sent waggon loads of timber and munitions which were unloaded by French civilian prisoners. Hidden among the trees and copses, men set to work. Trees were felled and immediately cut into logs, while efforts were made to conceal the destruction. Tracks were cut through the copses, sleepers laid, and the lengthening railway lines were immediately covered with camouflage netting to ward off spotting from the air. The earth was dug up in several places. At three points, the construction of concrete platforms was begun. All the areas laid bare by these gun emplacements were covered with wire mesh fitted to the tops of trees and overlaid with a net. Once the trees started to bud, the whole lot would be covered with foliage. This work was begun in November 1917 and continued actively in January and February of the last year of war.

Once the work had been completed, each gun was in position at the end of a railway spur laid just beyond the place where the railway line cut the road leading from Crépy to Couvron. There remained only to point each gun in the direction of Paris at an angle of 52 degrees and fire the projectile. After a three-and-a-half minute journey, it would land in the middle of the French capital. Such was the formidable German arrangement designed to demoralise Paris the day after the start of the great offensive on March 21. On the evening of March 22, an order from the General Staff was transmitted to the command post situated in the Crépy-en-Laonnois forest: 'Tomorrow morning, open fire on Paris'.

It was 0715 hours on the morning of March 23. Feverish activity had been in progress for several hours on the wooded eastern slope of the Mont de Joie. It was chilly that morning, and a thick fog covered the plateau and its vicinity. The sound of the guns could be heard from nearly all directions, and especially from the southwest. Two days previously at dawn, the formidable German offensive had been launched. Hidden in the middle of the Mont de Joie forest, an enormous tube lifted its muzzle to the sky. Brief orders were given and everyone got out of sight, sheltering in huge concrete shelters or in well equipped trenches close by. It was just after 0716 hours when a tremendous explosion shook the atmosphere, and the guns firing in the distance seemed to redouble in intensity as though in echo.

In Paris it was cold in the early morning of March 23. For some Parisians—and also some Parisiennes, for in the factories women were replacing the men who were at the front—it was time to go to work. The night had been quiet, no Gotha bomber had flown over the capital, and the last alert dated from the previous evening. At 0720 hours an explosion shook the 19th arrondissement. In front of No 6, Quai de la Seine, there was a hole in the roadway, some windows were broken and the walls in the vicinity had received splinters. Even though it passed practically unobserved, this explosion was a fantastic event: a shell fired over 74.12 miles away had just struck the capital. For just under three years, Paris had been untroubled by daylight bombardment. The few people who believed it was an attack thought it was an air raid, probably a single aircraft flying above the fog and dropping a bomb, without an alert being sounded.

Twenty-five minutes later, at 0745 hours,

be absolutely perfect, as each round required about 550 lbs (250 kgs) of powder packed into a chamber nearly 16 feet 5 inches (five metres) long. And then there remained to make a shell. With such a powder charge, erosion was considerable inside the tube every time a round was fired, and the artillery specialists rightly considered that the gun would wear out rapidly. In that case, accuracy of aim and range would decrease as the firing went on. They calculated that the gun could fire only 65 times, after which it would have to be rebored. They thus achieved the astonishing solution of numbering each round from one to 65 and increasing the weight of the shell by 33 lbs (15 kgs) between the first and the 65th round.

With little hesitation, they decided on railway transportation. It was not a simple business to move a piece of this size and

The incredible had happened: 'William's Gun', or 'Big Bertha' as she was unlovingly called by Parisians, could bombard Paris from 74½ miles away. Here a first model is being tested

a second missile fell, this time in the 10th arrondissement, striking a building in the Gare de l'Est. Now, people seriously believed it to be an air attack, but why had no alert been sounded? At last, at 0915 hours, the sirens sounded, warning the population to seek shelter. Three more projectiles had dropped and this time there were fatalities.

The first casualties

At 0815 hours at No 15, Rue Charles V, in the 4th arrondissement, a man had been killed—the first victim of this long range gun. A quarter of an hour later, again in the 4th arrondissement, Rue Miron, four people had been wounded. It was, however, at 0845, Boulevard de Strasbourg, the part facing the Gare de l'Est, that the greatest disaster of the day occurred: a shell killed eight people and wounded 13. As soon as this serious incident had occurred, and even though sightseers continued to scrutinise the sky looking for German aircraft, the trams of the Gare de l'Est-Montrouge line refused to take on any more passengers and returned to their depôts. A few minutes later, the alert was sounded. Public services were no longer functioning, people sought refuge, the shops closed and the streets became deserted.

Gradually, anxiety grew. People started asking questions about these strange detonations which were occurring practically every quarter of an hour, and the end of the alert was very long in coming. To complete the people's disarray, the official services of the War Ministry, which by 1000 hours still did not know what it was all about, or refused to recognise the truth, issued a communiqué which did not improve the situation. 'At 0820 hours, a few enemy aircraft flying at very high altitudes succeeded in crossing the lines and attacking Paris. They were immediately engaged by our fighter aircraft, both those from the entrenched camp and those from the front. Bombs are reported as having been dropped at several points and there are some casualties.' It was true that fighter aircraft had indeed taken off, but the explosions continued to take place every quarter of an hour.

After the fourth explosion, the rate of one shell every 15 minutes was maintained up to 1345 hours when the 21st round fell once again in the 19th arrondissement, at 57 Rue Riquet, without causing any casualties. Why was there a one-hour interval between the 21st and the 22nd and last round? A mystery! Possibly there may have been a technical difficulty with the gun or in the ammunition supply.

After midday, the thick fog which covered the Crépy-en-Laonnois region, began to disperse, its marvellous camouflage being replaced by the first rays of the sun, and consequently firing was interrupted because in the distance French observation 'sausage' balloons were rising on their cables, and it was unwise to divulge the site of the battery which must already have been approximately marked by sound despite every precaution. The last round was fired towards 1441 hours to burst at 1445 hours in the suburbs near Pantin station. On the slopes of Mont de Joie, the long tube which had been protruding into the sky since the morning returned to its horizontal resting position.

What was the opinion of the official services in the various ministries early that afternoon? The advocates of an air attack, as against shelling, were fast losing ground. It was 1430 hours when the telephone rang in the editorial office of the paper *Le Temps* which was due to be run off the presses within a few minutes. It was requested that publication should be delayed for a few minutes because an official communiqué was being prepared at the War Ministry. At 1500 hours—barely a quarter of an hour after the last shell had arrived—the telephone rang once again at the paper. The editor heard the duty officer tell him: 'Are you sitting down? Yes? Well, hold on to the table because the news is sensational!' And then he dictated

The bombardment of Paris was not restricted to guns alone: after an interval of three years, German bombers once again flew over the capital of France. *Below:* A statistical résumé of the German aerial effort over Paris. *Bottom:* The incredible velocity and trajectory of the shells fired by the original Paris gun

	Took off	Lost	Killed	Wounded	No.	Weight (lbs)
JANUARY 30	30	1	61	198	267	31460
MARCH 8	60	1	18	41	92	
MARCH 11	70	4	173	132		
MARCH 24	1				2	
APRIL 1	7				21	
APRIL 11	1		27	72		1543
MAY 22	1		1	12	22	
MAY 23	1		10		50	660
MAY 27	15	1			4	
MAY 30	6		3	3	17	
MAY 31	6					
JUNE 1	3			28	14	
JUNE 3-7	1					
JUNE 15	1		3	5	9	990
JUNE 26	1				6	660
JUNE 27	3	2		18	26	
SEPT 15	50	2	7	30	85	

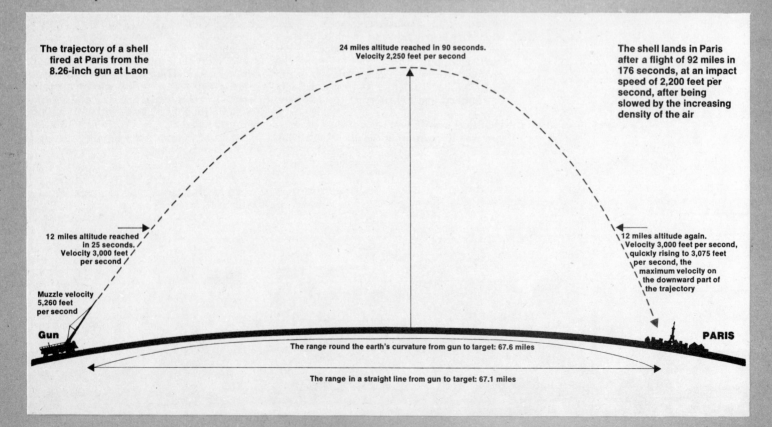

The trajectory of a shell fired at Paris from the 8.26-inch gun at Laon

24 miles altitude reached in 90 seconds. Velocity 2,250 feet per second

The shell lands in Paris after a flight of 92 miles in 176 seconds, at an impact speed of 2,200 feet per second, after being slowed by the increasing density of the air

12 miles altitude reached in 25 seconds. Velocity 3,000 feet per second

12 miles altitude again. Velocity 3,000 feet per second, quickly rising to 3,075 feet per second, the maximum velocity on the downward part of the trajectory

Muzzle velocity 5,260 feet per second

Gun

PARIS

The range round the earth's curvature from gun to target: 67.6 miles

The range in a straight line from gun to target: 67.1 miles

this communiqué: *The enemy has fired on Paris with a long range gun. Since 0800 hours this morning, every quarter of an hour, 9.5-inch (241-mm) shells struck the capital and its suburban area. There are about ten dead and 15 wounded. Measures to counter the enemy piece are now in course of execution.*

Thus was the fantastic exploit officially recognised. There was an error in this communiqué. The calibre of the shell was actually 8.3-inches (210-mm). As for the paper *Le Temps*, it thought it advisable to reassure its readers and avoid panic by adding that the front was still 62.1 miles (100 kms) away.

Twenty-two shells fell on this first day, 18 on Paris and four on its suburbs. They killed 16 people and injured 29. The following day, to the surprise of the Parisians who believed the gun had already been destroyed, firing was resumed and a first shell fell at 0645 hours in the Rue de Meaux (19th arrondissement), killing one person and injuring 14, most of them slightly. Twenty-two shells fell on Paris and its suburbs on that day, the last one (which wounded two people) at 1300 hours at Pré-Saint-Gervais.

Were there two guns?
This day was marked by a new event: shells started falling at very short intervals. For instance, it was noted that one shell fell at Pantin at 0917 hours and the next one at 0920 hours in Rue de la Lune in the 2nd arrondissement. These two rounds, only three minutes apart, could not have been fired from the same gun. From that moment, it had to be considered that there was not just one long range gun, but at least two. On that Sunday March 24, 1918 – Palm Sunday – a shell that dropped at 1145 hours near Blanc-Mesnil Church, killed four and injured seven churchgoers leaving after Mass. On Monday March 25, only six shells were fired on Paris, five between 0649 hours and 0740 hours and the sixth in the afternoon (at 1548 hours in the Père-Lachaise cemetery).

For some unknown reason, firing stopped for three days. No shells were fired on March 26, 27 or 28. Once again legends began to circulate among the Parisians

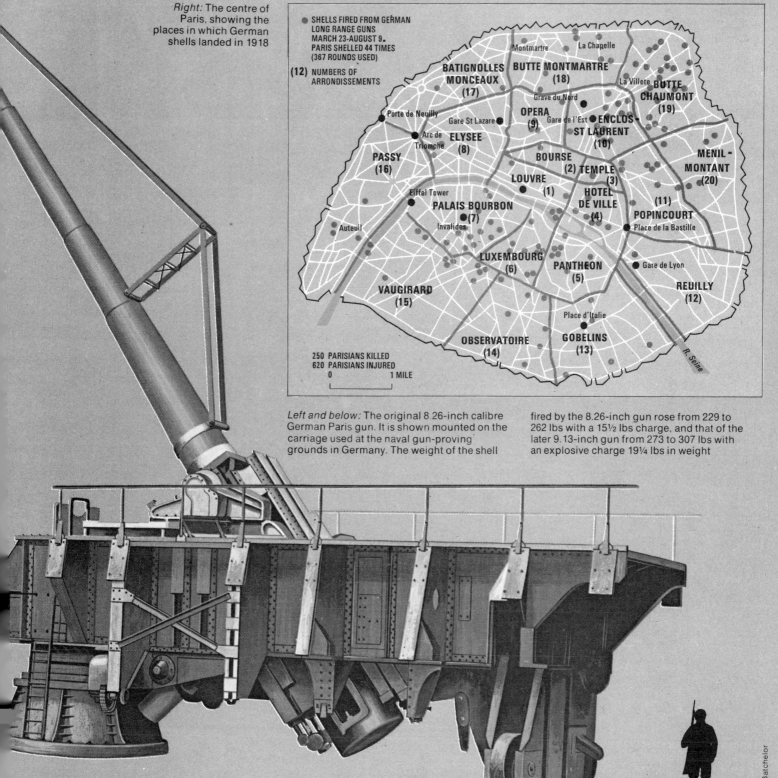

Right: The centre of Paris, showing the places in which German shells landed in 1918

● SHELLS FIRED FROM GERMAN LONG RANGE GUNS MARCH 23-AUGUST 9. PARIS SHELLED 44 TIMES (367 ROUNDS USED)

(12) NUMBERS OF ARRONDISSEMENTS

250 PARISIANS KILLED
620 PARISIANS INJURED
0   1 MILE

Left and below: The original 8.26-inch calibre German Paris gun. It is shown mounted on the carriage used at the naval gun-proving grounds in Germany. The weight of the shell fired by the 8.26-inch gun rose from 229 to 262 lbs with a 15½ lbs charge, and that of the later 9.13-inch gun from 273 to 307 lbs with an explosive charge 19¼ lbs in weight

John Batchelor

about the gun exploding or its destruction by French artillery, because it was true that the French artillery had begun firing in its turn in the direction of the Crépy-en-Laonnois forest, less than 30 hours after the firing of the first round.

On Friday March 29 – Good Friday – only four shells were fired by 'Bertha', and only one fell on Paris, but with dire results. On that day, a considerable crowd filled the old Church of Saint-Gervais, Rue Miron. There were no longer any seats, and many were the women and children who had come to pray for those falling in the terrible battle in progress for the past ten days. It was exactly 1633 hours, a priest who had given a short sermon had just descended from the pulpit and a concert given by the *Petits Chanteurs à la Croix de Bois* was going to end the ceremony, when suddenly, in an unnerving silence, the air was shaken by a dull explosion. A frightful shock followed by an appalling noise shook the church. A vast surface of the vaulted roof and part of the left hand side of the nave collapsed. Cries and shrieks of pain soon rose from the pile of rubble and masonry that entombed what remained of the congregation. A shell had just struck the church; there were 75 dead and 90 injured, some of whom subsequently died. An enormous crowd gathered outside and was with great difficulty pushed away from the building by a police cordon. This single shell accounted for more victims than any other day.

The firing continued with more or less intensity during April. The bad weather helped the Germans, as the French reconnaissance planes were greatly handicapped by a month of execrable weather. In the whole 30 days only April 12 was sunny.

On May 1 the shelling of Paris stopped:

Opposite page: Ruins of the church of Saint-Gervais in Paris, where 75 people died on Good Friday 1918. *Inset:* Strict attention to detail marked the installation of the 'Berthas'. This site, in the forest of Crépy-en-Laonnois, was heavily camouflaged by netting tied across the treetops: in spring leaves would spread over the netting, and mask the activity underneath. *This page:* The platform of the first gun. It consisted of a circular casing, mounted by a turntable on which the gun could traverse

French guns were causing casualties. French artillery continued its firing until May 3. For 26 days 'Bertha' was silent.

Before dawn on May 27, the Crown Prince launched his divisions towards the Chemin des Dames in a southerly direction. In less than ten days, on June 4, the Germans found themselves at Château-Thierry, and an advanced spearhead crossed the Marne. This was to be their last victorious offensive. While the German infantry overran the French positions on the dawn of May 27, a first shell fell on Paris at 0630 hours. Fifteen shells, seven for Paris, and eight for the suburbs, the last one of which fell at 1828 hours, fell on the Paris region killing four and injuring 20.

The 'Bertha' was back, and the 'psychological' firing was resumed at the same time as the new offensive was launched, but another mystery appeared: the shells were no longer coming from the Crépy-en-Laonnois forest but from much further to the west and slightly to the south, from the wood of Corbie, near Beaumont-en-Beine, 68.3 miles (110 kms) from Paris, giving an appreciable improvement in the wear and tear of the tubes. Another fact emerged with this resumed bombardment. The splinters picked up showed the calibre had been changed. It was no longer 8.3-inch shells that were falling on Paris but shells of a calibre of 9.5 inches. It has been suggested that during this period, the Beaumont-en-Beine gun was not the only one, but that a remaining piece on the Crépy-en-Laonnois slopes had also fired.

Five weeks were to elapse without a single shell dropping on Paris. It is certain that the wearing out of the guns was rapid and as soon as they could the Germans endeavoured to bring their pieces closer in order to reduce the powder charge. They had already gained 9.3 miles (15 kms) by installing their 'Parisian cannon' near Beaumont-en-Beine.

Since their last lightning offensive, the Germans were on the Marne at Château-Thierry. Why not place a giant cannon further to the south, especially since very great progress had been made in the field of installing platforms which allowed a relatively rapid erection? While the

establishment of the concrete platform on the slopes of Mont de Joie had required over seven weeks, the entirely metal one which the Germans had just built for firing in all directions gave them the opportunity of rapidly mounting a firing emplacement for their giant cannon very close to Paris and only 58 miles (93 kms) from the capital. The site was found in the wood of Châtel, not far from the Oulchy-le-Château – Château-Thierry highway, six miles (ten kms) to the north of the latter. When the gun arrived in the environs of Fère-en-Tardenois, the French artillery was already pounding the region. This was to be the third series of firings which lasted two days. On July 15, 15 shells exploded in Paris. They killed three and wounded four. The next day, July 16, firing was resumed at 1030 hours and ended at 1720 hours, but in seven hours only four shells were fired. The position was too untenable because of the French artillery, and the Germans evacuated their position to return to their original one.

For three weeks, since July 16, the 'Berthas' had been silent, when suddenly on August 5 at 1005 hours at No 19 Rue Danton in Vanves, a shell killed two and injured eight. The fourth and last series of firings lasted five days. The last shell of the 17 fired that day fell at 1930 hours in Avenue de la Grande Armée. From the rate of firing, it would seem that a single piece alone was in action, as the shells kept arriving in Paris every 15 to 20 minutes from the departure base in the wood of Corbie near Beaumont-en-Beine which the French artillery once again started to plaster. The next day, August 6, Paris and its suburbs received 18 rounds, the first one at 0857 hours, the last one at 1850 hours. On August 7, 12 shells were fired, killing five and injuring 40, while on August 8, six more shells were despatched, one man being reported killed.

But the end was near. On August 9, the first shell fell at 0916 hours in a field near Dugny. Ten other shells followed. They killed three and injured seven. On Friday August 9, 1918, the 367th and last shell fell in Rue Saint-Denis at Aubervilliers.

It was over. The advance of the Allied troops had overtaken the German gun, and it had returned home. The last of the giant cannons which killed 256 and wounded 620 had fallen silent for ever. No Allied Armistice Commission was ever able to find any trace in Germany of these guns. Everything had been demolished with oxy-acetylene cutting torches.

Further Reading
Hallade, J., *Mystérieux canons dans l'Aisne* (Paris 1968)
Poirrier, Jules, *Les bombardements de Paris* (Paris 1930)
Thierry, M., *Paris bombardé par avions, Zeppelins et canons* (Paris 1920)

JEAN HALLADE was born in 1922 and studied at the Institution Saint-Charles at Chauny. In 1943 he joined the Resistance and, after the Liberation, he volunteered for the army. He is passionately interested in history, especially in the naval and air aspects of the two world wars, and his publications include articles in specialised papers and three books, *Wings in the Storm; Mysterious Guns on the Aisne;* and *The Resistance was at the Rendezvous.*

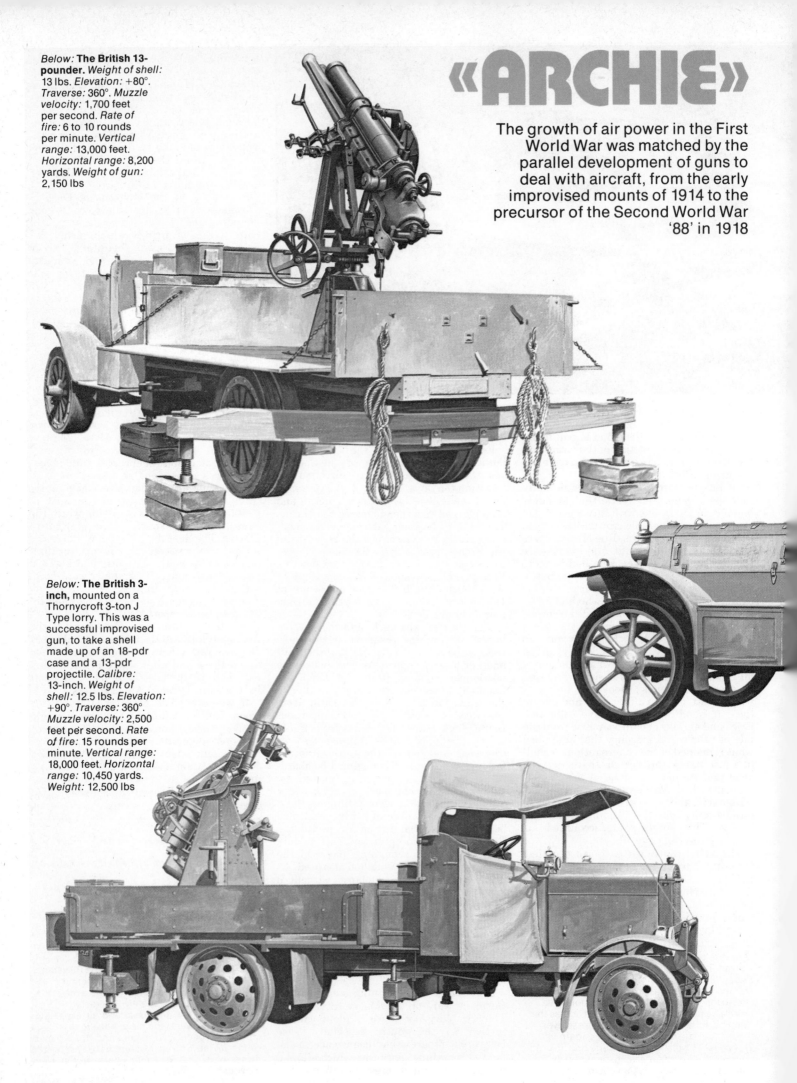

«ARCHIE»

The growth of air power in the First World War was matched by the parallel development of guns to deal with aircraft, from the early improvised mounts of 1914 to the precursor of the Second World War '88' in 1918

Below: **The British 13-pounder.** *Weight of shell:* 13 lbs. *Elevation:* +80°. *Traverse:* 360°. *Muzzle velocity:* 1,700 feet per second. *Rate of fire:* 6 to 10 rounds per minute. *Vertical range:* 13,000 feet. *Horizontal range:* 8,200 yards. *Weight of gun:* 2,150 lbs

Below: **The British 3-inch,** mounted on a Thornycroft 3-ton J Type lorry. This was a successful improvised gun, to take a shell made up of an 18-pdr case and a 13-pdr projectile. *Calibre:* 13-inch. *Weight of shell:* 12.5 lbs. *Elevation:* +90°. *Traverse:* 360°. *Muzzle velocity:* 2,500 feet per second. *Rate of fire:* 15 rounds per minute. *Vertical range:* 18,000 feet. *Horizontal range:* 10,450 yards. *Weight:* 12,500 lbs

ANTI-AIRCRAFT GUNS OF THE FIRST WORLD WAR

Right: **The German 7.7-cm.** *Calibre:* 7.7-cm. *Weight of shell:* 15 lbs. *Elevation:* +70°. *Traverse:* 360°. *Muzzle velocity:* 1,520 feet per second. *Rate of fire:* 6 to 10 rounds per minute. *Vertical range:* 14,000 feet. *Horizontal range:* 8,650 yards. *Weight:* 3,675 lbs.

Below: **The French 75-mm.** *Calibre:* 75-mm. *Weight of shell:* 15.8 lbs. *Elevation:* +70°. *Traverse:* 360°. *Muzzle velocity:* 1,735 feet per second. *Rate of fire:* 15 rounds per minute. *Vertical range:* 16,400 feet. *Horizontal range:* 9,850 yards. *Weight:* 8,800 lbs.

Left: **The German 8.8-cm,** forerunner of the famous anti-aircraft and anti-tank '88' of the Second World War. *Calibre:* 8.8-cm. *Weight of shell:* 21 lbs. *Elevation:* +70°. *Traverse:* 360°. *Muzzle velocity:* 2,575 feet per second. *Rate of fire:* 10 rounds per minute. *Vertical range:* 12,600 feet. *Horizontal range:* 11,800 yards. *Weight:* 6,700 lbs.

THE ARMOURED HERITAGE
The last tanks of the war

By the end of 1918, the design philosophy of the first tanks of 1916 had reached its culmination: its zenith, the Allied Mk VIII, was about to enter production. But the new designs about to see fruition were of more significance: track runs were lower, hull profiles more compact and the idea of a superimposed turret reintroduced. Thus the formula for future vehicles was laid down. It only remained to be seen if the tactical lessons had been learnt

Above: The last word in British tank design philosophy, the Anglo-American **Mark VIII** heavy tank, the International. This was the heaviest tank produced in the war, and also set new standards in crew comfort. Only four prototypes had been finished by the Armistice. *Weight:* 37 tons. *Crew:* 8. *Armament:* two 23-calibre 6-pdrs with 208 rounds and seven 3-inch Browning machine guns with 13,484 rounds. *Armour:* 16-mm front, 12-mm and 10-mm sides, sponsons and turret and 6-mm roof. *Engine:* Ricardo V-12 (British) or Liberty V-12 (American), 300-hp each. *Speed:* 7 mph max. *Step crossing:* 4 ft 6 ins. *Trench crossing:* 16 ft. *Range:* 50 miles. *Length:* 34 ft 2½ ins. *Height:* 10 ft 3 ins. *Width:* 12 ft 4 ins. *Left above:* The **Carro Armato Fiat 2000**, Italy's first venture into the design of heavy tanks. Only two were built. *Weight:* 40 tons. *Crew:* 10. *Armament:* one 65-mm gun and seven machine guns. *Armour:* 15-mm minimum, 20-mm maximum. *Engine:* Fiat 6-cylinder, 240-hp. *Speed:* 4½ mph.

Left below: The French **Char de Rupture 2C**. Designed in 1918, ten of these monsters were built between 1920 and 1922. *Weight:* 70 tons. *Crew:* 13. *Armament:* one 75-mm gun with 125 rounds and four machine guns with 10,000 rounds. *Armour:* 13-mm minimum, 45-mm maximum. *Engines:* two Daimler or Mercedes 6-cylinder, 250-hp each. *Speed:* 8 mph. *Range:* 107 miles. *Length:* 43 ft 2 ins. *Height:* 12 ft 8 ins. *Width:* 10 ft. *Below:* Germany's equivalent of the British Whippet, the **LK II**. *Weight:* 8¾ tons. *Crew:* 3. *Armament:* one 57-mm gun (another version had two machine guns). *Armour:* 8-mm minimum, 14-mm maximum. *Engine:* Daimler 4-cylinder, 55-hp. *Speed:* 8 mph. *Range:* 40 miles. *Length:* 18 ft 8 ins. *Height:* 8 ft 2 ins. *Width:* 6 ft 5 ins. The LK II was decidedly underpowered, and the limited traverse of its armament would have been a liability; but in the event, only prototypes had been built before the Armistice. Similar in design to the British medium tanks, it had no all-round protection

John Batchelor

THE SAMBRE

The Allied advances of August and September 1918 had by no means knocked the German armies out of the fight. But suddenly in October and November the German armies started to crumble. *Douglas Orgill. Above: Into battle 1918-style*

As the month of October opened in 1918, the shadows were falling fast across the whole of the German war effort. Yet on the Allied side of the lines, too, observers sensed a certain chill. The end of the war, to military eyes, was not yet clearly in sight; there were firm indications that the German armies were far from broken; and the dreadful possibility of a spring and summer campaign in 1919 was still heavy in the minds of the Allied staffs. In addition,

it was now autumn, and not the season in which to fight a dragging, bloody campaign. In September the Allied Commander-in-Chief, Marshal Ferdinand Foch, had planned a three-pronged offensive. In the south the Americans, with Gouraud's French Fourth Army on their left, would attack towards Mézières; the French centre would advance beyond the River Aisne; and in the north the British armies, supported by the French left, would drive forward to Cam-

CROSSINGS

brai and St Quentin. The forces available for this ambitious offensive were far from overwhelming, and at the British end of the front were barely adequate. Holding a line from St Quentin to Armentières was Haig's group of four British armies. These were Rawlinson's Fourth Army, Byng's Third, Horne's First, and Birdwood's Fifth. Between them, they comprised 52 divisions (including two American divisions and all reserves) and three cavalry divisions.

Opposing them were 63 German divisions, 30 of which were held in reserve – although the ratio in terms of man for man was not quite as high on the German side as it looked, since the prevalent influenza and heavy battle casualties had reduced some German nominal divisions to a shadow of their former strength. Further north still, holding the line from Armentières to Ypres, was King Albert's Flanders army group, consisting of Plumer's Second British

Army of 10 divisions, 12 Belgian divisions and 9 French reserve divisions.

On the battlefield Plumer faced the battered countryside of Flanders, where the tide of battle had ebbed and flowed for four years. This was a zone of deep devastation, which also embraced the front held by Haig's four armies. A senior officer who saw it that autumn reported later that . . . *from Vimy Ridge to the eastern outskirts of Amiens, and thence through St Quentin*

Südd Verlag

northwards by Cambrai to Douai, in an area of over 1,000 square miles, there was hardly a house to be found intact, no village which had not been gutted; the surface of the ground was torn and blasted by shell-fire, the vegetation withered by poison gas, the roads had been destroyed, the railways torn up, and all the bridges over the many rivers and canals blown down. In addition, the autumn rain turned the whole area, Passchendaele-style, into a desolate sea of mud. Behind the French forces in Champagne, too, there was another stretch of destroyed roads and desolated countryside, while the Americans had at their rear the old battlefield of Verdun. Armies need food, ammunition and stores: the delays inevitably entailed in the bid to rebuild an adequate communications system now made themselves felt on all the three fronts of Foch's offensive.

As October opened, King Albert's army group was about two miles from Roulers, yet it would be a fortnight before the town finally fell. Cambrai, too, had been enveloped to north and south by the First and Third Armies, but the Germans still clung to the greater part of the town. Further down the line, Gouraud had made an initial advance of about six miles, but had achieved little progress during the first nine days of the month. The worst delay of all was on the right wing of the triple offensive, where Pershing's Americans, crowded on to a narrow front where the staffwork was not good enough to operate them properly, were hopelessly wedged in the forest of the Argonne. The transport problem here was so bad that vehicles piled up in masses, unable to move forward or back, and troops in the line were going hungry and short of ammunition. The two wings of Foch's great attack, in fact, were stalled, while in the centre the French—who, like every other combatant, were none too keen to incur heavy casualties in what might be the last weeks of the war—waited willingly for the situation to sort itself out. These were disappointing days for Foch, as Haig was swift to note. Haig's diary entry for October 3 recorded that . . . *Foch's staff are terribly disappointed with the results of the American attacks on the west of the Meuse. The enemy is in no strength in their front, but the Americans cannot advance because their supply arrangements have broken down. The divisions in the front line are really starving, and have had to be relieved in order to be brought to where the food is.*

Haig himself, however, was well aware of the need to press on, and Foch had no need to repeat his stern principle of 'attaquez, attaquez, tout le monde à la bataille'. The British Commander met Foch at his headquarters at Mouchy le Chatel at noon on October 6, and announced that, provided enough guns and ammunition could be moved forward, he proposed to attack around Cambrai next day. He also asked for some more American troops to help push the assault along. 'I told F.', he said, 'that with three fresh American divisions it would be possible to reach Valenciennes in 48 hours.' Foch had already tried this gambit on Pershing, however, asking him for troops from the stalled American offensive to reinforce French armies. The answer from the American commander had been a brusque, unequivocal 'No'. Pershing was going to work out his own problems, and his mind was fixed exclusively on the problem of the Argonne.

Foch: disappointed in the results of the American operations on the west bank of the Meuse

Generals Byng and Currie. *Right:* Allied attacks that brought the Germans to breaking point

Haig, however, had grasped what needed to be done. The Allied delay had given the retreating Germans a little breathing space. Moreover, Ludendorff, who had seemed to be broken forever by the September storming of the Hindenburg Line, was now slowly recovering his nerve. He was fighting intelligently to hold the Allied armies out of Germany at least until the following spring. This meant a long stand on the line of the Meuse. Thus the American standstill in the Argonne played into his hands while he slowly withdrew his right flank from Flanders. The threat to his plan would be a British advance through Cambrai to Maubeuge, which would do much to cut off the German armies on both left and right flanks. This was precisely the advance which Haig now intended to make.

Meanwhile, the German army was told

of Germany's bid to secure an armistice, but in terms which emphasised the value of a stern continuance of the present struggle. An order of the day from Wilhelm II on October 5 reminded his troops: *For months past the enemy has been striving against your lines with powerful efforts, almost without a pause. You have had to resist by fighting for weeks on end, frequently without rest, and to show a front to a foe vastly superior in numbers. Therein lies the immensity of the task which has been given you, and you are fulfilling it . . . Your front is unbroken and will remain so. In agreement with my allies I have decided to offer peace once more to our enemies; but we will only stretch out our hands for an honourable peace.* To any German soldier who read beyond the bombast, this message must have had alarming undertones.

On the British front, Ludendorff was withdrawing grudgingly to a position partly along the line of the Selle which was known to the Germans as the *Hermann* position. In general, this line was intended to protect an area from the River Scheldt, south of Valenciennes, to the River Oise, just west of Guise, but it was very far from ready. Its chief strength had been given it by nature alone: the River Selle was an unpleasant obstacle for assaulting troops.

Second Cambrai: another tank 'duel'
At dawn on October 8, Byng and Rawlinson, with the Third and Fourth Armies respectively, launched the second battle of Cambrai, along an 18-mile front, and supported by 82 tanks. These were all the Tank Corps had managed to make fit for battle, after the extensive casualties and breakdowns which had somewhat inhibited its rôle since the great armoured stroke at Amiens on August 8. The troops had been skilfully assembled in the dark in difficult, shell-smashed and trench-ridden country. Not without difficulty, however, and the XIII Corps, on the left of the Fourth Army line, were badly harassed by shelling and air attack as they formed up. When they advanced, however, it was to success: some indication of gradually deteriorating German morale was given by ordinary line infantry who fought only briefly before running away. The machine gunners, however, as always stuck to their task, firing until the last moment.

One rare, noteworthy event marked the day. The Germans counterattacked with tanks—captured British Mark IVs—in groups of four or five. This happened on the left, near Awoingt, and two of the German-crewed machines were knocked out—one by a 6-pounder shell fired by a British tank, and the other, appropriately, by a round fired from a captured German field gun laid by a British tank section commander. The other German tanks then turned back into their own lines. This was the second tank-versus-tank action of the war, the first having been fought in April near Cachy. The German command, however, was never able to use tanks on anything like the scale of the Allies. At no time during the war did the German tank force exceed 40 machines—15 of their great slab-sided A 7 Vs, and 25 captured British models. At no time, too, did they succeed in using more than 25 in a single operation. For The British command was also unwilling to use tanks in an exploiting as well as a break-in rôle, though this, admittedly, might well have been beyond the mech-

Imperial War Museum

Bottom: Renault 37-mm tank. *Weight:* 14,300 lbs. *Armament:* 37-mm gun, with 237 rounds of ammunition. *Dimensions:* Length 16 ft 5 ins; width 5 ft 7½ ins; height 7 ft. *Engine:* 35 bhp. *Speed:* Cross-country 2.2 mph; on roads 4.8 mph. *Range:* 22 miles. *Armour:* Front 16-mm; sides 8-mm; top 8-mm; belly 6-mm. *Turret:* Either round (22-mm) or panelled (16-mm). *Crew:* 2 men. *Below:* A column of Renaults at Suvigney

National Archives

John Batchelor

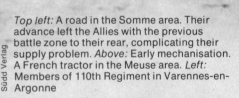

Top left: A road in the Somme area. Their advance left the Allies with the previous battle zone to their rear, complicating their supply problem. *Above:* Early mechanisation. A French tractor in the Meuse area. *Left:* Members of 110th Regiment in Varennes-en-Argonne

anical resources of the Tank Corps in October 1918. All that was offered for exploitation was cavalry—and it had already been demonstrated many times that cavalry on a modern battlefield was doomed; an easy victim for the ubiquitous wire and machine guns, and inhibited by the simple fact that it could not lie down and take cover. No decisive cavalry charge, in fact, could be made until the last machine gun had been captured. This was a prerequisite which could never be achieved; the British Official History observes sombrely that 'every effort was made by the higher commanders to pass the cavalry through to play havoc with the enemy's communications; but only a few patrols got much farther than the infantry's front line . . .'

However, with the tanks helping in what was now their traditional support rôle, the

XIII, II American, and IX Corps of Fourth Army moved forward to within a mile of Le Cateau by nightfall, while the Germans in front of Third Army finally pulled out of the battered city of Cambrai. The disciplined retreat planned by Ludendorff began to exhibit signs of panic, as troops and lorries and horse-drawn wagons jammed the roads and bridges in front of the River Selle. German battalion accounts of the day's fighting sound a note of despair. Battalions, it was said, averaged only 150 officers and men, and in some of them only 'severe measures' had held the formations together in the line. Strong measures were also taken rather higher up the line of command. The German army group commander, Boehm, was abruptly sent on leave, and was later reduced to the command of a single army. The army group itself was broken up, the German *Second*

Army going to Rupprecht's *Gruppe* based to the north, and the *Eighteenth* to the Crown Prince's *Gruppe* to the south. 'The troops,' said one official account, 'were completely used up and burnt to cinders . . .' There was, too, increasing evidence of the infection of revolution in the German army: the men whom Ludendorff had earlier noted as shouting 'blacklegs' at reinforcements moving up into the line were steadily growing in numbers and influence.

The next problem for the British command was to force the line of the Selle, the river on which the town of Le Cateau stood. Supply difficulties now dictated a pause in the onward movement: Rawlinson's Fourth Army, which was to make the assault, needed a week to establish its preparations on a proper footing. Rawlinson was covered on his right by Debeney's First French Army, and on his left by Byng's Third

Imperial War Museum

Army, with Horne's First Army on Byng's left. Between October 11 and 16 some small improvement was made in the British positions. Horne, in particular, battered his way up the Sensée Canal – an advance which was sure to inhibit the German defence of the Selle in the next few days. At 0530 hours on October 17, Rawlinson struck again, in mist so thick that it reminded Fourth Army of its earlier triumph in the fog on Ludendorff's 'Black Day' of August 8. His attack was on a ten-mile front, supported by almost 50 tanks. The programme set for Fourth Army took two days to achieve instead of the one which had been planned, for the Germans fought well, counterattacking at well-chosen points, and heavily shelling British assembly areas. Le Cateau was emptied of Germans by the evening of October 17, but the American 27th and 30th Divisions,

attacking a hillcrest railway line south of the town, met an unexpectedly ferocious response, and it was not until October 19 that Rawlinson's right wing closed up to the line of the Sambre Canal, and then stopped – five miles beyond the Selle. On October 20, also against desperate resistance, seven divisions of the Third Army and one of the First crashed their way across the Selle west of Le Cateau. The advance of all three armies began again three days later, and 15 miles were gained between the Sambre Canal and the Scheldt. A breach six miles deep and 35 miles broad had been punched into the German front, and thousands of prisoners were streaming into the cages. It had been a stunning victory, achieved in a strangely unspectacular manner, though at high cost. In the struggle for the Selle, the 24 British and two American divisions en-

gaged had taken 20,000 prisoners from 31 German divisions opposing them, and had captured nearly 500 guns. About 50 men from each British battalion in the assault had been killed or wounded.

The strong German resistance had, nonetheless, given Haig food for thought. He warned Wilson, Chief of the Imperial General Staff, that the Germans were not yet done, recording in his diary on October 19:

I visited the War Office soon after 10 am and saw Wilson. He gave me his views on conditions of armistice. He considers that the Germans should be ordered to lay down their arms and retire to the east of the Rhine. I gave my opinion that our attack on the 17th inst. met with considerable opposition, and that the enemy was not ready for unconditional surrender. In that case, there would be no armistice, and the war would continue for at least another year.

Later that day, to the Prime Minister, Lloyd George, Haig gave his considered view of the state of the Allied armies, summing it up later as:

French army: worn out and has not been really fighting latterly. It has been freely said that 'the war is over' and 'we don't wish to lose our lives now that peace is in sight'.

American army: is not yet organised: it is ill-equipped, half-trained, with insufficient supply services. Experienced officers and NCOs are lacking.

British army: was never more efficient than it is today, but it has fought hard, and it lacks reinforcements. With diminishing effectives, morale is bound to suffer.

And yet Haig was too pessimistic. The Battle of the Selle had seen the end of Ludendorff. His policy of retiring to a shorter holding line was manifestly being eroded by events, and Wilhelm II knew it. Ludendorff put his own head on the block when, in an astonishing act of insubordination, he issued an order saying that President Wilson's armistice proposals were 'a demand for unconditional surrender, and thus unacceptable to us soldiers'. On October 26 he was summoned to the presence of the Kaiser.

'There followed,' Ludendorff recorded later, 'some of the bitterest moments of my life. I said respectfully to His Majesty that I had gained the impression that I no longer enjoyed his confidence, and I accordingly begged most humbly to be relieved of my office. His Majesty accepted my resignation.' General Wilhelm Gröner, an expert on railway transport, replaced Ludendorff as First Quartermaster General.

As the British command prepared for what was to be the last great British thrust, the German slide was accelerating. The First Army reached Valenciennes, and the Fifth brushed aside feeble German rearguards to reach the Scheldt Canal. Plumer's Second Army, under King Albert's overall command in Flanders, had crossed the River Lys on October 19, preparing for a new move forward to the Scheldt. The next combined offensive was to be a drive by four armies, over a 40-mile front from Guise to Condé. The forces to be employed were the British Fourth, Third and First Armies, and Debeney's First French Army on the right. They were ordered to 'resume the offensive on or after November 3 with a view to breaking the enemy's resistance south of the Condé Canal and advancing in the general direction Avesnes-Maubeuge-Mons'. This advance would threaten Namur, the seizure of which would imperil all the German armies in western Belgium. Taken in conjunction with a resumed American advance through the Argonne to Mézières and Sedan, it would completely wreck any German attempt to establish a secure line on the Meuse. If it succeeded, the path into Germany would be growing broader and more inviting. Nevertheless, the attack would be made over difficult country, with an especially unpleasant feature in the shape of the Sambre Canal. This obstacle was 70 feet wide over a good deal of its length, and at least six feet deep. Along part of it the Germans had opened locks, flooding the low ground, though in general to a depth of only a few inches. Here the British Expeditionary Force had assembled four long years ago on its arrival in France. Mons was a charismatic word to every British soldier.

A little before dawn on November 4, the Battle of the Sambre began. 17 divisions of the three British armies attacked, with six in reserve. It was significant evidence of the inroads made by battle upon the limited resources of the Tank Corps at this stage of the war that only 37 tanks were available to support this mass of infantry. Rawlinson, on the right of the British line (with Debeney on his own right), was confronted by the canal at a point where it was both wide and commanded by good German defensive positions on the other side. He attacked with two corps—the XIII on the left and the IX on the right. The more difficult part of the canal was allotted to Braithwaite's IX Corps, which put the 1st and 32nd Divisions into the assault, with the 46th in reserve. Each division was allotted three tanks of the 10th Tank Battalion—an armoured support which can only be described as derisory.

The opening hours for Braithwaite's corps were grim. The assaulting battalion of the 1st Division was the 2nd Royal Sussex. Its

Above: German A 7 V tank. These were used alongside captured British Mark IV's. *Below:* François Flameng's painting of tank warfare

Musée de l'Armée

objective was a lock, with a group of houses on the far side, and a subsidiary stream a hundred yards short of the main canal. However, as the Royal Sussex battalion moved determinedly into the attack, a major setback was discovered—apparently the planks and bridging used in rehearsal were in fact a little too short for the actual obstacle of the subsidiary stream. Meanwhile, as the main body of the battalion began to pile up on the near bank, the German gunners found the range, and salvo after salvo, laced with streams of machine gun bullets from the houses beside the lock, tore into the struggling infantry. With tremendous courage and determination, the commanding officer of the Royal Sussex, Lieutenant-Colonel D. G. Johnson, led his infantry into the shell-storm to protect the bridge-carrying parties of the Royal Engineers, them-

selves commanded by Major G. de C. E. Findlay, of the 409th Field Company. While men dropped, hit and dying, into the stream, the bridges were eventually pushed and shoved into place, to be instantly crossed by the revengeful Royal Sussex, their machine gunners firing the heavy Lewis guns from the hip, cutting down German defenders running frantically from the houses beside the lock gate. Nearly 150 Royal Sussex and sappers lay dead or wounded behind them. For their part in this decisive action, Johnson and Findlay each received the Victoria Cross.

The other assaulting battalion of the 1st Division, the 1st Cameron Highlanders, were more fortunate, getting bridges across the canal in the first few minutes, and closing with the Germans on the far side in bitter hand-to-hand actions in the early morning mist. The 32nd Division, on the

John Batchelor

left of the 1st Division, had a bloody rebuff from a resolute German defence at Ors, but managed eventually to float a bridge across the canal on empty kerosene cans. The 32nd Division had lost 700 men in the morning's work; the 1st Division about 500.

Further still to the left the XIII Corps, with the 25th Division, pushed across the Sambre on rafts, and seized a trestle bridge which was capable of supporting the transit of field artillery. The 8th Royal Warwickshires, attacking Landrecies on the far side, were held up briefly when a German officer galloped up on horseback and man-

Below: A Belgian soldier in the uniform adopted in 1917. *Right:* Plumer. His Second Army fought alongside the Belgians. *Right:* German losses

Malcolm McGregor

marching back to the cages.

Further to the north, Byng and Third Army also had a day of success. Byng attacked with four corps in line, at slightly staggered times, so that his right-hand corps—Shute's V Corps—would be able to synchronise its attack with that of Fourth Army. Shute's corps and its next neighbour in line, Harper's IV Corps, had set an objective about five miles distant inside the Forest of Mormal. The forest itself imposed changes on the normal routine for attacking infantry. The usual protective barrage was kept 300 yards ahead of the infantry, since it was foreseen that shells striking the tops of high trees were liable to detonate in lethal airbursts hundreds of yards from where they had been intended to land. The Germans inside the forest fought with varying degrees of determination. The 13th Royal Welch Fusiliers picked up 60 prisoners without a shot being fired at Berlaimont, on the edge of the forest. On the other hand, in the corner of the forest near Englefontaine, the 9th Duke of Wellington's, of the 17th Division, ran into ground seamed with hedges and orchards, where the waiting German machine gunners cut loose wherever a glimpse of khaki was seen scrambling through the undergrowth. The battalion lost 13 of its 15 officers, and 226 out of 583 other ranks. In spite of this, the attacking battalions reached their objectives on time.

On other parts of the Third Army front the battle was a steady trudge onwards, marked by sporadic affrays on the edges of orchards or village cemeteries rather than

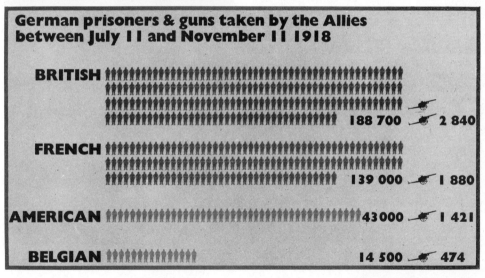

German prisoners & guns taken by the Allies between July 11 and November 11 1918

	Prisoners	Guns
BRITISH	188 700	2 840
FRENCH	139 000	1 880
AMERICAN	43 000	1 421
BELGIAN	14 500	474

aged to detonate the charges under the main bridge into the town, but soon succeeded in getting men across by two smaller wooden bridges, and by raft. They would have cleared the bank faster with more tanks: the three machines they had in support accounted for 40 machine guns during the morning, and only one of the tanks was knocked out. On the left of the corps line the 18th Division, fighting through a network of village houses, hedges, and orchards—each an ideal post for an ambushing machine gun—had to struggle until noon before a way could be cleared, but finally reached its objective inside the Forest of Mormal in the early afternoon. The Fourth Army bridgehead over the Sambre Canal was now 15 miles long and more than two miles deep, and the Mormal Forest had been penetrated. Four thousand German prisoners were

by large set-piece actions on an epic scale. The author Stephen Graham, a private in the Scots Guards in Haldane's VI Corps, remembered later the night of November 4, when his brigade was advancing on Amfroipet near the Belgian frontier:

In the afternoon we marched into Villers Pol, and most men, after sweeping and cleaning the billets, lay down and rested a few hours before the march to the line. Hot suppers and rum rations were dished out after midnight, and then at 2 am, with all the fighting impedimenta of shovels and bombs and sandbags, and what not, the battalion marched cautiously on, scouts reconnoitring each stretch of country in front and reporting all clear before we crossed it. It was a dark and windy night, and crossing the scenes of the day's fighting, we remarked here and there in the dark the vague shapes of the dead.

With dawn came the barrage, and Graham and the rest of his battalion, ducking and cursing under machine gun fire from the village cemetery, took Amfroipet. The Scots Guards pressed on to Bermeries, which was found to be already empty of Germans, and then ran into more opposition. This too was fairly quickly overcome. Graham describes this typical action:

An enemy line was located . . . in the low scrub alongside an orchard. At 1.30 pm a sharp encounter took place between one of our companies and a number of German machine gunners. The enemy was in deep slits, and his position cleverly hidden. It took about an hour to locate him certainly and dispose of him. Our men made a bayonet charge, and all the Germans were either killed, wounded, or taken prisoner . . . The men wallowed in mud all night, and it rained and rained, never ceased raining. The German artillery was very active, though firing largely at random . . . The last men to fall in the war fell, as it were, by accident; strolling back from the line towards headquarters; they were being brought back to Bermeries for a few hours' rest and were lighting cigarettes and chatting in little knots when two heavy shells came in their midst, tore one man's face off, ripped up another's stomach, and the like . . .

'Hostilities will cease . . .'
The whole British front was now on the move, and prisoners were streaming back in thousands. Behind the German front the new Commander, General Gröner, looked despairingly at the map. A Guards battalion, unopposed, took Maubeuge on November 9. On the 8th, the 32nd Division had taken Avesnes: Birdwood's Fifth Army crossed the Scheldt, and Horne's First Army was closing up to Mons. The Sambre was now 20 miles to the rear. Ahead lay, not so far away, the Rhine. It was time for the German command, at last, to call a final halt. At 6.50 am on November 11, Haig's headquarters gave the message that

so many had hoped for so long: 'Hostilities will cease at 1100 hours today . . . Troops will stand fast on the line reached.'

In Britain, New Zealand, South Africa and the other countries of the Empire, and in France and Belgium and Germany, too, the last telegrams were still going out, announcing on the very edge of joy that a father, a son, or a brother would never come home again. One of those who had scrambled to the assault on the Sambre on the morning of November 4 was the soldier-poet Wilfred Owen, whose *Anthem for Doomed Youth* still appears in every anthology of war poetry:

Above: Tanks again at Cambrai. Australian and American troops pass through the town in October 1918. *Below:* American troops using Renault tanks near Boureuilles, September, 1918

What passing bells for these who die as cattle?
Only the monstrous anger of the guns.
Only the stuttering rifles' rapid rattle
Can patter out their hasty orisons.
No mockeries for them from prayers or bells,
Nor any voice of mourning save the choirs,
The shrill demented choirs of wailing shells;
And bugles calling for them from sad shires.

There was a bugle for Wilfred Owen, too. He was killed on the Sambre crossing. His parents received their telegram on November 12. It was midday and all the church bells were ringing.

Further Reading
Blake, Robert (ed.), *The Private Papers of Douglas Haig* (Eyre & Spottiswoode 1952)
Blaxland, Gregory, *Amiens 1918* (Muller 1968)
Goodspeed, D. J., *Ludendorff* (Hart-Davis 1966)
Graham, Stephen, *A Private in the Guards* (Macmillan 1919)
Pitt, Barrie, *1918: The Last Act* (Cassell 1962)
Terraine, John, *Douglas Haig; the Educated Soldier* (Hutchinson 1963)
Military Operations: France and Belgium 1918, volume V (HMSO 1947)

DOUGLAS ORGILL was born in 1922 and was educated at Queen Mary's School, Walsall, and Keble College, Oxford, where he took an honours degree in modern history. During the war he served in the Royal Armoured Corps, being commissioned into the Lothians and Border Horse, an armoured regiment in 6th Armoured Division. He commanded a troop of Sherman tanks in Italy, and after the war served for a year in the Arab Legion in Jordan and Palestine. He is at present chief sub-editor of the *Daily Express*. His book *The Gothic Line*, a study of the autumn campaign in Italy in 1944, was published in 1967, and was followed by *The Tank*, studies of armoured warfare, in 1970. He is now working on an historical examination of cavalry.

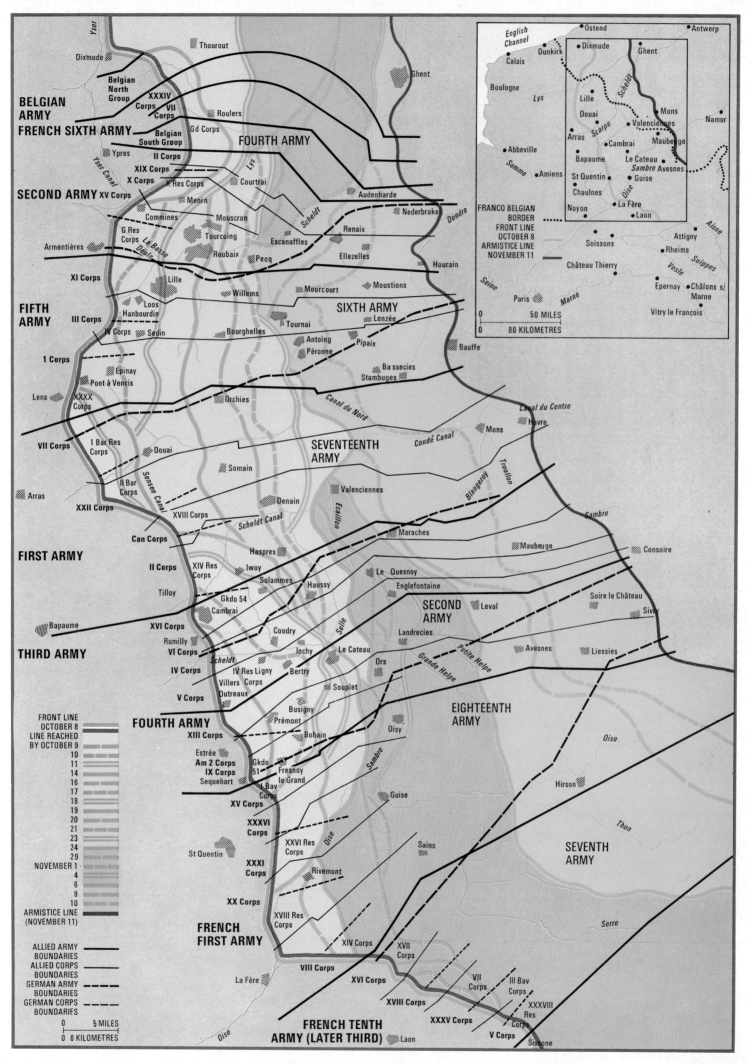

BELGIAN ARMY

FRENCH SIXTH ARMY

Yser

Dixmude

Thourout

Ghent

Belgian North Group

XXXIV Corps

VII Corps

Roulers

Gd Corps

FOURTH ARMY

Belgian South Group

Ypres

II Corps

XIX Corps

Lys

Courtrai

Audenharde

SECOND ARMY XV Corps

X Corps

X Res Corps

Yser Canal

Menin

Mouscron

Scheldt

Nederbrake

Dendre

Commines

Mouscron

Escanaffles

Renaix

G Res Corps

La Basse Deule

Tourcoing

Pecq

Ellezelles

Hourain

Armentières

Roubaix

Willems

Mourcourt

Moustions

XI Corps

Lille

SIXTH ARMY

FIFTH ARMY

Loos

Hanbourdin

Bourghelles

Tournai

Lenzée

III Corps

IV Corps

Sedin

Antoing

Péronne

Pipaix

Ba sscies

Stambuges

Bauffe

1 Corps

Epinay

Pont à Vencis

Orchies

Canal du Nord

Canal du Centre

Lens

XXXX Corps

Havre

Mons

VII Corps

1 Bar Res Corps

Douai

Somain

SEVENTEENTH ARMY

Condé Canal

Blangeroy

Trouillon

Sambre

II Bar Corps

Sensee Canal

Valenciennes

Arras

XVIII Corps

Denain

Ecaillon

XXII Corps

Scheldt Canal

Maraches

Maubeuge

Consoire

Can Corps

Haspres

Le Quesnoy

FIRST ARMY

XIV Res Corps

Iwuy

Englefontaine

Soire le Château

II Corps

Solammes

Haussy

SECOND ARMY

Leval

Sivry

Tilloy

Gkdo 54

Cambrai

Coudry

Landrecies

Avesnes

Liessies

Bapaume

Selle

THIRD ARMY

XVI Corps

Rumilly

VI Corps

Inchy

Le Cateau

Ors

Petite Helpe

Grande Helpe

IV Corps

Scheldt

IV Res Ligny Corps

Bertry

Souplet

EIGHTEENTH ARMY

Villers Outreaux

V Corps

Busigny

Oisy

Oise

FOURTH ARMY

Prémont

Bohain

XIII Corps

Estrée

Am 2 Corps

IX Corps

Gkdo 51

Fresnoy le Grand

Sambre

Guise

Sequehart

I Bav Corps

Hirson

SEVENTH ARMY

XV Corps

XXXVI Corps

XXVI Res Corps

Oise

St Quentin

XXXI Corps

Rivemont

Sains

Thon

XX Corps

XVIII Res Corps

Serre

FRENCH FIRST ARMY

XIV Corps

XVII Corps

VIII Corps

XVI Corps

VII Corps

III Bav Corps

La Fère

XVIII Corps

XXXVIII Res Corps

XXXV Corps

FRENCH TENTH ARMY (LATER THIRD)

Oise

Laon

V Corps

Sissone

Inset map (top right):

English Channel

Ostend

Antwerp

Calais

Dunkirk

Dixmude

Ghent

Boulogne

Lys

Lille

Scheldt

Mons

Namur

Abbeville

Douai

Scarpe

Valenciennes

Arras

Cambrai

Maubeuge

Somme

Bapaume

Le Cateau

Amiens

St Quentin

Sambre

Avesnes

Guise

Chaulnes

Oise

Aisne

Noyon

La Fère

Soissons

Attigny

Laon

Rheims

Suippes

Seine

Marne

Château Thierry

Vesle

Paris

Epernay

Châlons s/ Marne

Vitry le François

FRANCO BELGIAN BORDER ......
FRONT LINE OCTOBER 8
ARMISTICE LINE NOVEMBER 11

0   50 MILES

0   80 KILOMETRES

Legend (bottom left):

FRONT LINE OCTOBER 8

LINE REACHED BY OCTOBER 9

10
11
14
16
17
18
19
20
21
23
24
29

NOVEMBER 1
4
6
9
10

ARMISTICE LINE (NOVEMBER 11)

ALLIED ARMY BOUNDARIES

ALLIED CORPS BOUNDARIES

GERMAN ARMY BOUNDARIES

GERMAN CORPS BOUNDARIES

0   5 MILES

0   8 KILOMETRES

PART II: THE SECOND WORLD WAR
INFANTRY AT WAR

J. B. King

J. B. KING first wore uniform as a rifleman in a light infantry regiment during the Second World War. He retired thirty years later having attained staff rank and after experience in different branches of the Army. Like most soldiers he set about retiring into the countryside but, also like most soldiers, he found it too quiet and emerged once more to become a freelance writer and consultant on weapons.

Histories of warfare usually take account only of the broad ebb and flow of the tide of battle; objectives taken, positions held, battles won and lost. The ordinary soldier figures as no more than a statistic, on the strengths of opposing forces or the casualty lists.

Yet without the infantryman there would be no battle – and for him war is more than outlines and statistics. He is an individual, fighting his own fight, and all the complex machinery of war is devoted, ultimately, to getting him to the battlefield and supporting him while he is there. Ultimately, too, it is the infantryman who decides the outcome; who 'winkles the other bastard out of his foxhole and makes him sign the peace treaty!'

CONTENTS

Associated Press

US Marine at Okinawa, 1945

THE MEN WHO MEET THE ENEMY FACE TO FACE

There is an old joke which claims that the role of cavalry is to add tone to what might otherwise be merely an unseemly brawl; from this it might be inferred that the infantry's role in the battle is to promote the brawling. But there is a good deal more to it than that. Infantry is the basic element of the battlefield; tanks can roar back and forth like visiting Martians, guns can fire barrages for hours on end, aircraft can scream across the battlefield bombing and strafing, but when all this is done it requires a man with a rifle in his hand to walk across the ground on his two feet, occupy it, rule it, deny it to the enemy, and, in the words of an American General, "winkle the other bastard out of his foxhole and make him sign the peace treaty". There have been attempts to replace manpower with firepower, but the experiment has never succeeded. As long as there are wars, there will be infantry.

Another common misconception is that the infantry are no more than cannon-fodder, somewhat dense individuals unfitted for anything better or more mentally demanding than walking about with a rifle. This too is highly erroneous; the modern infantryman is a highly trained specialist, competent to handle several weapons and skilled in many arts. He is expected to master two or three types of rifle, machine-guns, submachine-guns, mortars, light artillery, radio, radar, infra-red devices and drive a variety of vehicles; he must have a good knowledge of first aid, he must be able to cook and look after himself in conditions verging on the appalling, find his way in featureless country, survive an atomic bomb, find direction by the stars, have a knowledge of water purification and sanitation, be able to use his enemy's weapons as well as his own, have a knowledge of minor tactics, be self-reliant in emergency . . . the list can go on and on.

The infantryman is the army's jack-of-all-trades, but with it he is master of one – the trade of fighting the enemy. He is a member of the world's most exclusive club – the men who actually meet the enemy face to face, eyeball to eyeball, and decide the issue on the spot. Artillery, tanks and aircraft can break an enemy's spirit, demoralise him, inflict casualties; but unless this is followed up by determined men on foot the advantage is lost and the effort wasted. Infantry are the army's backbone.

Organisation

The organisation of infantry units is much the same in all countries, since the tasks to be performed are the same. Starting at the bottom, the individual soldier forms part of a **section**, ten or a dozen men including one or two junior NCOs, the basic tactical unit. The section will contain one machine-gun and possibly one or two submachine-guns, the rest of the men carrying rifles.

In British practice the section split into two groups, the rifle group and the machine-gun group, the latter consisting of just three men with the light machine-gun. Thus it could move in 'bounds' when confronted with an enemy, the machine-gun taking up position to give covering fire while the rifle group moved forward. They in turn would take up position and give covering fire to allow the machine-gun group to move up and leapfrog them, taking up a fresh position to cover another move of the rifle group. This alternate movement would go on until the rifle group were close enough to come to grips with the enemy when, after a bombardment of grenades, the riflemen leapt up and charged at the enemy, firing as they went, assisted by the machine-gun's fire until they closed with the enemy, when the machine-gun would stop and move up to consolidate the position which had been taken.

This basic movement was called 'Battle Drill', and it was actually taught as a drill, on the parade ground, before it was practised on the ground. As a result it became ingrained in the individual soldier so that he understood the basic tactical move and, in emergency, could take command and continue the attack in the knowledge that he was doing the right thing. Moreover, the same basic premise governed the application of larger forces: one unit covering while the other unit moved forward – or as one contemporary instruction termed it, "one leg on the ground and one in the air". With minor variations it was the basic move of the infantry of any army.

The section formed part of the **platoon**; this usually contained three or four sections, plus a small headquarter element consisting of radio operator, officer in command, sergeant and one or two men. Thus the platoon could split up into sections or work as a whole, one section covering while the others moved, an enlarged version of the same Battle Drill.

The platoon, in its turn, formed part of the **company**; three or four platoons, plus a headquarter element, plus, perhaps, an anti-tank platoon or mortar platoon to provide specialist weapon support for the whole. Again, the company could act in its component parts or it could act as a complete entity, one platoon covering the movement of the others.

Three or four companies went to make up the **battalion**. Once more this would have a headquarter company, which might include further heavy weapons, radio, pioneer elements, transport elements and all the service troops needed to make the battalion a completely self-sufficient military unit. It could move itself, feed itself, had medical facilities and was really the lowest sub-unit capable of going into battle independently. The average soldier identified with his battalion; ask a private what unit he belonged to and the answer would rarely take account of the section, platoon or company – he was in the "Second Battalion of the so-and-so's"

The battalions formed part of the **regiment** – normally three to a regiment. The regimental structure was similar to that of the battalion in that it carried all the administrative services necessary to be

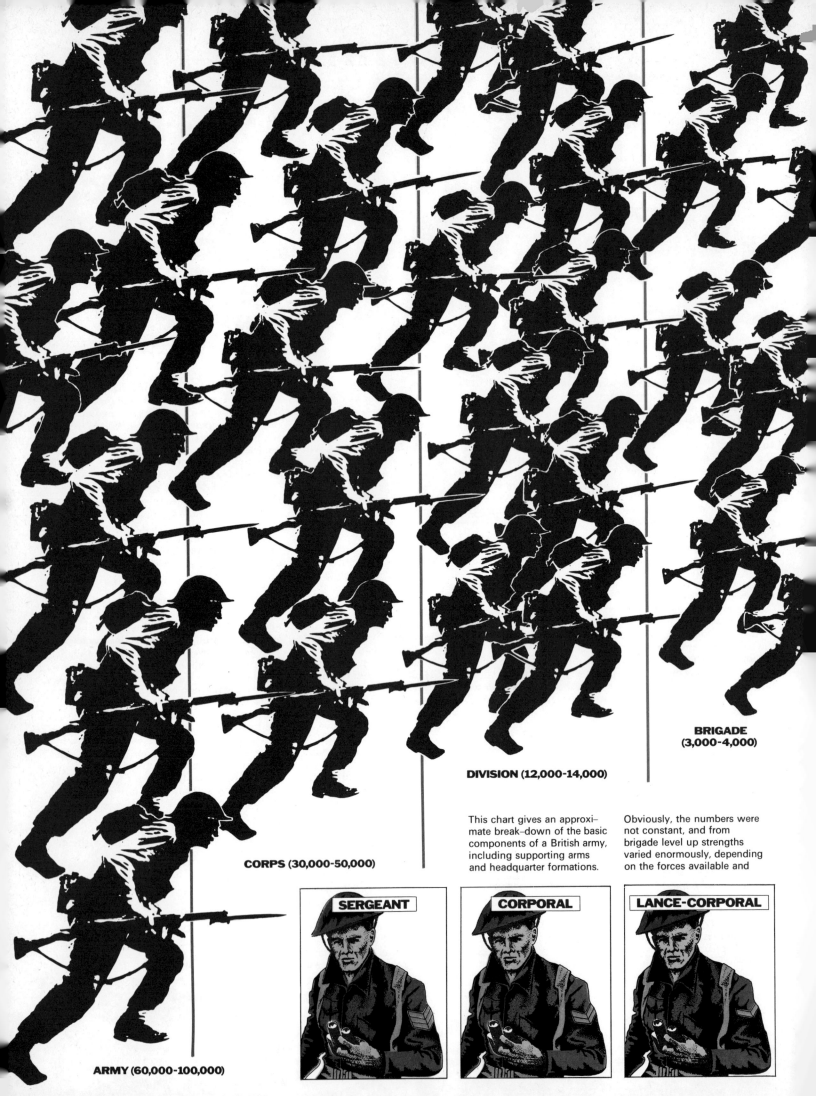

BRIGADE
(3,000-4,000)

DIVISION (12,000-14,000)

CORPS (30,000-50,000)

This chart gives an approxi–
mate break–down of the basic
components of a British army,
including supporting arms
and headquarter formations.

Obviously, the numbers were
not constant, and from
brigade level up strengths
varied enormously, depending
on the forces available and

SERGEANT

CORPORAL

LANCE-CORPORAL

ARMY (60,000-100,000)

HOW THEY FALL IN

BATTALION (700-800)

COMPANY (120)

PLATOON (32)

SECTION (8)

FIELD-MARSHAL

GENERAL

BRIGADIER

COLONEL

the tasks to be carried out.
The illustrations show
officers and men of the
British Army wearing uni‑
forms of 1940

MAJOR

CAPTAIN

LIEUTENANT

Drawings: Faulkner/Marks

self-sufficient, plus more heavy weapons and communications, transport and supply echelons. But at this level there were many differences between different nations, depending on how they felt their role could best be accomplished.

The German Infantry Regiment, for example, had a very mixed composition: headquarters; signal section; motor-cycle despatch rider section; a mounted infantry platoon (though this was rarely implemented during war); a pioneer platoon to attend to such details as minefields, bridge-building and so on; an infantry gun company armed with 75-mm and 150-mm guns and howitzers; an anti-tank gun company armed with 37-mm or 50-mm anti-tank artillery; a transport column; an anti-aircraft gun section armed with 20-mm automatic guns; and three battalions each containing a headquarters, signal section, three infantry companies and a machine-gun company. This totalled 95 officers and 2,993 soldiers, with a firepower of 112 machine-guns, 144 submachine-guns, 36 heavy machine-guns, 27 anti-tank rifles, 45 mortars, 8 field guns, 12 anti-tank guns, 4 anti-aircraft guns and about 1,500 rifles.

But the regiment rarely operated on its own; it formed part of a **division** and, keeping to our German example, this became a very potent fighting force. The division

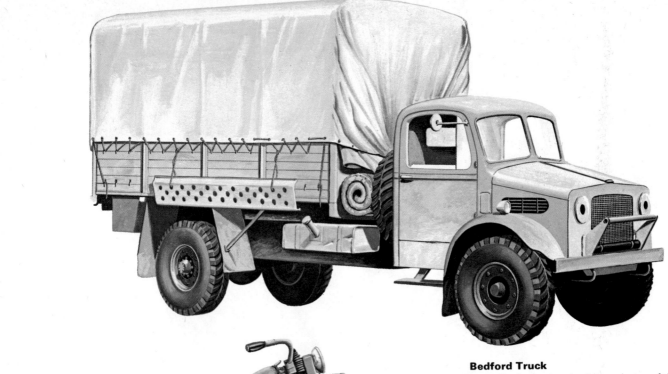

Bedford Truck
The Bedford was the absolute mainstay of the British Army as far as transport was concerned, especially for the infantry. It could carry loads of up to 3 tons, and was used for transporting men, weapons, ammunition, petrol and supplies

US Army Harley-Davidson Motorcycle
Known officially as the 'Chain Drive Solo Motorcycle', this was the civilian Harley—Davidson Model WLA with combat tyres and a bracket for mounting a Thompson submachine-gun. Widely used by messengers, reconnaissance scouts and Military Police

consisted of a headquarters; a divisional reconnaissance unit of armoured cars and cyclists; a signals regiment; three infantry regiments; an artillery regiment; an anti-aircraft battalion; an anti-tank battalion; an engineer battalion; a medical company; a veterinary company (for the German Army still relied very heavily on animal transport) and a service company which provided cooks, armourers and other specialists.

It must be said that this was the theoretical line-up: the actual composition of an infantry division varied depending on what was available, where the unit was going to operate, what sort of battle was envisaged, what sort of weapons were considered necessary and so forth. None of the world's armies produced infantry divisions in the field which exactly matched the paper composition. But the theoretical composition of the German infantry division given above produced a strength of 497 officers, 14,953 men, with 432 submachine-guns, 430 light machine-guns, 116 heavy machine-guns, 81 anti-tank rifles, 84 anti-aircraft guns, 75 anti-tank guns, 141 mortars, and 74 pieces of artillery.

There are, however, different kinds of infantry in some armies. The British Army distinguishes between infantry and light infantry, though the difference today is largely one of tradition. The light infantry in days of old were expected to be just that – light, and quick to move, ready to take

advantage of the enemy while the slower-moving 'heavy' infantry followed up behind to consolidate. Their role was principally scouting and skirmishing, but the different nature of war in the present century has diminished their special role; they operate on the same lines as other infantry today.

Nevertheless, they still have their own idiosyncracies which stem from their old role; for example, they march at a faster pace than the rest of the army – 160 paces a minute against the normal 120 – which leads to confusion when they form part of a ceremonial parade with other troops; their arms drill is different, since they always carry their rifles at the 'trail' in one hand, instead of on their shoulders; they never refer to bayonets but always to 'swords' in some regiments; and during the Second World War they always asserted their individuality by wearing their field-service caps perfectly vertical instead of tipped over one ear like the rest of the army.

Different types
Other types of infantry which have appeared from time to time in different armies include 'mounted infantry' – mounted on horses to allow them rapidity of movement, but expected to dismount and fight in the usual infantry manner after contact with the enemy; 'lorried infantry', the same sort of thing but using petrol instead of horseflesh; 'motorised infantry',

like lorried infantry but usually provided with armoured transport and backed up by attached armoured car units; and 'airborne infantry', which could be either another name for parachute troops or could be ordinary infantry ferried into battle by aircraft or glider. There are also 'mountain regiments' of infantry which differ in that they operate more in the vertical plane than the horizontal, and usually include such esoteric specialists as ski troops and expert rock-climbers, backed up by mule-pack artillery. And there have even been such things as 'cyclist regiments' of infantry mounted on bicycles – used in the Second World War by the Japanese and Germans – and even 'motor-cyclist battalions' of the German Army which roared into action on BMW sidecar combinations.

It must not be thought that only infantry divisions contained infantry; armoured divisions also carried their quota although, of course, the predominant arm was the armour. The German Panzer Division, for example, carried with it a lorried infantry regiment of 24 officers and 637 men, plus a motor-cycle battalion of 18 officers and 288 men. Not very much, considering the overall strength of a Panzer division, but the tanks were there to do the fighting; the infantry were taken along to consolidate and hold the ground which had been taken. Which brings us back to the point where we came in.

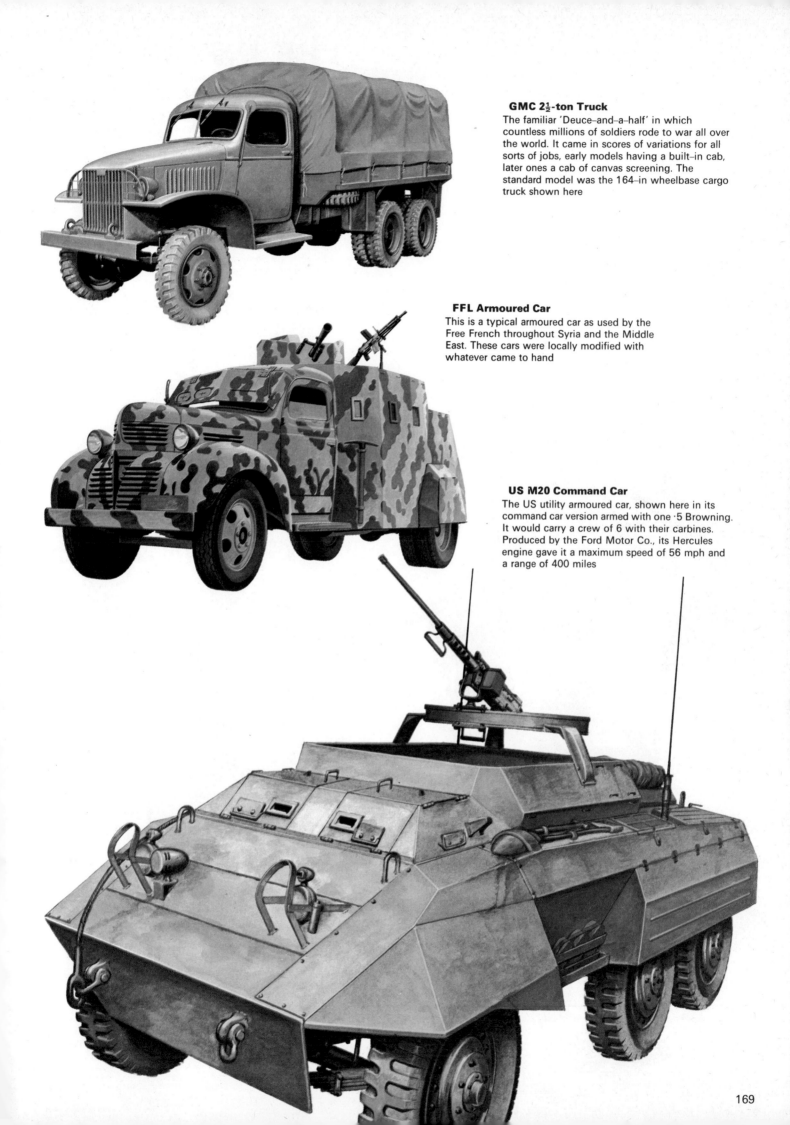

GMC 2½-ton Truck
The familiar 'Deuce–and–a–half' in which
countless millions of soldiers rode to war all over
the world. It came in scores of variations for all
sorts of jobs, early models having a built–in cab,
later ones a cab of canvas screening. The
standard model was the 164–in wheelbase cargo
truck shown here

FFL Armoured Car
This is a typical armoured car as used by the
Free French throughout Syria and the Middle
East. These cars were locally modified with
whatever came to hand

US M20 Command Car
The US utility armoured car, shown here in its
command car version armed with one ·5 Browning.
It would carry a crew of 6 with their carbines.
Produced by the Ford Motor Co., its Hercules
engine gave it a maximum speed of 56 mph and
a range of 400 miles

THE INFANTRY
READY FOR ACTION

A British soldier equipped for anti-tank work, with trenching tools, PIAT, and carrying his Sten submachine–gun. The PIAT was an alarming weapon to use, but effective enough

An American soldier carrying his rifle — the famous Garand ·30 M1, described, with good reason, by General Patton as the 'best battle implement ever devised'

BRITISH

AMERICAN

A German soldier armed with MP40 machine–
pistol and with a stick grenade in his belt.
Often, the Germans stuck their grenades in the
tops of their jackboots

A Japanese soldier with his 6·5–mm Arisaka
rifle, fitted with its sword bayonet. Japanese
soldiers were not particularly well equipped:
the first Arisaka was produced in 1897

GERMAN

JAPANESE

Arthur Gaye

171

RIFLES
A SOLDIER'S BEST FRIEND

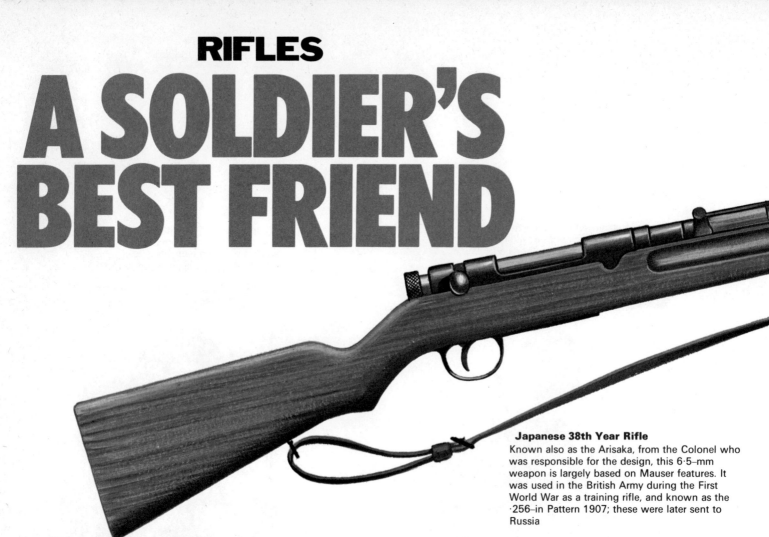

Japanese 38th Year Rifle
Known also as the Arisaka, from the Colonel who was responsible for the design, this 6·5-mm weapon is largely based on Mauser features. It was used in the British Army during the First World War as a training rifle, and known as the ·256-in Pattern 1907; these were later sent to Russia

The basic weapon of the infantryman is, of course, his rifle; so much so that one normally speaks of 'rifle battalions' or 'rifle companies', though not, in British parlance, of 'Rifle Regiments' since this means something else again – a regiment of the Kings Royal Rifle Corps. Another common expression when speaking of the strength of an infantry unit is to refer to it as having "so many bayonets", reflecting the other standard infantry weapon. The rifle is generally represented to the recruit as potentially his best friend, though in the early days of training he rarely sees it in this light, thinking of it more as a somewhat demanding mistress, always in need of cleaning and polishing, and containing all sorts of pitfalls for the unwary in the way of unnoticed corners in which dust can accumulate, only to be detected, with dire results, by the eagle eye of the inspecting sergeant-major. But once on campaign, the truth becomes blindingly obvious, and from then on the rifle's care and protection is always foremost in the soldier's mind.

With the exception of the American Army, the standard rifle with which all combatants began the Second World War was an elderly bolt-action dating from the last years of the previous century. In those days rifles were built to last, constructed by traditional gunsmithing techniques of the finest materials, assembled carefully, and in the hands of a trained soldier deadly at a mile range. It is customary for target-shooters to look down their noses at military rifles as being inaccurate, but any military rifle, when properly zeroed and handled by a soldier who knows what he is doing, can put five shots out of five into a man-sized target at a thousand yards without much trouble.

Unfortunately, this is a technique which rarely needs to be exercised on the battlefield, and it took a long time for this fact to sink in, as will be seen. As a result, all the standard bolt-action rifles used powerful ammunition, were heavy, recoiled somewhat forcefully, and reached out to over 2,000 yards. But withal, they were robust and reliable, simple to use, and deadly.

The oldest design was that in service with the Soviet Army, the Moisin-Nagant of 1891, known originally as the 'Three-Line Rifle'. This came from the calibre, three of the old Russian measurement known as 'lines', approximately one-tenth of an inch; hence three lines meant a calibre of 0·30-in. It was long (51·25 in) and heavy (9¾ lb) and mounted a socket bayonet around the muzzle. This was intended to be fitted all the time, and the rifle sights were graduated to take the weight and balance of the bayonet into account. Certainly, one rarely sees pictures of the Tsar's soldiers with this rifle without the bayonet being fitted.

Vast numbers

Although this was a sound enough rifle, it was too cumbersome, and in 1930 it was re-designed to a more compact length (40 in) and less weight (just under 9 lb). The sights were improved and it took the place of the 1891 model as the standard infantry rifle. However, there were such vast stocks of the 1891 that it remained in use as a training weapon and in the hands of reserve troops throughout the war. A number of them were also fitted with telescopic sights and used for sniping.

Next in order of age came the Japanese Arisaka, which first saw the light of day in 1897. This was chambered for a 6·5-mm cartridge and was more or less a copy of a

Mauser design but with some refinements of dubious value added by the Japanese. In 1905 the design was overhauled to do away with some of the less useful ideas and bring it more or less into line with an ordinary Mauser, and this stayed the standard rifle throughout the Second World War. However, experience during the Manchurian and Chinese wars in the 1930s led the Japanese to the conclusion that a 6·5-mm bullet was not sufficiently powerful, and after some experimenting they adopted a 7·7-mm cartridge, more or less copied from the British ·303 model except that the cartridge case was rimless.

The next step was to produce a rifle to fire the new round, and this was the Type 99, issued in 1939. Except for its different calibre and a reduction in length, it was simply the 1905 model over again, but the changing times were reflected in the fitting of a complicated anti-aircraft sight to allow the troops to shoot at attacking planes. It is open to question whether this was worth much in practice. In any case the production of this rifle never reached great numbers, and the majority of the Japanese Army continued to use the 6·5-mm model throughout the war.

An interesting development of the Type 99 rifle was a collapsible model for parachute troops, in which the barrel could be separated from the rest of the weapon by unlocking an interrupted thread joint, so that the two sections could be conveniently carried by a parachutist. Another arm specially developed for parachute troops was a short carbine version of the 6·5-mm rifle in which the stock was hinged behind the trigger and thus the butt could be folded alongside the action, again for convenience in carriage. Since the Japanese

French MAS36CR39 Rifle
This weapon was the last bolt–acton rifle ever to be adopted by any nation, the French taking it into use in 1936, the same year that the Americans were adopting the Garand. It is notorious for not having a safety catch

Italian M1891 Rifle
A Mannlicher–Carcano design, though the only Mannlicher feature was the clip–loaded magazine, the rest of it being more or less copied from Mauser. In 6·5–mm calibre, it was a reliable enough design, representative of its times

Italian Carbine Model 38
A design with some unfortunate features; this weapon has a sight which cannot be adjusted for range and some had folding bayonets underneath. Their war record was not particularly impressive, but the weapon goes down in history as the gun that killed President Kennedy in 1963

German Gewehr 98
The Mauser rifle which, in one form or another, armed many of the world's armies during the 20th century. The Gewehr 98 was the German Army's standard model, but similar versions under other names could be found all over the world. A cross–section of the mechanism is shown above

used their airborne forces very little, these weapons were not produced in large numbers.

Next in order of age comes the famous Mauser, standard rifle of the German Army since 1898, and always known by its year of introduction as the 'Gewehr 98'. The Mauser is, of course, the standard to which all other rifles are referred, and there is no doubt that more countries have been armed with Mauser rifles, or copies of Mauser rifles, or adaptations of Mauser rifles, than any other design. There is some room for the belief, however, that this was as much due to Mauser's salesmanship as to the intrinsic virtues of the rifle. While it was certainly accurate and robust, the bolt action was awkward to operate and difficult to clean, two features which argue against it as a combat rifle.

The original rifle version was, in common with most other designs of the 1890s, long and fairly heavy, and in the early years of the twentieth century it was supplanted by a short version, called the Karabiner 98, or Kar 98. Until 1903 it had generally been the custom throughout the world to produce two versions of the standard bolt action rifle, one long one for use by infantry and a short one for use by cavalry, the shorter version being called a 'carbine' or 'dragoon rifle' or some similarly descriptive term.

In 1903 the British Army decided that this

was unnecessary and produced a 'short rifle', shorter than the usual rifle and longer than the usual carbine, so that one weapon could be used to arm all troops. This example was soon followed by others; the Kar 98 was the German version and it soon became their standard rifle. The basic design was never altered from that day forward, the only changes being in the sights, to cater for improved ammunition, and in minor details which permitted the weapon to be mass-produced more easily. Other nations, after the First World War, took the Kar 98 as their model and produced their own versions, such as the Czechoslovakian Model of 1924, and the Belgian Model 1922, but a soldier trained on any one of them could easily find his way around any of the others.

Ten rounds in ten seconds
The British Army had introduced their first Lee-Enfield rifle in 1895, a long rifle in the same style as its contemporaries, and, as already related, in 1903 they produced a short rifle, known as the 'Short, Magazine, Lee-Enfield' and always abbreviated to SMLE by the troops. The Lee bolt action was, in theory, less efficient than the Mauser since the lugs which locked the bolt were at the rear of the bolt and not at the head; this meant that the body of the weapon had to be stronger and also that

when the rifle was fired the bolt was slightly compressed backwards. This, it was averred, led to inaccuracy. It may have done, if the firer was trying to take the pip out of the ace of hearts at a thousand yards, but for all practical battlefield purposes it made no difference.

Where the Lee-Enfield scored was in the ease of operation of the bolt, due to those same rear-end locking lugs. This came to the fore in a technique developed and taught during the Second World War for house-to-house fighting, in which the rifle was held at the hip with the thumb and forefinger of the right hand grasping the bolt and the middle finger inside the trigger guard; a quick flip of the wrist and the bolt was operated, and as the hand came to rest the middle finger fell onto the trigger and fired the rifle. It was possible to get off ten rounds in ten seconds very easily by this method; it may not have been accurate, but it kept the other man's head down until you could get close enough to throw a grenade at him.

At about the same time as the British Army were adopting the Lee-Enfield, the United States Army were looking for something to replace their Krag-Jorgensen rifle. After examining a wide range of alternatives, they came to the conclusion that the best answer was a Mauser. For $200,000 they obtained a licence from Mauser to

British Lee-Enfield Rifle
The British Lee—Enfield was criticised when it was first produced, but millions of soldiers proved the critics wrong. Probably the best combat bolt–action rifle ever invented, no less than 27 different versions were produced. It is still in service today, rebarrelled for the 7˙62–mm NATO cartridge, as a sniping rifle

Russian Simonov Model 1936 AVS
Although not the first Russian automatic rifle, this was the first successful one. Produced in 1936 in 7·62–mm calibre, in automatic and semi–automatic versions, it was widely used by the Soviet army

cover certain of Mauser's patents and set about developing what came to be known as the 'Springfield' rifle. It was designated the Model 1903 although the first versions did not appear until 1905.

In a similar manner to the British, they moved away from the long rifle and carbine combination to produce a short rifle which could be used by any service. By the time of the Second World War it was obsolescent, being in the process of replacement by an automatic rifle, but since the supply of the automatic model took some time to get organised, the Springfield stayed in service throughout the war for training and in the hands of reserve and home defence troops. Indeed, it was still in service as a sniper's rifle in the Korean War in 1952. It saw a few modifications during its life, principally in order to make it easier to produce in quantity, but the design was basically unchanged after 1903.

MacArthur's decision
The automatic rifle which was to supplant the Springfield was, of course, the US Rifle M1, or, more familiarly, the Garand. John C. Garand had begun work on an automatic rifle in 1920, and in 1929 his design was the only one to survive some searching tests at Aberdeen Proving Ground. Development and perfection continued, and in 1936 it was officially adopted as the standard service rifle. Responsibility for this far-reaching decision rested with General MacArthur, then Chief of Staff, and whatever MacArthur did or did not do in later years, this single decision ought to earn him the everlasting thanks of his country. The Garand wasn't perfect (the perfect automatic rifle isn't with us yet) but it showed that a reliable automatic rifle was possible, that troops armed with automatic rifles were not liable to blast all their ammunition off in the first two minutes of battle – a possibility which had haunted staff officers ever since the first inventor called on them with an automatic rifle design – and that the mechanics of an automatic rifle were not beyond the mental powers of the soldier – another belief of the traditional officers. General Patton said once that the Garand was "the best battle implement ever devised" and he wasn't far wrong.

The operation of the Garand was deceptively simple; on looking at it one wonders why it took so long to develop and why it wasn't in production fifty years earlier. But like all simple designs, it took time to get it simple and reliable. The bolt is a rotating bolt, much the same as that on an ordinary bolt-action rifle, but is operated by a rod which hooks over a lug on the bolt and then disappears beneath the woodwork of the rifle. The rod passes to a gas cylinder beneath the barrel and also carries the bolt return spring. This spring also performs the task of placing pressure on the rounds in the magazine, to keep them feeding up to the bolt as it moves.

When a shot was fired, the gas, pushing the bullet through the barrel, was tapped off through a tiny hole near the muzzle and led into the gas cylinder. Here it pushed on the head end of the operating rod, which was driven back and, by its connection to the bolt lug and by suitably shaped cam surfaces, rotated the bolt and drove it back to extract and eject the fired cartridge case. At the end of the bolt's stroke, the return spring, which had been compressed by the movement of the operating arm, expanded again and pulled the bolt forward, stripping the top round from the magazine and chambering it. As the operating rod made its final movement it turned and locked the bolt once more, ready for the next shot. As the bolt moved to the rear, it also cocked the hammer, so that all the firer had to do was take aim and press the trigger.

The magazine was loaded by a clip of eight cartridges, and this, in the eyes of some authorities, was the only drawback to the Garand rifle. The clip was inserted complete – the rounds were not stripped out of it as were the charger-loaded rounds of bolt action rifles – and a follower arm in the magazine pushed the cartridges up as

A US paratrooper, armed with a Garand M1 rifle, takes a prisoner at bayonet point during the Allied advance in Normandy, 1944

Associated Press

US Army Carbine ·30 M1

Developed to provide a lightweight weapon for men who did not need a full-sized rifle, the carbine became a highly popular gun. Later versions for paratroops (with folding butt), and with the ability to fire full automatic were developed. Over seven million carbines were manufactured

they were used. As the last round was fired and the bolt moved back, the empty clip was automatically ejected and the bolt held to the rear ready for another clip to be dropped in. But it had to be a whole clip or nothing; unlike most other designs, it could not be 'topped-up' by inserting one or two loose rounds.

The other defect was the ejection of the clip; it sailed through the air for several feet and when it fell on hard, particularly frozen, ground it gave out a distinctive ring. An alert enemy often heard this and then knew that the owner of the rifle was busy pushing a new clip in as hard as he could go; which gave the alert enemy a chance to stand up and get a shot in without much fear of being shot at himself.

Once production got into its swing, the Garand was turned out rapidly, and over four million were made by the end of the war. After the war another 600,000 were made and it was also built under licence by the Beretta Company in Italy, who subsequently re-designed it and used it as the foundation for a full line of automatic rifles. One of the impressive things about the Garand is that it saw practically no modification throughout its life; the design had been so thoroughly worked out before it went into production that it was not necessary to make any changes.

However, a number of experimental variations were tried out at one time and another: changes in, for example, the angle of the bolt-opening cam, changes in the gas piston design in order to try and reduce recoil, and so on. But the only variations on the standard pattern which ever got into production were the T-26, a shortened version intended for jungle warfare in the Pacific, and the various sniping rifles.

The T-26 was requested in July 1945 and a contract for 15,000 rifles was let out; but the war was over before they could be delivered and the contract was cancelled after a small number had been made. They were evaluated by the Army but turned

down as being too violent in action (like all shortened standard rifles) and were eventually sold on the surplus market, usually being advertised as the 'Short Tanker's Model' for some unknown reason. The sniper rifles were known as the M1C and M1D, the differences being in the model of telescope fitted; except for being selected for their shooting capabilities and fitted with a flash hider on the muzzle, they were exactly the same as any other M1 rifle.

The cooks' rifle

The success of the M1 led to a request in pre-war days from the Infantry School for a lightweight semi-automatic rifle to arm such specialists as mortar crews, machine-gun crews, radiomen, cooks, bakers, and all the other members of a battalion who were normally engaged in other duties and found the carriage of a full sized rifle inconvenient. Most of them were provided with pistols, but with experience of the First World War behind them, the infantry knew quite well that there were never enough pistols to go round, and that moreover a pistol is not the easiest weapon to shoot without a good deal of practice. Their request was turned down, but when the war loomed closer they repeated it, and this time it reached more sympathetic ears.

The first step was to produce a suitable cartridge, since a lightweight rifle firing the normal ·30 rifle round would be too violent to handle. A ·30 calibre cartridge of smaller dimensions, using a straight-sided cartridge case and a round-nosed bullet was developed, and with this before them numerous manufacturers were invited to design a suitable weapon. As it happened, the Winchester Company had developed a full sized rifle in 1940 and this had been offered to the US and British Armies but had been turned down; not on account of any deficiency but because both armies were committed to their existing designs and could not afford to try changing horses in mid-stream. Indeed, the British report

on the trial of the rifle said "with slight modifications this rifle is suitable for adoption into the British service".

The Winchester design used a unique form of gas operation in which the gas tapped from the barrel entered a small chamber beneath and drove a captive piston back for about one-tenth of an inch – enough to drive an operating rod to the rear and open and close the bolt. This gas piston design was now combined with an operating slide and bolt similar to that used on the Garand rifle to make a short carbine, and after testing alongside numerous others which had been submitted, the Winchester design was selected as being the best. It was put into production as the US Carbine M1 in October 1941. Shortly afterwards an M1A1 variation was produced in which the stock was a folding steel framework and a pistol grip was fitted, this version being for use by airborne troops.

Both these weapons were highly successful and popular, and after some experience with them in action, the infantry asked whether it would not be possible to arrange matters so that the carbine could be fired as an automatic – that is, keep firing as long as the trigger was pressed. In this way it would double as a submachine-gun. The request was considered reasonable, and a small modification was fitted, turning the design into the carbine M2.

Since this modification meant more rapid expenditure of ammunition, a 30-round magazine was developed to replace the earlier 15-round model. Firing at 750 rounds per minute, the M2 soon became the standard model, the M1 and M1A1 being relegated as 'limited standard'; over seven million carbines of all models were manufactured during the course of the war.

Russia's trial run

The only other army to have any automatic rifles at the outbreak of war was Russia's. They had been experimenting with various types from the early 1930s and in 1940 they had two Tokarev models in service in small numbers, largely as a trial run to see whether it was a good idea or not. The first was the Model 1938, the second the Model 1940, and as far as operation went they were almost identical. A gas cylinder lay above the barrel and carried a piston rod which protruded above the rifle chamber where it made contact with a 'bolt carrier', a heavy steel unit which surrounded the bolt. On firing, the gas pressure drove this piston rod back so that it struck the bolt carrier and drove it backwards; movement of the bolt carrier, through cams, lifted the bolt and unlocked it from the gun body, then drew it back to eject the fired case. At the same time, a return spring was compressed by the carrier and the hammer was cocked. When the carrier came to rest, the spring returned it and the bolt was taken forward, picking up a fresh round from the 10-round magazine and placing it in the chamber. The final movement of the bolt carrier as it reached the end of its travel drove the rear end of the bolt down to lock it securely.

Several thousands of the Tokarev were made, and they were used extensively in the Russo-Finnish war of 1939–40, but they were never entirely successful. The weight had been kept low by excessive paring of metal, leading to a degree of fragility, and the old rimmed 7·62-mm round was not well suited to use in an automatic weapon. One

US Rifle M1
John Garand's greatest effort, and the weapon which allowed the US Army to be the only one to enter the Second World War with an automatic rifle as standard. Five and a half million were made, and the design lives on in the present–day US Rifle M14 and Italian Beretta BM59 models

problem was extraction of the fired case, and the designers went as far as making the chamber with flutes milled in it so that gas could wash around the outside of the cartridge case and thus make it less liable to stick – which rather suggests that the design was basically unsound. The rifle apparently remained in production until 1944, but it never completely replaced the bolt-action Moisin-Nagant and appears to have been largely used as a sniping rifle. There was also a selective-fire version, ie. one capable of either single shots or automatic fire, but this was uncommon and was probably made simply to see if the idea was any good.

Not their best work

The British Army was alive to the advantages of a self-loading rifle, but although they had tested several specimens they had never found one which came up to their specification. Indeed, the specification was so stiff that it is doubtful if any automatic rifle of the present day could meet it in its entirety, and it was not until political pressure forced the modification of their demands that they were able to find a rifle with which to re-equip the army. But once the war was on, although various models were tested, there was no chance to re-equip, since the production capacity necessary to turn out the required millions of rifles would never be available. Moreover, one drawback to a successful rifle was the rimmed ·303 cartridge. The Army were well aware of this and had been considering a change to rimless ammunition for many years, but again the production problem involved, the immense logistic problem of changing over ammunition in the middle of a war, and – not least – the question of what was to be done with the millions of rifles and billions of cartridges already in existence, were enough to deter anybody but a lunatic from changing.

In Germany however, most of these same conditions applied except the ammunition problem, and numerous attempts at producing an automatic rifle were made before some successful designs appeared. The rimless 7·92-mm Mauser cartridge was well suited to an automatic weapon and after seeing the success of the Garand and the apparent success of the Tokarev, the Germans decided that they had to do something to equalise matters. In 1940 the hunt for an automatic rifle began; the following year saw two designs on trial, the Gewehr 41 (Walther) and the Gewehr 41 (Mauser).

Quite frankly, to anyone accustomed to the clever and ingenious designs emanating

from German gunsmiths, these two weapons come as a horrible shock, and one has the feeling that the designers were clutching at straws and not really feeling at their best just then. The system of operation in both designs relied on a cup-shaped muzzle attachment trapping some of the emerging gas behind the bullet, turning it back on itself and using it to operate a piston which encircled the muzzle of the rifle. It was a variation of an idea, evolved by a Dane called Soren H. Bang in the early years of the century, which had never been made to work satisfactorily. The Mauser rifle used the gas piston to operate a rotating bolt; the Walther design used a bolt which moved in a straight line and did not turn but had two flaps let into its side. These flaps were moved out to lock into recesses in the gun body and moved in to unlock the bolt by the action of the gas piston.

The Mauser design was dropped in favour of the Walther model and many thousand were made and issued to units on the Eastern Front in 1942–3, but it was never a very successful weapon. It was ill-balanced, difficult to make, easily jammed and difficult to clean and maintain. As soon as something better came along, production was stopped, but the weapons in the hands of the troops continued in use until the end of the war.

In view of what has already been said about the difficulty of changing ammunition during a war, what happened next is incredible; in fact it turned out to be the smartest thing the Germans did. During the pre-war years a number of senior German officers had applied themselves to a careful analysis of the infantry's use of the rifle, and they had come to the conclusion that the normal pattern of weapon, firing a powerful cartridge to 2,000 yards or so, was no longer needed. Except for specialists such as snipers, it was rare for an infantryman to have to fire at ranges greater than about 400 yards, and for this a much less powerful cartridge would be suitable.

They therefore set about laying down specifications for a short cartridge and a selective fire rifle to fire it. If the cartridge were shorter, then the mechanism of the rifle would have less distance to move, and so the rifle itself could be shorter and lighter. Also, if the round were less powerful, then the recoil shock would be less and again the rifle could be lighter and shorter. If the ammunition were smaller, then it would weigh less, and the soldier could carry more of it.

All these factors and many others were taken into account, and the more they

thought about it the more attractive it sounded. But by the time the whole thing had been hammered out and an experimental cartridge designed, war had broken out, and the arguments outlined previously were applied here – change of weapon, change of ammunition, what to do with the millions of rounds in stock, and so forth.

But the demands of the war showed that it was feasible after all, because new rifles were going to have to be made and fresh stocks of ammunition too, and they might just as well be in a more practical form. In order to make things a little simpler, the original cartridge design was scrapped and a new one drawn up which used a bullet of standard 7·92-mm calibre and a cartridge case based on a shortened version of the standard one, so that much of the work could be done on existing machinery. With this clear, the Haenel and Walther companies were asked to produce prototypes of an 'assault rifle', one of the most important features being that the design had to be suited to rapid mass-production.

The designs were ready in 1942 and became known as the Maschinen Karabiner 42(H) and 42(W). The Walther design used the same barrel-encircling piston as their G41 but drove it by a more conventional gas port in the barrel. The bolt was locked by tipping it upwards at the rear to lock into a recess in the gun body. The Haenel design used almost the same bolt mechanism but drove it by a more conventional gas piston moving in a cylinder mounted above the barrel. About five thousand of each were made for trials on the Russian front, as a result of which it was decided to drop the Walther design and put the Haenel model into volume production.

Outwitting the Führer

But now trouble arose. Weapon production had to be approved by Hitler, and when the project was put before him he refused to allow it to continue. As far as he was concerned, what an infantry rifle needed was range and a powerful cartridge – this, he said, he had realised from his own experience in the First World War – and all the explanations of the advantages of the new weapon were lost on him. Nothing doing.

The designers and promoters of the new weapon were somewhat put out by this, but being experienced in the politics of the time they knew what to do. They re-christened it the 'Machine Pistol 43' and put it into production; henceforth when anything appeared on paper about the project, Hitler

German MP43
Although called a Machine Pistol, this was a bit of political camouflage to get the weapon into production behind Hitler's back. It was actually the first 'Assault Rifle', firing a shortened cartridge and capable of single shot or automatic fire

NAME	CALIBRE	WEIGHT (lb)	MAGAZINE CAPACITY	ACTION	VELOCITY (F.P.S.)
Germany					
Kar 98K	7.92 mm	8.6	5	Bolt	2,450
G41(W)	7.92 mm	11.0	10	Gas	2,550
FG 42	7.92 mm	9.9	20	Gas	2,500
MP 43	7.92 mm	11.25	30	Gas	2,125
Great Britain					
Lee-Enfield No. 1 Mk III	.303	8.4	10	Bolt	2,400
United States					
Garand M1	.30	9.5	8	Gas	2,800
Springfield M1903A1	.30	8.0	5	Bolt	2,800
Italy					
Mannlicher-Carcano	6.5 mm	7.6	6	Bolt	2,300
Mannlicher-Carcano	7.35 mm	7.5	6	Bolt	2,500
Japan					
Ariska Type 38	6.5 mm	9.5	5	Bolt	2,400
Type 99	7.7 mm	9.1	5	Bolt	2,400
Soviet Russia					
Mosin Nagant 1930	7.62 mm	8.7	5	Bolt	2,660
Tokarev 1940	7.62 mm	8.6	10	Gas	2,725
Simonov M 1936	7.62 mm	8.9	15	Gas	2,519
France					
MAS 1936	7.5 mm	8.3	5	Bolt	2,700

assumed it to be a replacement for the existing submachine-guns, and as he was all in favour of these weapons, it all went past him unquestioned.

The weapons were being turned out from three factories and issued to the Russian Front, where every German unit was clamouring for the new gun when, at a conference at Hitler's headquarters, the whole affair was 'blown' by some divisional commanders from Russia asking Hitler point-blank when they were going to get the new weapons.

Hitler was furious and instituted an enquiry, but fortunately for the officers who had defied his ruling the report of the enquiry was so encouraging that Hitler changed his mind, gave the weapon his blessing, and announced that henceforth it would be called the *Sturmgewehr* or Assault Rifle. It is doubtful if any of the officers concerned felt like pointing out to him that they had already invented the name in the specification they issued three years earlier; there was no sense in asking for more trouble.

Official string-pulling
The German Parachute troops were part of the *Luftwaffe*, and when the first moves towards an assault rifle were reaching some sort of result, the promoters tried to enlist the *Luftwaffe's* aid by offering them the new assault rifle as a likely weapon for the paratroops. Unfortunately for this plan, the paratroopers had been the victims of some unfortunate experiences in the invasion of Crete, when they were brought under long range rifle fire from British troops. This convinced them that nothing but a full-power cartridge was of any use to them and they were quite firm in their refusal of the new weapon. They asked the Army if they would not reconsider their decision and develop a similar weapon using the standard cartridge, but the Army refused.

The parachutists, being soldiers attached to the Air Force, were able to use both channels when they wanted anything. Whatever the Army refused them the *Luftwaffe* would provide, and vice versa, and when things got too difficult they could always enlist the aid of Goering, Marshal of the *Luftwaffe*, who regarded the paratroops as his private army, for whom nothing was too good. So through *Luftwaffe* channels they now drafted a specification for a new rifle and gave it to Rheinmettal-Borsig who, by coincidence, were part of the Hermann Goering Werke combine.

The resulting weapon came into service in 1942 as the '*Fallschirmgewehr* 42' (Parachutist's Rifle Model 1942) but it was never

Luger Pistol
The Luger Pistol was the standard German Army pistol from 1908 until it was officially replaced by the Walther P–38 in 1938; even so, thousands remained in service throughout the war. Well made and accurate, it was not however the ideal pistol for combat, being difficult to make and with an action too dependent upon ammunition quality

perfectly right and modifications continued to be made throughout its life. Little more than 7,000 were made, one of the reasons being that the parachute troops in the German Army became less and less important as the war drew on.

The FG 42 was a remarkably good design. It was built in a 'straight-line' configuration, with the butt in prolongation of the barrel axis, which stopped the tendency to rise when fired at automatic; when fired at single shot it fired from a closed bolt – that is to say, the bolt closed and locked behind the cartridge without the weapon firing, and the firer then pressed the trigger to drop the hammer and fire the round. This gave greater accuracy because there was no movement going on at the moment of discharge. On the other hand, a machine-gun firing from a closed bolt is dangerous, since it means that when the finger is taken off the trigger the mechanism loads one last round but does not fire it, thus leaving it in what is probably a very hot chamber, in which it is liable to 'cook off' or explode from the heat, usually surprising the operator and catching one of his comrades unexpectedly.

When firing automatic, therefore, the FG 42 worked from an open bolt, so that

Webley Revolver
The British Army's revolver from 1887 to 1963 was the Webley. Hard–hitting, accurate and reliable, it was issued in ·455 calibre until the early 1930s when it was replaced by a ·38 model which was easier to shoot accurately

French Lebel Revolver
This is the Model 1892, standard French Army pistol until the Second World War. Although of old design and using a relatively small bullet (8–mm calibre) it was a reliable weapon

when the finger was taken from the trigger the action stopped with the bolt held back and an empty chamber; air could thus pass down the barrel and cool the weapon before the next burst. When the trigger was pressed again, the bolt went forward and chambered a round, firing it automatically as soon as the bolt was locked. All this, moreover, was accomplished in a weapon weighing less than 10 lb.

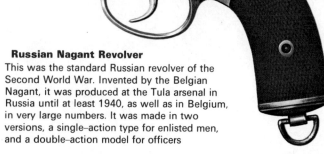

Russian Nagant Revolver

This was the standard Russian revolver of the Second World War. Invented by the Belgian Nagant, it was produced at the Tula arsenal in Russia until at least 1940, as well as in Belgium, in very large numbers. It was made in two versions, a single-action type for enlisted men, and a double-action model for officers

Polish Radom Pistol

The Radom was a Polish combination of the Browning Colt model 1911 and the 1935 high-powered FN Browning. Made at the Radom arsenal in Poland, it was used extensively by the German forces after 1939

Nambu Automatic Pistol

The standard Japanese Army pistol, this 8-mm automatic, although resembling the German Luger, has a much simpler action

Astra Pistol

Manufactured in Spain, this 9-mm pistol is unusual in being able to accept and fire almost every type of 9-mm and .38 automatic round

Walther P-38

This became the German Army's standard pistol in 1938, replacing the Luger. It was easier to manufacture and a more reliable combat pistol. It was unusual in that it used a double-action trigger lock, so that it could be carried with the hammer lowered on to a loaded chamber and fired by simply pulling the trigger. It is still in service with the German Army today

PISTOLS

NAME	CALIBRE	WEIGHT (lb)	MAGAZINE CAPACITY	VELOCITY (F.P.S.)	Remarks
Germany					
Pistole '08 (Luger)	9 mm	1.9	8	1,150	Auto
Pistole '38 (Walther)	9 mm	2.1	8	1,150	Auto
Great Britain					
Pistol No 2 (S&W)	.38	1.5	6	650	Revolver
Pistol No 2 (Enfield)	.38	1.7	6	650	Revolver
Pistol Mk 4 (Webley)	.455	2.37	6	620	Revolver
Browning GP35	9 mm	2.2	13	1,100	Auto
United States					
Colt M1911A1	.45	2.5	7	860	Auto
Colt Army M1917	.45	2.5	6	860	Revolver
Smith & Wesson 1917	.45	2.25	6	860	Revolver
Italy					
Glisenti 1910	9 mm	1.8	7	1,050	Auto
Beretta 1934	9 mm	1.5	7	825	Auto
Japan					
Nambu 4	8 mm	1.9	8	1,100	Auto
Nambu 14	8 mm	2.0	8	1,100	Auto
Type 94	8 mm	1.75	6	1,000	Auto
Type 26	9 mm	1.9	6	750	Revolver
Soviet Russia					
Tokarev 33	7.62 mm	1.8	8	1,375	Auto
Nagant 1895	7.62 mm	1.75	7	1,000	Revolver
France					
Lebel	8 mm	1.9	6	625	Revolver
Spain					
Astra	9 mm	2.2	8	1,100	Auto

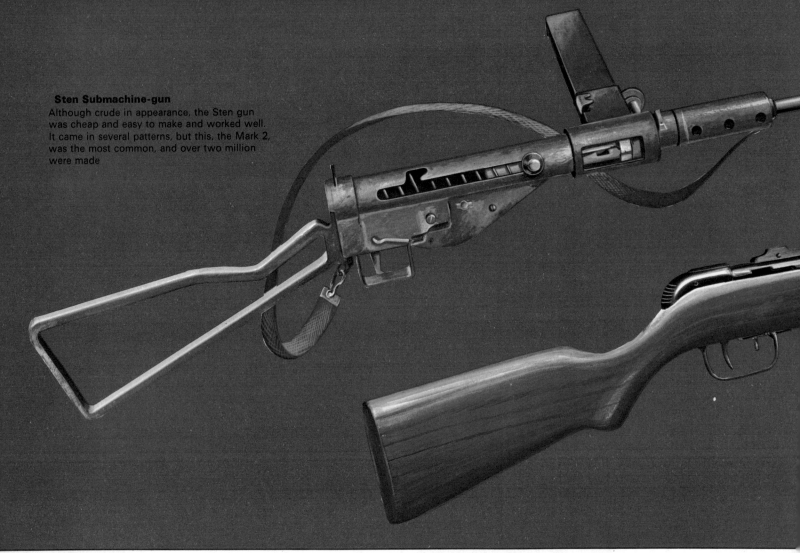

Sten Submachine-gun
Although crude in appearance, the Sten gun was cheap and easy to make and worked well. It came in several patterns, but this, the Mark 2, was the most common, and over two million were made

The next category of weapon we should consider is the submachine-gun, and in this class the Germans and Russians had a clear lead over the rest of the world since they had been able to do some field research during the Spanish Civil War. Both of these countries, and Italy, sent troops, thinly disguised as popular volunteers, and an assortment of weapons they wished to have tried out, and as a result the superiority of fire of the submachine-gun was noted and studied.

The Germans already had a variety of submachine-guns in service in small numbers, but the reports from Spain led them to think seriously about a standard model to be produced by the million. After looking at the current models, the German Army sent for the director of a company called the *Erfurter Maschinenfabrik B. Geipel GmbH*, who had produced a number of designs under the name 'Erma'. They handed him a specification and told him to produce a weapon to suit and be quick about it.

As it happened, the company had been working on a fresh design and with a little modification this was produced and offered for the Army's approval. Approval was given and in the middle of 1938 the Erma company began producing the submachine-gun which has almost become a German Army trademark – the Maschinenpistole 38.

The MP-38, for some unknown reason, attracted the name "Schmeisser" during the war, and it has generally stuck. Although Hugo Schmeisser did design many machine pistols and submachine-guns during his life, he had nothing whatever to do with this one; it was designed by one Heinrich Vollmer. The nearest Schmeisser ever got

to it was to manage a factory which was making them during the war.

The MP-38 was a leader in its field, and some of the ideas seen in it have been copied many times over in later designs. It was the first to adopt a folding stock which was successful, the first to utilise stamped steel and plastic in its construction and the first weapon to have no wood of any sort. Like almost every other submachine-gun, the MP-38 fires from an open bolt and operates on the system generally called 'blow-back' because the firing of the cartridge causes the case to be blown back against the bolt and thus force it to the rear to start off the cycle of operation. One might think this would lead to the case being blown out at dangerous speed, but since the weight of the bolt is enormous compared with the weight of the bullet, the bullet has left the barrel and the pressure in the barrel has dropped to safe limits before the inertia of the bolt has been overcome and it starts to move.

German grumbles
Though the simplicity of this system has a lot to commend it, it also has one or two drawbacks. For example, if the weapon has a magazine fitted and is dropped sharply on its butt, then the bolt will slam back against the spring and, going forward, will collect a round from the magazine and fire it – since the firing pin is part of the bolt and as soon as the bolt closes it fires the cartridge.

This was one of the defects of the MP-38, as it was of many other submachine-guns, and in 1940 this, plus the fact that the MP-38 was not easy to manufacture, led to it being re-designed as the MP-40. The difference

between the 38 and 40 is not easy to see, but the best clue is that the body of the 38 is made of corrugated steel with a plain magazine housing, while the 40 has a plain body and corrugated magazine housing. The other differences are largely concerned with easier mass production, but one important feature was the alteration of the cocking handle so that it could be moved inwards to lock the bolt securely in the forward position, removing any danger of bouncing the bolt and accidentally firing a round.

One of the complaints of the German soldiers on the Russian Front was that the Soviets were using submachine-guns which had 71-round drum magazines, while the Germans were using weapons with 30-round magazines. An interesting attempt to try and even things up was a modification of the MP-40 which changed the magazine housing into a sliding assembly. Two standard 30-round magazines could be clipped in and the assembly pulled over to one side so that one magazine was lined up with the bolt. After emptying this magazine, the firer simply pushed the sliding assembly across and brought the other magazine into line. It was not particularly good, because the all-up weight of the machine-gun now rose to over 12 lb, which is something of a handful. It seems that very few were made.

The Russians also learned from the Spanish Civil War. They had adopted a submachine-gun in 1934, the PPD designed by Degtyarev, a name which frequently crops up when discussing Russian firearms. This had been used with considerable effect by the various Communist 'volunteers' and as a result the Russians decided to put it

CRUDE BUT EFFECTIVE

Soviet PPSh Submachine-gun
One of the simplest and crudest submachine-guns ever made, the PPSh Model of 1941 was none the less effective. Firing 900 rounds per minute, most models were built so that they only fired automatic; over five million were made before the end of the war

into volume production. The first model was known as the PPD34/38 and after a few thousand had been made it was improved into the PPD40 model, which used a rather better design of magazine. This was the 71-round drum which the Germans found such a nuisance in later years. Its mechanism was a simple blowback, and it was recognisable by its wooden stock and fore-end, and the slotted cooling jacket round the barrel. The PPD40 however was not simple to make, being machined from solid steel, and when the German Army invaded, the cry was for a weapon which could be rapidly produced by semi-skilled labour. Fortunately the Red Army had been testing a gun which fitted this demand, and in 1941 it was approved for production, the PPD40 being dropped.

The new weapon was the PPSh41, designed by Georgi Shpagin, another well-known Russian arms designer, and it was, compared to any other submachine-gun previously known, crude beyond belief. The entire body and jacket were welded up from steel stampings, the parts of the firing mechanism were stamped out, and the only pieces made with any sort of precision were the bolt and barrel. The jacket extended beyond the muzzle and folded over to form a primitive muzzle deflector which helped to keep the muzzle down when fired, since – like most of its type – the recoil tended to force it up.

In 1942 the German Army beseiged Leningrad, and the occupants of the city were desperate for weapons. Supply from the rest of the Soviet Union being almost impossible, a local engineer, A. I. Sudarev, designed a submachine-gun which was

built in local factories and sent straight out to the nearby Front. This became standardised as the PPS-42. He later improved it into the PPS-43, though the changes are largely concerned with even simpler production and make little difference to the operation or appearance.

All or nothing

The PPS-42 and -43 models were probably the most crude devices ever put into the hands of troops. They were entirely put together from stamped steel components, welded and riveted together; the bolt is as simple as can be, and only that and the barrel demand any sort of precision in manufacture. The stock folds over to the top of the gun, and as with the PPSh the barrel jacket runs round the front of the barrel to form a primitive muzzle brake or deflector. The magazine was a curved box holding 35 rounds and there was no provision for firing single shots; it was all or nothing with this weapon. Its manufacture continued until after the war was over, but it never reached the vast quantities of the PPD or PPSh models.

The vast number of submachine-guns produced by the Soviets is a reflection of the particular attraction this weapon had for them. In the first place it was simple to teach and simple to understand, a great factor when you are confronted with the problem of turning millions of peasants into soldiers overnight. There is also a good deal of logistic sense in arming whole battalions with the same weapon, from the commanding officer down; it simplifies ammunition supply and weapon replacement immensely. The production of such crude

and simple weapons by the use of semi-skilled labour is much easier. And the submachine-gun lends itself to the Soviet way of fighting. It is no good sitting in a hole with a submachine-gun and hoping to snipe at somebody three hundred yards away. The only way to use it is to get out of the hole and go looking for the enemy.

The best employment of these weapons is shown by the peculiar Soviet innovation of Tank Rider Battalions. These were 500-strong units, armed solely with PPSh submachine-guns and carrying as many spare magazines as they could. They rode into battle on the back of tanks, and when the tank found some enemy they leapt off and went for him, covered by the tank. If the tank was hit, those who were still capable leapt off and grabbed the next tank along. Their training was very simple, since the only thing they knew was the assault; they never retreated and they left the consolidation to others. From all accounts their life expectancy was short.

The British Army, along with the US Army, had long resisted the submachine-gun, referring to it in disparaging terms, but when the war began in 1939 the British soon realised that these despised weapons were a necessity in modern war. They had, it is true, tested almost every type ever made over the years and thus were in a position to be able to say what they wanted, but by that time it was more a question of what they could get. The best, they considered, was the Finnish Suomi, but the Finns were taking all the weapons the factory could produce, so they turned to the next best, went to America and bought the Thompson submachine-gun.

Thompson Submachine-gun
The original Thompson model was the 1928 which could be fitted with a drum magazine; this was not required for military work, since it was too heavy and too noisy, so when the weapon was simplified into the M1, shown here, only a box magazine was used

The Thompson had certain drawbacks. It was heavy, it was difficult to make, and it was expensive. But it had one virtue which over-rode all that – it was reliable. And even when cheaper and newer submachine-guns came along, a lot of people refused to part with their Thompsons for just that reason, that they could rely on them in a tight corner. The first models were provided with the 50-round drum magazine so beloved of the G-Men and their opponents, but what goes down in a Chicago alley doesn't always work in war, and the British infantry patrols on the Saar front in 1940 reported that the rounds in the drum slapped back and forth, making a noise at night and drawing enemy fire. The drums were replaced with box magazines and no British submachine-gun ever again had a drum.

But the expense and difficulty of supply of the Thompson led the British to think about manufacturing a gun of their own, and the events of 1940 led them to speed up their deliberations. After much consideration of various designs it was decided that the best thing to do was to make a British copy of the German MP-28, one which was designed by Schmeisser during the First

World War. It was a blowback gun with a wooden stock and box magazine at the side, and plans were drawn up to manufacture this under the name of Lanchester, the engineer responsible for modifying the design to suit British engineering standards.

Just as this was about to go into production, two designers from the Royal Small Arms Factory at Enfield came forward with a weapon they had put together – the Sten gun. In appearance it was almost as bad as some of the Russian weapons, all metal, with a tubular stock and made up of stamped and welded sheet metal. But it worked very well and it was cheap and easy to make, and without much more argument the Sten went into production.

The first model, the Mark 1, had a spoon-like flash hider-cum-deflector on the muzzle and a wooden foregrip, and about 100,000 were made. Then the Mark 2 – the most common version – was introduced. This did away with the foregrip and flash hider, had a simple perforated tube doubling as a barrel locking nut and hand grip, and a single-tube butt. Two million of these were

made, and untold numbers of them were dropped over Occupied Europe, where they still turn up from time to time. A silenced version was also made for clandestine operations. Then came the Mark 3, with a long welded jacket completely covering the barrel; this was made in vast numbers in Canada and Britain. The Mark 4 was an experimental model for paratroops which never saw service; finally came Mark 5.

Did it really work?
This was somewhat more elegant than the earlier models, in an attempt to produce a weapon which would inspire its owners with more confidence It had a pistol grip and a wooden butt and fitted a bayonet on the muzzle, but for all that it was still the same old Sten gun at heart. It was first issued to Airborne units but later replaced all the other marks as standard issue and remained in British service until the 1950s.

The US Army also took the Thompson in 1941 when the rapidly expanding army showed the need for such a weapon. Prior to this a small number had been bought for evaluation, and the US Marines had used them since 1928. The Thompson as adopted by the US Army was known as the M1928 or 1928A1, and was the original Thompson design; this was somewhat different to the other submachine-guns we have discussed

A US Marine sights in on a Japanese sniper on Okinawa with his Thompson submachine–gun

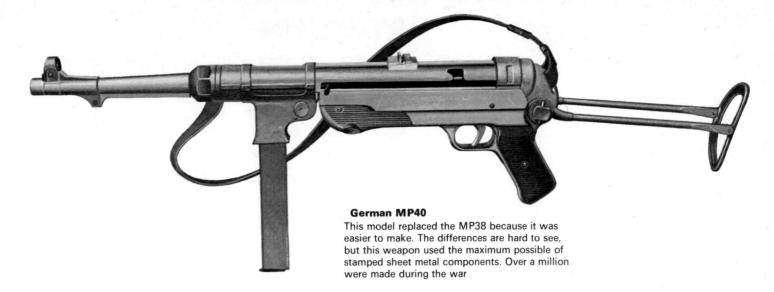

German MP40
This model replaced the MP38 because it was easier to make. The differences are hard to see, but this weapon used the maximum possible of stamped sheet metal components. Over a million were made during the war

because it was of a type of action known as the 'delayed blowback'. In an attempt to keep the breech closed until the pressure had dropped, the bolt of the Thompson had an inclined slot cut in it, in which rode a separate locking piece. This locked the bolt to the gun body at the moment of firing. The angle of this slot was carefully calculated so that under extremely high pressure, as when the round was fired, the slot and locking piece jammed securely together and into the body slot. Then, as the pressure dropped, so the friction lessened until the locking piece could ride up the slot and unlock the bolt from the gun body and allow it to move back to complete the unloading and loading cycle.

There has always been considerable argument as to whether this system did or did not achieve any locking at all, but one thing it did do was make the gun that much more difficult to manufacture. As a result, the US Army demanded a simpler version so that production could be speeded up, and the lock system was dispensed with, turning the weapon into a simple blowback with no ill-effects. Sundry other small changes were made and the new weapon was known as the M1, and was adopted in April 1942.

In spite of this the Thompson was still a manufacturing problem and tests were now done on a variety of other designs which were put forward for possible adoption. None was entirely successful and eventually the Army itself designed a cheap and simple weapon, more or less the American equivalent of the Sten gun, known as the M3 submachine-gun; due to its shape it was instantly called the 'grease gun' and the name stuck.

Weaknesses dealt with

The M3 was a simple blowback gun made entirely from steel stampings except, of course, for the bolt and barrel. Cocking was done by a lever on the side, the stock was a collapsible wire form, and – one unique feature – it could be adapted to either the US Army standard ·45 cartridge or the European standard 9-mm cartridge by simply changing the barrel and bolt; when used with the 9-mm cartridge, an adapter was slipped into the magazine housing and Sten gun magazines were used. But although this design was simple and cheap, it soon became apparent that it could be made even more simple and even cheaper.

After complaints from units in the field drew attention to weaknesses in the cocking device, the weapon was redesigned, doing away with the cocking handle and bringing the cocking procedure down to the simplest possible level. A large flap in the gun body was opened, a finger poked into a hole in the bolt, and the bolt hauled to the rear, and that was that. Closing the lid entered a steel pressing into the hole in the bolt and thus acted as a safety catch. One or two other small changes were made and the result entered service as the M3A1. All these guns were made by the Guide Lamp Division of the General Motors Corporation, and they produced no less than 646,000 of both types before the war ended.

We have not mentioned Italy so far in this narrative, since by and large their weapons were not particularly noteworthy, but in the submachine-gun field they deserve mention since, although little known, they possessed one of the best submachine-guns of all time – the Beretta. The Italians were, in fact, in the forefront of submachine-gun development, as they had a submachine-gun in the hands of their infantry as early as 1918, and in the intervening years the company of P. Beretta had worked on improving their design. The Model 1938 was a 9-mm weapon, wooden stocked, with a perforated barrel jacket, a 40-round box magazine and a bayonet. Large numbers of this excellent weapon were provided for the Italian infantry units and subsequent modifications were made in order to simplify production. It remained in production until the 1950s, many thousands being sold to other countries in the postwar years.

SUBMACHINE GUNS

NAME	CALIBRE	WEIGHT (lb)	MAGAZINE CAPACITY	RATE OF FIRE (R.P.M.)	VELOCITY (F.P.S.)	REMARKS
Germany						
Bergmann MP 34	9 mm	8.9	24 or 32	650	1,250	Largely used by SS Troops
Erma	9 mm	9.1	25 or 32	500	1,250	
MP 38 or 40	9 mm	8.75	32	500	1,250	
MP 3008	9 mm	6.5	32	500	1,250	Copy of Sten-Gun
Great Britain						
Sten Mark 2	9 mm	7.75	32	450	1,000	Commonest model
United States						
Thompson M1	.45	10.6	20 or 30	700	900	M1928 similar; also 50-round drum
M3A1	.45	8.2	30	400	900	M3 similar data
Reising 50	.45	6.75	20	550	900	Used only by US Marines
Italy						
Beretta 1938A	9 mm	9.25	30 or 40	600	1,350	
Japan						
Model 100, 1941	8 mm	8.5	30	450	1,100	1944 model fired at 800 rpm; few made
Soviet Russia						
PPD 40	7.62 mm	8.1	71	800	1,600	Drum Magazine
PPSh 41	7.62 mm	8.0	35	900	1,600	71-round drum also available
PPS 43	7.62 mm	7.4	35	700	1,600	

MACHINE-GUNS
COVERING FIRE FOR THE RIFLEMEN

So far we have considered the personal weapons of the infantryman, those he actually carries from day to day and uses for both offence and for his own personal defence. Let us now move up to the weapons used for supporting the movement of the troops – the machine-guns. These, it will be remembered, formed in most cases the pivot of action; the machine-gun covered while the riflemen moved and vice versa. The only army in which this basic principle was less well defined was the US Army, since, with every man having an automatic rifle, an immense firepower could be brought to bear by rifle fire alone, and the squad machine-gun was less important. This was just as well, since the US Army never had anything which could have been charitably called a squad machine-gun.

What passed as the light machine-gun with the US Army was the Browning Automatic Rifle. This had been developed in 1917-18 as a weapon to be fired from the hip as the infantry advanced across No Man's Land to the German trenches, the idea being to keep the enemy's heads down until the troops were closed in. Since the war was over before many Brownings could be got across to France, the correctness or otherwise of this theory never had a chance to be proven, but with thousands of BARs coming off the production line, something had to be done. So they were fitted with a bipod and called the squad automatic.

They were not the best choice for the job. The design was such that they were violent in action, which led to inaccuracy. The magazine held only 20 rounds and was fitted beneath the weapon, making it difficult to change when laid on the ground; there was no provision for changing the barrel when it got too hot, so sustained fire was out of the question. But like it or not, that was all there was, and the US Army soldiered through the Second World War with it and even used it in Korea. Many ex-GIs swear that it was the greatest, but one feels that if they had ever had the chance to use a real light machine-gun in action they

Bren Gun Mark 1
Designed in Czechoslovakia, the Bren was adopted by the British Army in 1938. Probably the best light machine–gun of all time, it is gas–operated and has a quick–change barrel

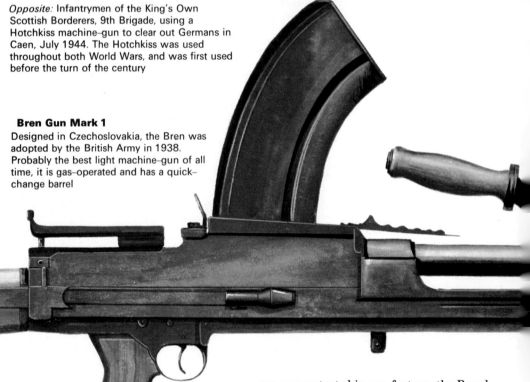

would have rapidly changed their opinion.

In an attempt to provide something better, the Infantry Board grafted a butt and bipod on to the venerable Browning air-cooled medium machine-gun and called it a light machine-gun, but they weren't fooling anybody. It was four-and-a-half feet long and weighed 32½ lb, and that isn't a light machine-gun by anybody's standards.

The British Army undoubtedly had the best light machine-gun of the war in their Bren gun. This was a Czechoslovakian

design which they had examined in 1935 and for which they had bought the manufacturing rights. Gas operated, air-cooled, with a quick-change barrel, fed by a 30-round box magazine on top, it was simple to use, accurate, extremely reliable and trouble-free; it formed the base of fire for the British infantry section from 1938 onwards and, re-worked to the NATO calibre, continues to do so today.

One of the odd features about the Bren gun was that the entire British production

was concentrated in one factory, the Royal Small Arms Factory at Enfield, and the fact that it was never severely damaged by an air raid is nothing short of miraculous. But the War Office was well aware of the possibility and asked the famous Birmingham Small Arms Company to develop a simplified version of the gun which could have been built rapidly in any light engineering shop as an insurance against the disruption of Bren production. Harry Faulkner, BSA's Chief Designer, rapidly produced the Besal Gun which although simple in construction was nevertheless reliable. It was made in prototype and proved, then the drawings and jigs were put to one side to await the day they would be needed; they never were, and the Besal – or as it later became known, the Faulkner

machine-gun – never came into production.

Another insurance against enemy interference was to have the gun made in Canada, and eventually about 60 per cent of all Bren guns were made there both for the British Army and also in 7·92-mm calibre for the Chinese Nationalist Army. The Bren was probably the most widely used of all light machine-guns, seeing service with the British, Canadian, Australian and New Zealand Armies, as well as with the free forces of France, Czechoslovakia and Poland, the Indian Army, the Chinese Army and even, in its original Czechoslovakian form, with the German Army.

For heavier support the British relied, as

they had done for years, on the well-tried and faithful Vickers medium machine-gun. This was, in essence, the Maxim Gun of 1887 improved by Vickers in 1912, and it was a water-cooled and tripod mounted weapon which could fire at 450 rounds a minute for hours on end. It was normally issued to machine-gun companies to support attacks or deploy in defence, but each battalion had an allocation of them for organic use. Some regiments – the Manchester Regiment for example – became purely machine-gun regiments, supplying gun companies across the front when needed.

Maxim's original design also appeared in Russia, their standard heavy machine-gun

being the Maxim 1910 model. For some unknown reason, probably connected with ensuring reliability in Arctic cold, the Russian version weighed far more than any other pattern of Maxim and, with its peculiar wheeled mounting, tipped the scales at over 160 lb, which was probably why it had wheels. It could also be fitted to a type of sledge for dragging across the snow, and was frequently towed into action by squads of ski-troopers hauling on lines. But a water-cooled gun was not the best answer for the conditions of warfare in Russia, though it took the Russians a long time to realise this. Eventually they introduced the Goryunov SG43, an air-cooled weapon with a heavy barrel, gas operated and also fitted to a wheeled carriage. Like that of many other Soviet weapons, the interior of the barrel was chromium-plated, giving much longer life and allowing sustained fire for long periods without excessive wear.

The standard light machine-gun of the Soviet Army was one of Degtyarev's designs

and known as the DP (for Degtyarev Infantry). This was a very successful design and remained in first-line service with the Soviets for many years. It used a 47-round flat drum mounted on top, was gas operated, and was the foundation of a series of guns for tank and ground use, all based on the same basic mechanism.

Doing the most damage

The German Army, after its experiences in the First World War, had some very definite ideas on machine-guns; they regarded the machine-gun as the prime tool of the infantry, with the riflemen there merely to back it up. The machine-gun was the killer, and all efforts were devoted to putting it into a position where it could do the most damage. After the Armistice in 1918 the Allied Disarmament Commissions made short work of the various gunmakers and placed severe restrictions on manufacture, so that when Hitler came to power in 1933 and began to re-equip the army, there was no problem of what to do with the huge stock of weapons still on hand from the war, as there was in other countries. Thus the Army were ready for a new machine-gun, and due to their views on the tactical use of the weapon and some clever designing, they were to be provided with some of the most effective machine-guns developed during the war, guns which considerably influenced both design and handling in the years which followed.

One of the problems for German gun designers in the post-First World War years was the problem of keeping busy; they were not allowed to produce some types of weapon, and only limited quantities of the permitted types. So rather than fire all their skilled designers they resorted to various subterfuges to keep them working. The famous Rheinmettal company secretly bought control of a small Swiss engineering firm and set them up in business as *Waffenfabrik Solothurn*. The rest of the world thought "What clever designers these Swiss are", but in fact every design which

German Machine-gun 42
Designed by Mauser to replace the MG34, this was intended for mass–production and introduced a number of novel features. It survived the war, and modernised versions of it are still in service with the German Army

The Vickers Machine-gun
The medium machine–gun of the British Army from 1912 to 1966 and, like all Maxim designs, a model of reliability; 10,000 rounds an hour of sustained fire has been recorded in tests with this weapon

came from Solothurn originated in Dusseldorf. Through Solothurn they were also able to share in the control of an Austrian company, the *Waffenwerke Steyr*, so that weapons were designed in Germany, made in prototype and tested in Switzerland, and then put into production and sold commercially from Austria. One of the weapons which was put forward by the Solothurn Company was a new machine-gun called the MG30.

The MG30 was a highly ingenious design, deriving its operation from the recoil of the barrel. As the barrel went back, the bolt was unlocked by two rollers working in tracks in the gun body; the barrel then returned to the firing position while the bolt travelled back and forwards once more to extract and re-load. It was fed from a fifty-round belt and fired at 800 rounds per minute. Due to careful design it was very stable and accurate, but for some unknown reason the German Army did not take to it. After some improvements had been made it was re-submitted and accepted but then passed to the famous Mauser Company for some improvement. This they did, and the final result was issued as the MG34 and became the standard German machine-gun. As might be expected from such a company, the weapon was excellently made and finished, but this meant that it was not easy to produce, and before the war was over five factories were hard at work turning out more and more of them, trying to keep up with demand.

Great ingenuity

Because of the production problem, and because the fine tolerances of the MG34 led to trouble when firing in dust and dirt, a fresh design was begun which resulted in the MG42. Work started in 1941, and although the 42 looks very much like the 34 from a distance, both mechanism and construction are vastly different. The main point about the 42 was that it was designed for mass-production, utilising stamped metal components and paying less attention to high quality finish. The mechanism was changed so that the bolt, instead of rotating, unlocked in a straight line by rollers which moved in and out of recesses in the gun body. The belt feed mechanism was highly ingenious and built into the top cover of the gun, and has since been widely copied in other designs. The rate of fire was very high – over 1,200 rounds a minute – and this made the gun rather hard to control,

Italian Machine-gun Breda Model 30
An awkward–looking weapon, with a peculiar hinged magazine, fixed to the gun, which had to be filled from rifle clips. Notice that there is no carrying handle; the gunner had to sling it on his shoulder or carry it in his arms

French Machine-gun 24/29
Known also as the Châtellerault, from its place of origin, this was a gas–operated gun with a mechanism very similar to that of the Bren. Numbers were captured and used by the German Army after 1940, but it remained in French Army service until the 1950s

Browning ·50-in Machine-gun
The ·5 Browning was a heavy machine–gun widely used for anti–aircraft work, as well as for infantry support, while the ·30 was the infantry's standard medium machine–gun. Both used the same recoil-operated mechanism and can be found in service with armies all over the world

... up to 1300 rounds per minute ...

A Bren gunner of the 6th Durham Light Infantry, 50th Division, in action in the ruins of a house in Normandy, June 1944

Japanese Machine-gun Type 96
An improvement over an earlier model, this weapon used a normal box magazine instead of a cartridge hopper and a rear sight based on the Bren design. Notice the finned barrel, a common Japanese feature, designed to improve cooling

as it only fired at full-automatic. In spite of this it was well liked by the Army. It was hoped that it would entirely replace the MG34 in service, but the demands of the war were so great that it never did, both guns remaining in use throughout the war.

One interesting thing about these two guns was their manner of use. The Germans issued each type with a bipod, so that it could be used as a light machine-gun, and with a tripod and optical sight so that if required it could work as a medium machine-gun; it was all things to all men. Although slightly heavy for a light gun, it was a good deal lighter than most medium guns, so

that it could and did fulfil both roles quite happily. A quick-change barrel allowed the guns to be fired for long periods, changing the barrels every 300 rounds or so.

This multiple-role idea took hold during the war; in the early days of the war there was still a distinction between light and medium machine-guns, the MG34 usually acting as the medium gun. To fill the role of light gun there were a number of different designs, most of which had been adopted in the late 1930s simply in order to have some machine-guns – any machine-guns. The MG15, for example was originally designed as an aircraft gun, but by the addition of a bipod and butt it became a light machine-gun; another was the Swedish-designed Knorr-Bremse, a cheap and nasty affair which had the distressing habit of falling to pieces while it was being fired. But these were only regarded as stop-gaps until there were sufficient MG34s to go round, and although many of the light weapons re-

mained in service throughout the war, they were gradually relegated to reserve and occupation troops, with the better designs going to the combat units.

But when it comes to poor designs, the Italian and Japanese Armies were in the forefront; indeed, one wonders how they managed at all with some of their machine-guns. Take, for example, the Italian Model 30. This was an awkward-looking weapon, full of pieces sticking out and corners to catch the dust. It worked on the blowback system, which may be very well in a sub-machine-gun but which is stretched to its limits when used with a powerful rifle cartridge in a machine-gun. The problem is that the pressure inside the chamber causes the cartridge case mouth to expand and seat firmly in the chamber; but since it is a blow-back gun, at the same time the bolt is opening and trying to extract the case. This usually tears the case in half and jams the gun.

To overcome this the Italian designers fitted an oil pump and tank inside the gun which squirted oil on to the cartridge as it was being inserted into the chamber, so that it would slip out without ripping to pieces. This sort of idea works well in a nice clean trials environment, but put the gun down in a sandy desert and it becomes a different story. The dust settles on the oily cartridge, and every round acts like a valve-grinder as it goes into the chamber, scoring and wearing the steel away to a rough surface

Japanese Machine-gun Type 92
Known widely as the 'Woodpecker' from its peculiar stuttering noise, this was the Japanese Army's medium machine–gun. It was little more than a remake of an earlier design, using a larger calibre

Hotchkiss FM 1922
A French 7·5–mm light machine–gun, developed in 1921, and widely issued to French infantry companies in the years leading up to the war

which eventually grips the cartridge worse than ever and promotes even more trouble.

Another masterpiece from the same designer was the Model 37, the standard medium machine-gun. An air-cooled and gas-operated weapon, it would have been an excellent gun had the design of the bolt allowed a less violent extraction. As it was, it meant that once again the rounds had to be oiled before they went in, with all the attendant troubles. But in addition to this the Model 37 had one of the most peculiar feed and ejection systems ever devised. The cartridges were clipped into a 20-round tray and inserted into the left side of the gun. The mechanism extracted one cartridge from the tray, loaded it, moved the tray across, fired the round, extracted the empty case . . . *and put it neatly back into the tray* before extracting the next round. The tray came out of the right side of the gun filled with empty cartridge cases, which the unfortunate gunner's mate had to strip out by hand before he could start re-loading.

The Japanese also had some odd ideas on feed systems. Their Model 11 light machine-gun had a large square hopper on the left side, into which six ordinary rifle clips were dropped, and the mechanism then stripped the rounds from the clips one at a time, throwing out the clip every fifth round. The original idea was that any rifleman could contribute ammunition to the gun in emergency, but this was ruined by the discovery that the gun did not work very well with the standard rifle ammunition and had to be supplied with special low-powered rounds. To add to the problems, this was another gun where the extraction was violent and the cartridges had to be oiled before they went in.

In 1936 the gun was redesigned to try and do away with these problems; an ordinary box magazine was fitted and the oiler removed, but the extraction trouble persisted and so the machine issued for filling magazines was fitted with an oiler. If anything this was worse, because it now meant that instead of an oily round being open to collect dust for the fraction of a second before it was loaded, the magazines full of oily cartridges were carried around by the gunners, sometimes for several days, collecting dirt and grit all the time.

Eventually the Japanese saw the light and developed a new gun using 7·7-mm ammunition and based very much on the Czechoslovakian ZB26 design which was the basis of the British Bren gun. It was no longer necessary to oil the rounds or provide special ammunition, but production of this gun never reached great numbers and the earlier models were still used throughout the war.

MACHINE GUNS

NAME	CALIBRE	WEIGHT (lb)	MAGAZINE CAPACITY	ACTION	RATE OF FIRE (R.P.M.)	VELOCITY (F.P.S.)	REMARKS
Germany							
MG 15	7.92 mm	28.0	75	Recoil	850	2,500	Ex-aircraft gun
MG 34	7.92 mm	26.7	75	Recoil	850	2,500	Also belt fed
MG 42	7.92 mm	25.5	Belt	Recoil	1,200	2,500	
Knorr-Bremse	7.92 mm	22.0	20	Gas	500	2,600	Principally used by SS units
Great Britain							
Bren Mk 1	.303	22.3	30	Gas	500	2,450	Water-cooled medium
Vickers Mk 1	.303	40.0	Belt	Recoil	450	2,450	Water-cooled medium
Vickers-Berthier	.303	22.0	30	Gas	600	2,450	Indian Army only
United States							
Browning Auto Rifle	.300	16.0	20	Gas	500	2,800	Squad automatic
Browning M1919A6	.300	32.5	Belt	Recoil	500	2,800	Squad automatic
Browning M1917A1	.300	32.6	Belt	Recoil	500	2,800	Water-cooled medium
Browning .50	.50	84	Belt	Recoil	550	2,930	
Italy							
Fiat-Revelli 1914	6.5 mm	37.5	50	Blowback	400	2,100	Water-cooled medium
Breda 1930	6.5 mm	22.5	20	Blowback	475	2,000	Squad automatic
Breda 1937	8 mm	43.0	20	Gas	450	2,600	Air-cooled medium
Japan							
Taisho 3	6.5 mm	62.0	30	Gas	400	2,400	Air-cooled medium
Taisho 11	6.5 mm	22.5	30	Gas	500	2,300	Squad automatic; hopper feed
Type 99	7.7 mm	23.0	30	Gas	850	2,350	Squad automatic
Type 92	7.7 mm	122	30	Gas	500	2,400	Weight includes tripod
Type 96	6.5 mm	20	30	Gas	550	2,400	
Soviet Russia							
Maxim 1910	7.62 mm	52.5	Belt	Recoil	550	2,800	Water-cooled medium
SG 43	7.62 mm	30.25	Belt	Gas	600	2,800	Air-cooled medium
DP	7.62 mm	20.5	47 drum	Gas	550	2,750	Squad automatic
DShK	12.7 mm	78.5	Belt	Gas	550	2,800	Heavy support and anti-aircraft
France							
MG 24/29	7.5 mm	24.5	25	Gas	600	2,590	
Hotchkiss FM1922	7.5 mm	19.25	N.A.	Blowback	550	2,250	Approximate figures

Before leaving the individual weapons of the infantry soldier, there is one last type which ought to be considered: the grenade. Few infantrymen would dream of going into battle without a few grenades in their equipment, for nothing was so lethal at short range. How they were carried was a matter of personal preference; the drill books said one thing, but the soldiers did what they felt was most convenient and carried them in whatever fashion allowed them to be reached quickly. The German soldier was fond of carrying his stick grenade tucked into the top of his jackboot; the Russian had his tucked into his belt or in the pockets of his smock jacket; the conservative Englishman kept them discreetly in his pouches, largely because the design did not lend itself to being carried any other way; the American usually hung them on his equipment by their handles, which was a trifle hazardous.

The range of grenades available was limited when the war began, but as more and more peculiar problems arose, more and more grenades were designed to deal with them, until by the end of the war the number of different grenades in use ran into hundreds. We can only afford space to consider some of the better known and some of the more unusual models.

Broadly speaking the grenades with which the war began were the same ones which had seen the end of the First World War, largely because many authorities considered they were only suitable for trench warfare. This was soon found to be wrong, but since the old designs were perfectly sound, they stayed in service throughout the war and in many cases are still in use with their respective armies. The British 'Mills Bomb' is a case in point; this was patented in 1915, went through a number of minor modifications during the following three years, and has remained the standard British hand grenade ever since, although from time to time there have been attempts to replace it with something a little more up to date.

Action stations

The Mills is the classic 'pineapple' shape, with the cast iron body deeply grooved to allow it to break into lethal fragments. A central striker is held up by the hand lever and locked by a safety pin. The thrower grasps the grenade so as to hold the hand lever firmly, pulls the pin and throws. As the grenade leaves his hand, the lever flies off due to the striker spring power, and the striker hits a cap, lighting a short length of fuse which burns through to a detonator which in turn explodes the grenade. A good thrower should be able to drop the grenade where he wants it up to a range of about thirty yards, but since the explosive was capable of scattering the fragments much further, once he had thrown it he had to be quick to get behind some cover.

In order to reach out further, it was possible to screw a flat disc on to the bottom of the Mills Bomb and fire it from a special cup fitted on the end of the service rifle, using a blank cartridge to launch it. This gave a range of up to 200 yards but, since the flight over this distance took longer, it was necessary to have a fuse in the grenade which burned longer. The standard grenade fuse which catered for both hand and rifle

German hand grenades in action during the second Kharkov offensive, 1942

GRENADES
THROW AND DIVE FOR COVER

Robert Hunt Library

launching burned for seven seconds and in peacetime everybody had been happy with that. But in 1939 when the British Army went to France and began patrolling against the Germans, they found that the seven second fuse had some disadvantages in a hand grenade – the Germans picked them up and threw them back before they had time to explode. Some urgent messages were sent back to England and a new four second fuse was developed for use when the grenade was hand thrown. There were some very surprised Germans on the French frontier soon afterwards.

Eggs and Potato Mashers

The Germans, for their part, were using their stick grenade, commonly known to millions of soldiers as the 'Potato Masher' from its shape. This too had first emerged in 1915 and had been somewhat improved over the years. Its mechanism was a system widely used by the Germans but rarely by other nations, a 'friction igniter'. The handle of the grenade was hollow, with the explosive-filled head screwed on to the end. Into the head went a detonator assembly, from which a string ran down the hollow handle, ending in a porcelain bead. The spare string and bead were tucked into the handle and held there by a screw cap. When the time came to use it, the screw cap was removed so that the bead fell out on the end of its string. Holding the handle in one hand, the thrower pulled the string with the other and threw the grenade. Pulling the string pulled a roughened steel pin through a sensitive chemical, causing it to ignite; this lit the five second fuse, which in turn fired the detonator and exploded the grenade.

An alternative issue to the German Army was the 'Egg Grenade', an oval light steel device about the size of an orange. Screwed into the top was a dark blue button; unscrewing this revealed a string, and the ignition was just the same as the stick grenade – pull the string and throw. But the Egg Grenade lent itself to some unpleasant modifications; the blue igniter could be removed and replaced with a red-capped version which had a delay of only one second, which was of little use as a hand grenade, but if left lying around a position after a retreat often gave unpleasant surprises to any new arrival who didn't know the colour code.

Even more devastating was an unofficial modification dreamed up on the Russian Front. The standard blue igniter set was removed and, using two pairs of pliers, dismantled and the safety fuse removed. The detonator was now screwed back directly into the friction igniter and the grenade was left lying around for a Russian soldier to pick up when the Germans left the position. Of course, since the blue cap looked normal, Ivan was likely to pick it up to throw at the retreating Hermann; he unscrewed the cap, pulled the string, and there was an instantaneous explosion which usually killed him.

Anti-tank warfare

One of the greatest problems of the war was the question of how the infantry soldier could deal with an enemy tank. As we shall see later he was given a number of missile weapons at various times, but there was always a demand for some sort of grenade which every soldier could use if the need arose. An ordinary hand grenade was useless – after all, this was the sort of thing

which tanks were intended to be proof against – so it had to be some special type. The answer to this problem had been discovered just before the war in the phenomenon called the 'Hollow Charge'.

It had been found, many years before, that if a charge of explosive were hollowed out, and the hollowed face placed against a piece of steel, the shape of the hollow was reproduced in the plate. After much experiment and research it was found that if the hollow were made into a regular cone or hemisphere and thinly lined with certain metals, the explosion of the charge would convert the liner into a stream of molten metal which was, in effect, focused by the shape of the hollow and directed at a point on the target armour at such speed and pressure that it blasted straight through. Once this was appreciated, numerous grenades were developed to utilise the principle, for now a small charge of explosive was capable of piercing four or five inches of hard armour. Once the armour was pierced, a jet of flame and hot gas went into the tank and, if it struck a fuel tank, ammunition or one of the crew, it did considerable damage.

The only problem was how to make sure that the hollow portion was in correct alignment with the tank armour, and to do this numerous tricks were tried. One of the most effective was the German Magnetic Anti-tank grenade. This was a cone charge with a handle similar to that of the stick grenade and with the same friction igniter.

Around the cone were three powerful magnets; the soldier could either run up to the tank, stick the grenade firmly in place and then pull the string and run for it; or he could pull the string and throw the grenade so that the magnets struck the armour and adhered. Either way the tank was as good as dead. An alternative, forced on them by the shortage of the special metals needed to make good quality magnets, was to coat the edge of the cone with a very sticky adhesive to hold the charge in place.

Finally they developed the *Panzerwerfmine* – literally translated as the 'thrown anti-tank charge'. This did away with the time action of the fuse and used a simple impact fuse which set off the charge as soon as it struck, so that it was no longer necessary to worry about magnets or sticky stuff. To make sure it landed with the hollow charge pointing in the right direction, a canvas tail was fitted.

But nevertheless, it sometimes happened that the German grenadier was confronted with a Russian tank when he had no special anti-tank grenades handy. He developed his own solution to this problem – the 'Bundle Charge'. He took eight or ten stick grenades and unscrewed the explosive heads, removing the fuses. Then, using a complete stick grenade as the foundation, he strapped the spare heads around it with insulation tape or cord. When a Russian tank appeared, he lay quiet until it went past him, then jumped up, pulled the string

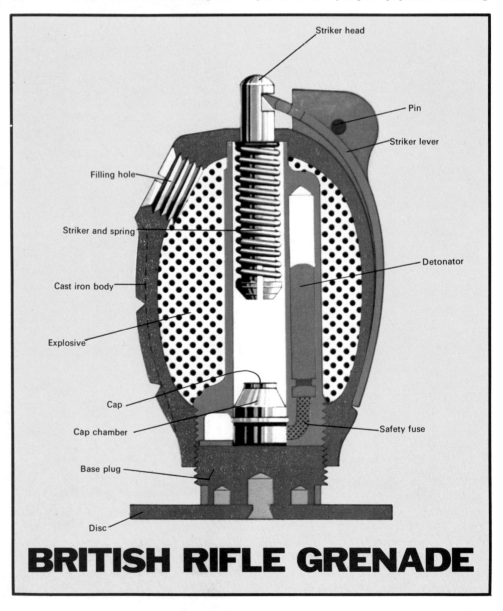

Striker head

Pin

Striker lever

Filling hole

Striker and spring

Detonator

Cast iron body

Explosive

Cap

Safety fuse

Cap chamber

Base plug

Disc

BRITISH RIFLE GRENADE

of the central stick grenade and threw the bundle on to the engine deck of the tank and ducked. The central grenade acted as a primer for the rest, and the explosion which followed was enough to wreck the engine of any tank and stop it dead.

The British Army also had a hollow charge anti-tank grenade, but this was to be rifle-launched from the same cup attachment as was used with the Mills Bomb. Called the 'Number 68' it had four small fins and an impact fuse and was quite effective against two or three inches of armour at ranges up to about 100 yards. But the most way-out British designs came about after Dunkirk, when everybody was trying to produce weapons to repel the forthcoming invasion. From this period came the 'Sticky Bomb', a glass flask filled with nitroglycerin and coated with cloth soaked in bird-lime. The handle carried a striker and lever very similar to that of the Mills Bomb. The thrower pulled the pin and flung the bomb at a tank; the sticky coating adhered to the armour and the nitroglycerin, though un-likely to blow a hole in armour, would severely damage tracks or engine compart-ments so that the tank would be stopped and then dealt with by some other weapon.

Another fearsome anti-tank weapon was the 'Self-Igniting Phosphorus Grenade' which was a high-sounding name for a lemonade bottle full of phosphorus and benzene. When thrown at a tank the bottle would break open and the phosphorus would ignite spontaneously on coming into contact with the air; this lit the benzene and the liquid fire would run into the various apertures in the tank.

As with Britain and Germany, the Ameri-can standard hand grenade originated in the First World War, but not with the Ameri-cans. It had begun as a French grenade and was adopted by the US Army as being a quicker solution than trying to design and produce one of their own. The ignition system was a trifle erratic, so during the inter-war years the US designers improved it into the well-known 'mousetrap' system, so-called because the lever held down a small spring-loaded arm with a firing pin which, when the lever was released, snapped across like a mousetrap and fired the fuse.

Unlike the Mills design, the lever of the US grenade stood away from the grenade body, so that it was easy to hook it through loops on the clothing or equipment and carry several grenades festooned around the chest ready for instant use. If they were used fairly quickly this was harmless enough (though there was always the danger that some sniper might hit one with a bullet), but if they were carried for several days there was the danger that the handle would

be deformed and, when the pin was pulled, although the lever was still held down its other end could slip free and allow the striker to swing over and light the fuse. If the thrower didn't notice and get rid of it immediately, the results were drastic.

For anti-tank work the US Army had a hollow charge rifle grenade which was used rather differently to other people's designs. Instead of having a cup on the muzzle of the rifle, into which the grenade fitted, the American grenade launcher was an exten-sion to the barrel. The grenade had a long hollow tail which slipped over the extension and the grenade was shot off the end by a blank cartridge. It was highly effective and was later adopted by the British Army.

The cast-iron pineapple grenade used by the US Army was thought to be old and inefficient – though in reality it was no worse than many others still in use – and

efforts were made to try and produce some-thing better. One of the designs proposed was the 'Beano' grenade; the theory here was that, since all Americans could throw baseballs, then the grenade ought to be shaped like a baseball and weigh the same amount. Then there would be no need to train men to throw grenades, since they merely had to use the same technique they had been using since their schooldays.

Unfortunately, it took a long time to get the grenade working satisfactorily, since instead of a time fuse it was decided to use an impact fuse; and since there was no way of telling how the grenade would land, it had to be a fuse which would work no matter which way the grenade fell. By the time the design was perfected the war was over and the 'Beano' never got into service.

The Italians also developed grenades with 'all-ways' impact fuses, which had a

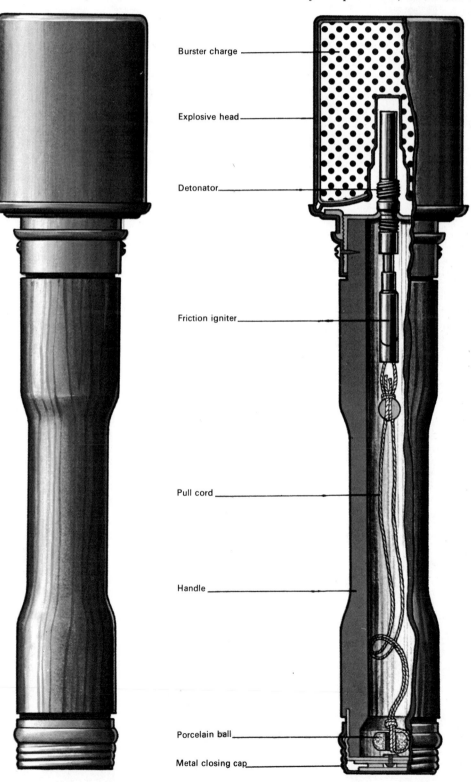

Burster charge

Explosive head

Detonator

Friction igniter

Pull cord

Handle

Porcelain ball

Metal closing cap

British Grenade No. 36M (left)
Last of the 'Mills Bomb' family, the 36M was developed in 1918 but has remained in service ever since. Extremely reliable, it could be fired from a rifle discharger cup by screwing a flat plate onto the base

German Stick Grenade
This was the German Army's favourite grenade from 1915 to 1945, and it changed very little in those years. The friction igniter is screwed into the base of the head and its operating string runs down the hollow handle. On pulling the string a roughened wire is drawn through a sensitive chemical to light the 5–second fuse

195

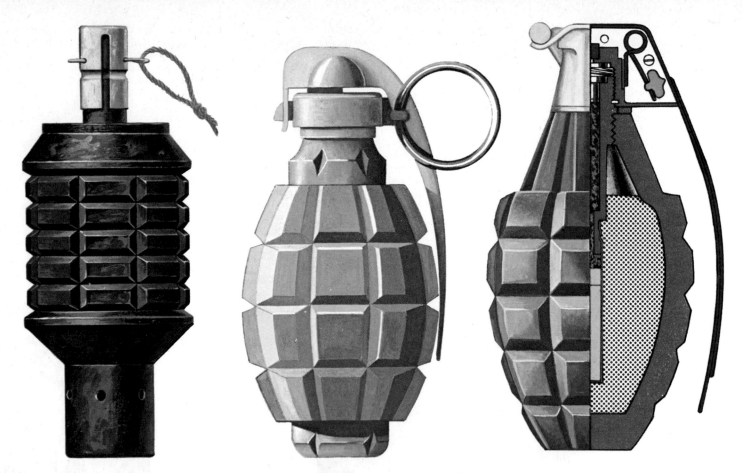

distressing habit of not always working when they landed, but working very effectively if anybody disturbed them afterwards. They were all painted a vivid red, and their violent habits led to them being nicknamed "Red Devils" by the British troops in the North African campaign.

Perhaps the most astonishing grenade was one developed in the latter days of the war by the Japanese. They had tried one or two anti-tank grenade designs without much success and finally, relying on the suicidal desperation of the Japanese soldier, they produced what the Allied troops called the 'Lunge Mine'. This was a powerful hollow charge unit mounted on the end of a bamboo pole and fitted with an instantaneous fuse operated by a string tied to the pole. On the approach of an Allied tank the Japanese volunteer would leap from hiding, holding the lunge mine in front of him, rush up, and jam the explosive charge against the tank. He then said a quick prayer and pulled the string. The resultant explosion pierced the tank with the hollow charge – and the soldier with the bamboo pole.

When it came to stretching the range of the grenade just that little bit further than

the soldier could throw, the rifle grenade was the usual technique, but the German Army came up with a new idea. Every infantry section was provided with a signal pistol; this was the usual one-inch bore single-shot weapon which was loaded with a cartridge containing coloured lights or an illuminating star on a parachute. But the Germans developed a number of small grenades which could be loaded into the signal pistol and launched as offensive weapons. Their size was not very great but they made a satisfying bang and were capable of being aimed fairly accurately.

The next step was to give it a longer rifled barrel, turning it into what they called the 'Battle Pistol'. This could still fire signal cartridges, but also had a range of rifled grenades, similar to the pattern fired from the rifle discharger cup. Some of these were anti-tank grenades, using the hollow charge principle, and in order to give a worthwhile effect at the target, the heads of the grenade were much larger than the bore of the pistol, so that they had to be muzzle loaded with a small-section stem entering the pistol barrel. It took a fairly determined and courageous man to attack a tank with this device, but the grenade was highly effective if it was directed at a part of the tank where the armour was thin.

Japanese Model 91 Grenade (top left)
A fragmentation grenade which could also be fired from a small mortar. In the grenade role the pin was removed and the striker smacked against the boot–heel to start the fuse burning before throwing it. For firing from a mortar a small propelling charge unit could be screwed on to the base

French 'Pineapple' Grenade (top centre)
The French Army's 'Defensive' grenade, which used a mechanism slightly different to that of the British Mills bomb but handled in the same way – hold the lever, pull the pin, and throw

US Army Grenade Mark 2 A1 (top right)
This was developed from the French model after the First World War. The firing mechanism was improved to the 'mousetrap' system, but the operation is the same. It was generally filled with smokeless powder instead of high explosive in order to give larger and more lethal fragments

South African troops in action in the Western Desert, January 1942

ANTI-TANK WEAPONS
FROM BULLETS TO BOMBS

After loading a Bazooka, an American soldier takes cover as the weapon is fired. The firer kept the Bazooka on his shoulder while it was being loaded, but the loader had to stay well clear of the rear of the weapon as it was fired, because of the flash from the projectile's rocket motor

 *(vertical caption, left side)* Robert Hunt Library

When it came to the business of attacking tanks at a distance there were some radical changes in technique as the war progressed. At the outbreak, all the major nations with the exception of the United States provided their infantry with an anti-tank rifle. This was no more than a very powerful and heavy rifle firing a special hard slug which could penetrate about half an inch of armour plate. The propelling cartridges were large and the calibre usually about half an inch, so that the weapon had to be heavy to try and soak up some of the recoil. In spite of that, they still kicked like mules and they were far from popular with the men who had to carry and use them. Moreover, after the first few months of the war, the tanks became thicker and harder to stop and the anti-tank rifle soon fell so far behind in the race that most armies got rid of them. The only army to retain them throughout the war was the Soviet Army, though it is doubtful whether they saw a great deal of use after 1943, but since the Soviets never developed any other anti-tank weapon for the infantry to use, they had to keep the rifle for lack of anything better.

In the British Army the anti-tank rifle was replaced by a device called the Projector, Infantry, Anti-Tank, which was shortened to PIAT for convenience. This was a 'Spigot Discharger', the invention of a Lieutenant-Colonel Blacker, who was, in fact, an artillery officer. The PIAT had no barrel; the projectile was a hollow-charge bomb with a long tail and fins, and the tail section was a hollow tube with a cartridge inside it.

Enter the Bazooka

This bomb was laid on a tray at the front of the PIAT, and when the trigger was pressed, a steel rod was driven by a powerful spring into the tail tube of the bomb, where it struck and fired the cartridge. The resulting explosion blew the bomb off the spigot and out of the tray, to send it about 150 yards or so through the air and, hopefully, hit a tank. The explosion also blew the spigot back into the body of the weapon, re-cocking it ready for firing the next shot. It was unpleasant to fire, and it was a trifle alarming to look up after firing it to see the bomb sailing slowly through the air, but when it hit a tank it was highly effective.

The Americans had no anti-tank rifle to replace, but having seen the result of the *Blitzkrieg* in 1940 they realised that some sort of infantry anti-tank weapon was necessary and set about organising one. Back in 1918 a Doctor Goddard had offered them a rocket tube which could be fired by one man, but since the war was over before much could be done with it, and since there seemed to be very little use for it, the idea was turned down. Now it was revived, and the rocket bomb given a hollow charge head. The weapon was a simple smooth-bore tube which the soldier could lay on his shoulder, taking aim through an open sight at the side. His assistant loaded a finned rocket into the tube and connected an electric firing wire; the firer pressed the trigger which connected two torch batteries in the pistol grip and fired the rocket motor by electricity. The rocket launched itself from the tube, giving no recoil to the weapon since the blast simply shot straight through the end of the tube, and there was a very potent anti-tank weapon.

The question then arose of what to call it. Officially it was the 2·36-in Rocket Launcher,

but that was a bit of a mouthful. Some unknown bystander remembered a radio show he had just heard on which a comedian called Bob Burns produced a peculiar musical instrument of his own invention as a comic prop; he called this thing his "Bazooka". And from that moment on the 2·36-in Rocket Launcher was the Bazooka. It was first used in the North African campaign where, according to legend, a German tank commander surrendered after one of his tanks had been hit by a rocket, complaining that he saw no future in going on when he was being shot at by 155-mm guns.

And the Stovepipe

The German Army, in 1942, were beginning to find out a lot of things about Soviet tanks that they didn't know before, such as how hard they were to stop and what a lot of them there were. In desperation the infantry cried for something better than their anti-tank rifles; as it happened an American Bazooka had just been captured in North Africa and, with some rockets, flown back to Germany for examination. The Germans were so impressed with this weapon that they immediately designed a copy, increasing the calibre to 3·5 inches so as to get a more effective warhead. It was officially known as the *Raketenpanzerbuchse* or 'Rocket tank rifle', but the German soldiers, like the Americans and British, preferred to call things by their own names, and this device was christened "Stovepipe" from the flame and smoke which came from it when it was fired.

Although Stovepipe was good, the army wanted something easier to carry and more powerful in its effect, while the production people in Germany were concerned at the amount of propellant that a rocket device used. To try and solve all three problems at once sounded impossible, but an ingenious German scientist produced, as his solution, one of the most famous anti-tank weapons of the war. This was the *'Panzerfaust'*, a simple but deadly weapon which was produced in tens of thousands and could stop any tank in existence.

'Tuck it under your arm'

The weapon consisted of nothing more than a mild steel tube about three feet long and an inch and a half in diameter, carrying a primitive backsight and a trigger. Inside the tube was a small charge of gunpowder. Into the front of the tube went a hollow charge bomb; the head was six inches in diameter and carried a charge of three and a half pounds of explosive, powerful enough to pierce seven inches of armour. The tail stem of the bomb had four flexible fins which wrapped around the stem so that it could slip into the firing tube. The foresight was a pin on the edge of the bomb. All the firer had to do was tuck the tube under his arm, take aim, pull the trigger and throw the empty tube away. Pressure on the trigger fired a cap which lit the black-powder charge, this exploded, and a portion of its energy drove the bomb out of the front of the tube, while the remaining blast went backwards and out of the rear. The blast passing out of the rear balanced the recoil due to the bomb being launched, and thus the *Panzerfaust* was a recoilless gun and not, as has often been said, a rocket launcher. It was, moreover, the first 'expendable' weapon; fire once and throw away, a policy which has caught on in postwar years with a number of one-man anti-tank weapons.

Panzerfaust
One of the most remarkable anti–tank weapons developed during the war, the Panzerfaust was turned out in the hundreds of thousands in a variety of sizes. Basically it was a small recoilless gun firing a large over–size bomb. Its hollow–charge warhead could penetrate any tank in the world – and did

Projector, Infantry, Anti-tank

Better known as the PIAT, for obvious reasons, this was the British Army's lightweight anti-tank weapon. It was a spigot launcher, using a heavy steel rod to fire a hollow–charge bomb to about 100 yards. It took nerves of steel to wait until the tank was that close, but the bomb was certainly effective

US Rocket Launcher 2·36-in

Better known as the Bazooka, this shoulder–fired hollow-charge rocket launcher could deal with tanks in a surprising fashion. There were various versions, the later models coming apart for easier carriage. A 3·5–in 'Super–Bazooka' was designed but was considered unnecessary, and it did not see service until the Korean War

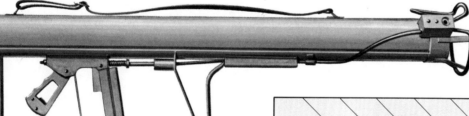

Panzerschreck

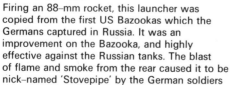

Firing an 88–mm rocket, this launcher was copied from the first US Bazookas which the Germans captured in Russia. It was an improvement on the Bazooka, and highly effective against the Russian tanks. The blast of flame and smoke from the rear caused it to be nick–named 'Stovepipe' by the German soldiers

PROJECTORS

NAME	CALIBRE	WEIGHT (lb)	LENGTH (in)	TYPE OF SHELL	WEIGHT SHELL (lb)	PENETRATION –ALL RANGES (in)	REMARKS
Germany							
Panzerfaust 30*	1.9 in	2.0	31.5	Ho Chg	7.0	7.0	Recoilless
Panzerschreck	88 mm	20.0	64.0	Ho Chg	7.25	4.0	Rocket launcher
Great Britain							
Piat	–	32.0	39.0	Ho Chg	3.0	4.0	Spigot discharger
United States							
Bazooka	2.36 in	13.25	54.0	Ho Chg	3.4	6.0	Rocket launcher

General note: Penetration figures refer to all ranges, since, as the hollow charge bomb operates by the explosive inside it and its special shape, it is independent of range and velocity.

*Note: Panzerfaust came in other models–60 and 100–which differed only in the propelling charge and consequently in the maximum range. The number indicates the fighting range in metres.

ARMOURED SUPPORT

French Hotchkiss 35 Tank

This 11–ton tank was actually intended for cavalry use, but since it was better than most of those intended for infantry support, the infantry came to lean on it more and more. Its 37–mm gun was moderately effective, but like all French tanks it suffered from having a one-man turret, giving the occupant too many jobs to do at once

British Sexton

Another example of the artillery support always available to the infantry is this British Army 25–pounder self–propelled gun. The chassis came from a Canadian tank (the Ram) and the whole equipment weighed 25½ tons. The gun fired a 25-lb shell to 13,400 yards, as well as smoke, flare, star, anti-tank and incendiary shells

Italian Light Tank CV33/5

An Ansaldo design, this vehicle appeared in 1933 and was slightly improved in 1935. It is a little more than an infantry–accompanying machine–gun carrier, and in this role was quite useful. When wrongly used as a fighting tank, it invariably came off worst

German Panzer 1
This Krupp design was the first tank delivered to the re–arming German Army in 1934. Like the Italian vehicle, it was little more than a machine–gun carrier and had no real fighting ability

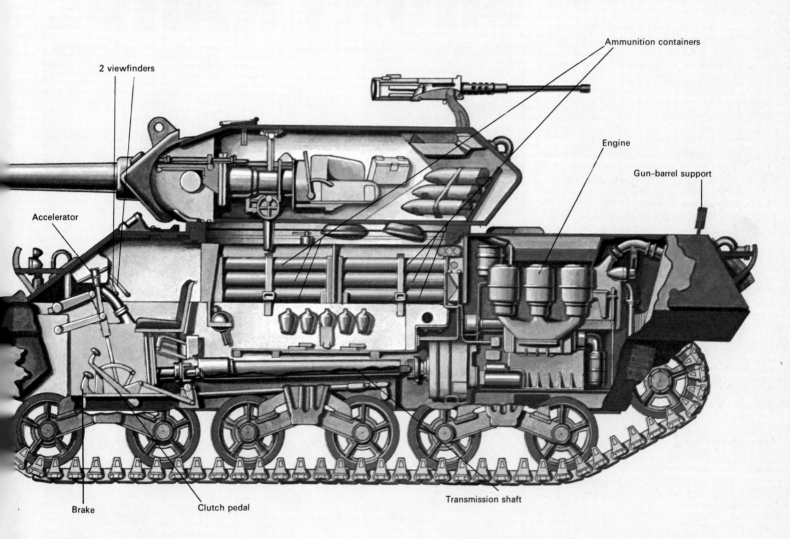

2 viewfinders

Accelerator

Ammunition containers

Engine

Gun–barrel support

Brake

Clutch pedal

Transmission shaft

US M-10 Tank Destroyer
This vehicle was built up from normal tank parts, but sacrificed top and side armour for thicker front armour and a heavier gun. The US version mounted a 76–mm gun; the British one, shown here, the more powerful 76·2–mm 17–pounder

Czech 38-t Tank
A good design which saw very little use as a tank since it was ageing when the war began, but it served as the foundation for innumerable designs of self-propelled infantry assault guns used by the German Army during the war. Slightly modified, one of these designs continues in service today with the Swiss Army

MORTARS
BOMBARDING AT SHORT RANGE

While the infantry's task is generally accepted as being close-quarter fighting, the fact remains that they need to reach out to more than arm's length or grenade-throwing range in order to deal with long-range enemy weapons or to bombard a hostile strong point so that they can get closer to it. For this the standard weapon is the mortar, and a wide variety have been developed since the original Stokes mortar of 1915, though few have strayed very far from Stokes' original idea.

The typical mortar is a very simple contrivance; a smooth-bore barrel with a fixed firing pin at the bottom, resting in a baseplate which sits on the ground, with the barrel cocked up in the air by a bipod with some form of screw gear for elevating and traversing the muzzle. The bomb fired from such a weapon is usually tear-drop shaped, with fins at the rear and fuse at the front. The propelling charge is in two parts, called the 'primary' and 'secondary' cartridges; the primary is a cardboard and brass shotgun shell with a charge of fast burning powder which slips into the tail tube of the bomb, at the centre of the fins. The secondary charges, up to six in number, are celluloid containers of powder or sheets of explosive clipped between the fins.

When the bomb is to be fired, it is inserted tail first into the muzzle of the mortar and dropped. It slides down the barrel and the cap in the primary cartridge strikes the firing pin at the bottom of the bore with sufficient force to fire it; the powder in the cartridge fires, and flashes through holes in the tail tube to light the secondary charges; these explode and the rush of gas lifts the bomb and throws it from the muzzle.

It will be obvious from this description that the prime asset of a mortar is simplicity and cheapness; the thing can be made by almost any competent engineering shop without having to call on specialised gun makers, and the bombs are usually made of cast iron with sheet metal fins and one of the cheaper and less efficient explosives inside. As a result, accuracy and long range are not to be expected, but for what they are asked to do, they are ideal. The other virtue of the mortar is that it can be dismantled and carried by a three-man team to any place that a man on foot can go. Generally, they are taken on long marches by some form of transport, but once battle is joined, man-carry is most often the only way of moving them about.

For the simple soldier

Mortars fall into two groups. Firstly there are the small models which throw a light bomb for about 500 yards, mortars which some armies, with more honesty, call grenade dischargers; secondly the medium mortars, about three inches in calibre and having a range of about three thousand yards. There are mortars larger than this, but in most cases they are treated as artillery weapons, requiring the fire control and observation back-up usually found with artillery organisations.

Of the light mortars, the only ones of interest came from Russia and Japan, and they are of interest because of their system of operation. The Russian 50-mm mortars were designed with the idea of making them as simple to operate as possible, so that a few simply-memorised settings could produce almost anything needed. Instead of being adjustable to any elevation between

202

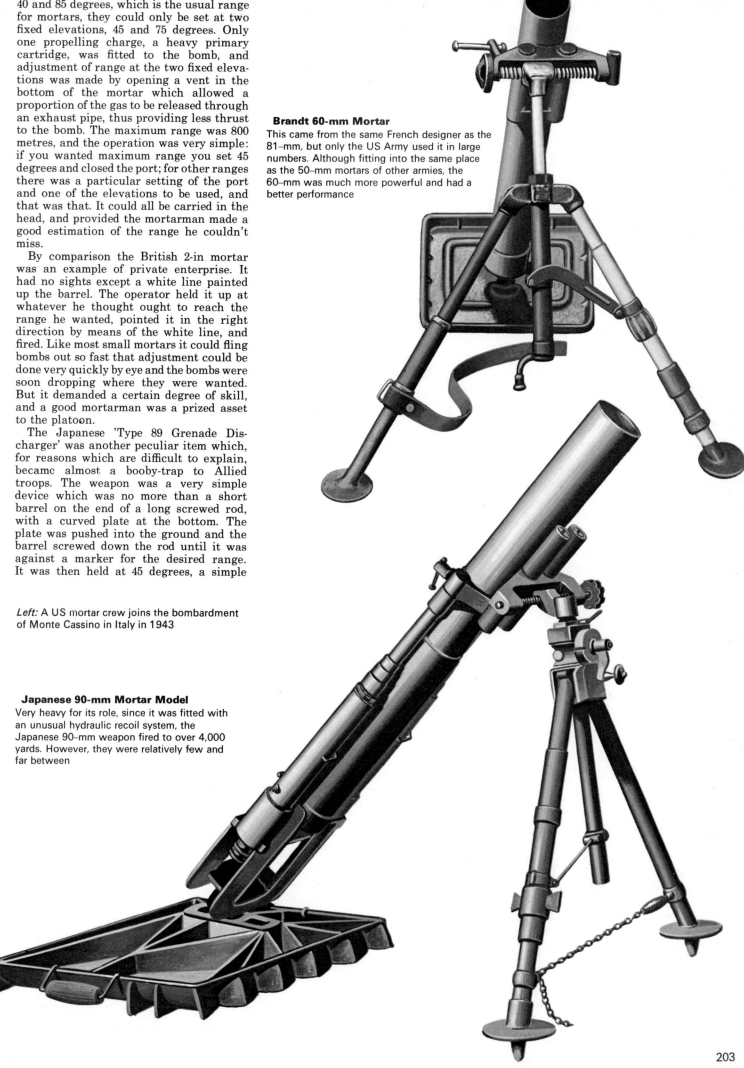

40 and 85 degrees, which is the usual range for mortars, they could only be set at two fixed elevations, 45 and 75 degrees. Only one propelling charge, a heavy primary cartridge, was fitted to the bomb, and adjustment of range at the two fixed elevations was made by opening a vent in the bottom of the mortar which allowed a proportion of the gas to be released through an exhaust pipe, thus providing less thrust to the bomb. The maximum range was 800 metres, and the operation was very simple: if you wanted maximum range you set 45 degrees and closed the port; for other ranges there was a particular setting of the port and one of the elevations to be used, and that was that. It could all be carried in the head, and provided the mortarman made a good estimation of the range he couldn't miss.

By comparison the British 2-in mortar was an example of private enterprise. It had no sights except a white line painted up the barrel. The operator held it up at whatever he thought ought to reach the range he wanted, pointed it in the right direction by means of the white line, and fired. Like most small mortars it could fling bombs out so fast that adjustment could be done very quickly by eye and the bombs were soon dropping where they were wanted. But it demanded a certain degree of skill, and a good mortarman was a prized asset to the platoon.

The Japanese 'Type 89 Grenade Discharger' was another peculiar item which, for reasons which are difficult to explain, became almost a booby-trap to Allied troops. The weapon was a very simple device which was no more than a short barrel on the end of a long screwed rod, with a curved plate at the bottom. The plate was pushed into the ground and the barrel screwed down the rod until it was against a marker for the desired range. It was then held at 45 degrees, a simple

Left: A US mortar crew joins the bombardment of Monte Cassino in Italy in 1943

Brandt 60-mm Mortar
This came from the same French designer as the 81-mm, but only the US Army used it in large numbers. Although fitting into the same place as the 50-mm mortars of other armies, the 60-mm was much more powerful and had a better performance

Japanese 90-mm Mortar Model
Very heavy for its role, since it was fitted with an unusual hydraulic recoil system, the Japanese 90-mm weapon fired to over 4,000 yards. However, they were relatively few and far between

spirit bubble indicating the correct angle, and the bomb popped in and fired by a trigger and striker. Screwing the barrel down caused some of the rod to move into the barrel and reduce the space inside, so that the bomb was further up or down when fired and thus the gas pushed on it for a longer or shorter time; this governed the range of the bomb.

It was customary for the Japanese mortarman to carry this mortar strapped to his leg; and probably due to a faulty translation of a Japanese document, the weapon became known in Allied circles as the 'Knee Mortar'. As a result, when specimens were captured and examined, one or two soldiers came to the conclusion that the shaped spade at the bottom was just the right curve to fit a man's leg and the approved method of using it was to kneel down and place the mortar on the thigh. After several Allied soldiers entered hospital with compound

fractures of the femur, word got around.

In the medium mortar class there was much less scope for individuality; most mortars adhered strictly to the typical pattern, so much so that two nations on opposite sides had the same mortar firing the same design of bombs. In the 1930s both Italy and the USA had been in the market for a useful medium mortar and both had gone to the same place – a French designer named Edgar Brandt – and purchased manufacturing rights to his design of 81-mm mortar. There were very small differences between the two mortars when they finally reached the Italian and US Armies, largely due to modifications to suit their own style of use and their national manufacturing standards, but nevertheless the ammunition was based on the same designs and was interchangeable.

The heaviest infantry mortar of the war was the German 120-mm, a powerful weapon

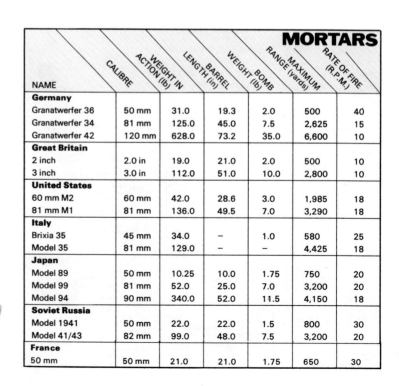

British 2-in Mortar
The standard British mortar, one of these was issued to each platoon. It could fire a 2–lb bomb to 500 yards

NAME	CALIBRE	WEIGHT IN ACTION (lb)	BARREL LENGTH (in)	WEIGHT BOMB (lb)	MAXIMUM RANGE (yards)	RATE OF FIRE (R.P.M.)
Germany						
Granatwerfer 36	50 mm	31.0	19.3	2.0	500	40
Granatwerfer 34	81 mm	125.0	45.0	7.5	2,625	15
Granatwerfer 42	120 mm	628.0	73.2	35.0	6,600	10
Great Britain						
2 inch	2.0 in	19.0	21.0	2.0	500	10
3 inch	3.0 in	112.0	51.0	10.0	2,800	10
United States						
60 mm M2	60 mm	42.0	28.6	3.0	1,985	18
81 mm M1	81 mm	136.0	49.5	7.0	3,290	18
Italy						
Brixia 35	45 mm	34.0	–	1.0	580	25
Model 35	81 mm	129.0	–	–	4,425	18
Japan						
Model 89	50 mm	10.25	10.0	1.75	750	20
Model 99	81 mm	52.0	25.0	7.0	3,200	20
Model 94	90 mm	340.0	52.0	11.5	4,150	18
Soviet Russia						
Model 1941	50 mm	22.0	22.0	1.5	800	30
Model 41/43	82 mm	99.0	48.0	7.5	3,200	20
France						
50 mm	50 mm	21.0	21.0	1.75	650	30

MORTARS

which could throw a 35-lb bomb to 6,600 yards, and it was heartily disliked by everyone who came under fire from it. Yet this, too, had originated in France; the original model had been bought for evaluation by the Soviets and then modified to suit their purposes. It went into service with the Soviet Army in 1938, but so many were captured by the Germans in 1941–42 that they found it worth while to issue them to their own units. It proved so effective that eventually it was re-designed slightly to make it a little stronger and went into production in Germany as the *Granatwerfer* 42. In the Soviet Army it had been manned by artillery units, but the German infantry preferred it to their own 81-mm mortar because of its longer range and more lethal bomb. Although heavy, it was easily moved since each mortar was provided with a transporter which could be attached to the mortar in seconds.

French 50-mm Grenade Launcher
This weapon fired a bomb weighing 1·75 lb to a maximum range of 650 yards. It was not used a great deal during the war

Brandt 81-mm Mortar
This mortar was designed in France and bought by Italy, the USA and Japan, among others. As a result, they all had similar weapons during the war. Firing light or heavy bombs, its maximum range was generally about 3,500 yards

THE SILENT BATTLEFIELD

German Pot Mine

This simple anti–personnel mine was developed late in the war; it is basically a metal can holding about six ounces of explosive with a chemical igniter screwed into the lid. The igniter was simply a soft metal container with flash powder and a glass ampoule of acid inside; stepping on it crushed the ampoule and allowed the acid to contact the powder. The reaction was sufficient to set off the mine

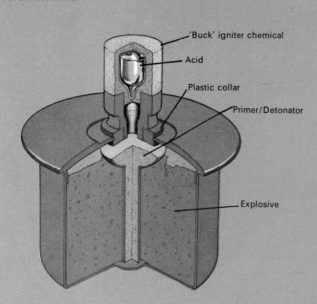

'Buck' igniter chemical

Acid

Plastic collar

Primer/Detonator

Explosive

Russian TM-41 Mine

This was a simple pressure–operated anti–tank mine; a weight of 350 lb would set off the 9–lb explosive charge

Operation of the S-Mine

S-Mine stood for 'Schrapnell-Mine', and this was the favourite German anti-personnel mine. It was buried so that only the prongs of the igniter were above the ground; a pressure of only 7-lb on these would cause the igniter to fire. This lit a propelling charge which fired the shrapnel container into the air; as it was launched, a delay fuse was lit which acted when the container was about 4 ft above the ground and detonated the charge of explosive in the container. This blasted the surrounding steel balls in all directions

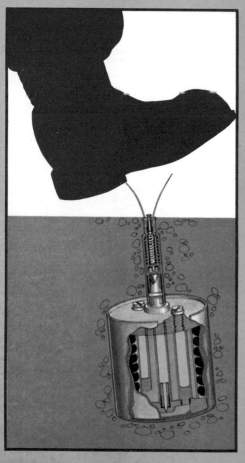

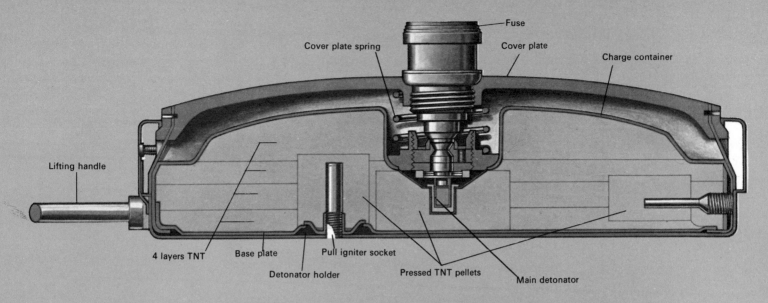

Fuse

Cover plate spring

Cover plate

Charge container

Lifting handle

4 layers TNT

Base plate

Pull igniter socket

Detonator holder

Pressed TNT pellets

Main detonator

Tellermine

The standard German anti–tank mine, this contained about eleven pounds of TNT and was activated by pressure on the fuse set in the centre. About 350 lb would set off the mine, and the blast would rip the tracks off any tank

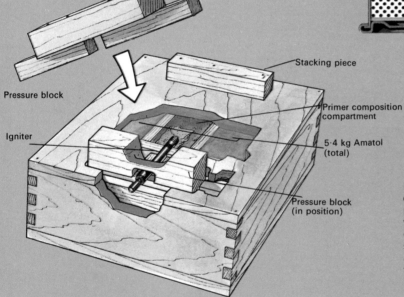

Pressure block

Igniter

Stacking piece

Primer composition compartment

5·4 kg Amatol (total)

Pressure block (in position)

British Anti-tank Mine Mark 7

Developed by Britain after the Second World War, this resembles the German Tellermine but contains almost twice as much explosive

German Holzmine 42

In order to defeat Allied magnetic mine detectors, the Germans began using wood, plastic and glass for their mines. This is a wooden anti–tank mine filled with 10 lb of high explosive. A 200–lb pressure on the upper surface sheared two wooden dowels inside and allowed the firing pin to be released, detonating the mine

German Riegel Mine 43

This anti–tank mine contained 8·8 lb of Amatol, and required a pressure of 440 lb on the ends or 880 lb on the centre to set it off. The explosive rested on the shear wires; when these were broken the striker was freed, and the detonator ignited

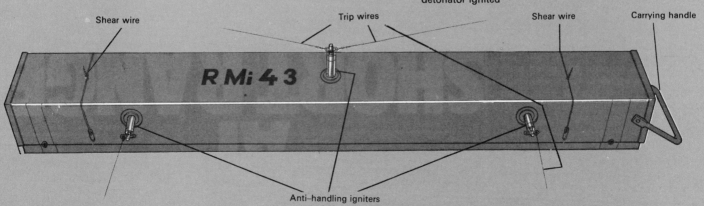

Shear wire

Trip wires

Shear wire

Carrying handle

R Mi 43

Anti-handling igniters

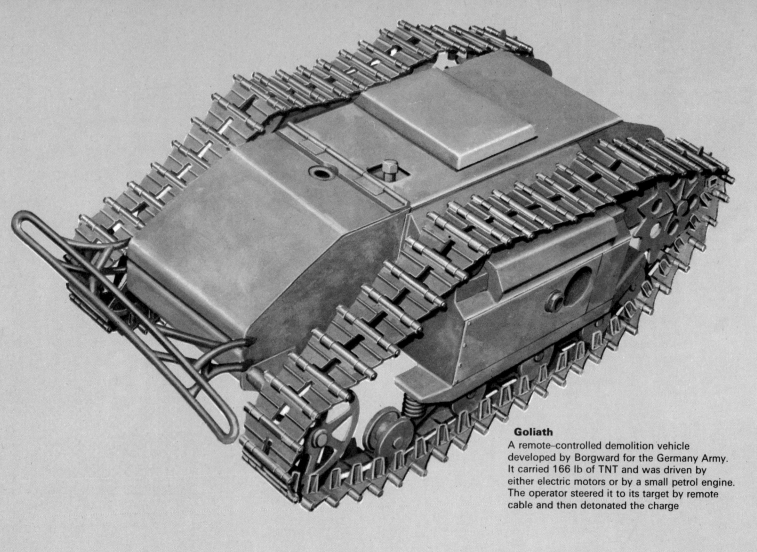

Goliath

A remote–controlled demolition vehicle developed by Borgward for the Germany Army. It carried 166 lb of TNT and was driven by either electric motors or by a small petrol engine. The operator steered it to its target by remote cable and then detonated the charge

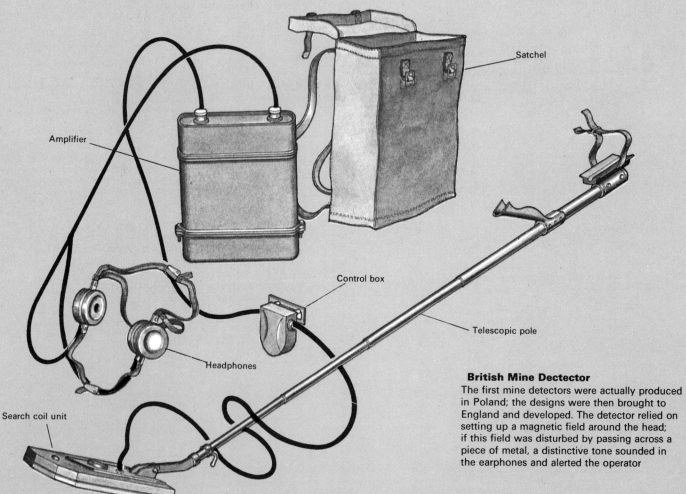

Satchel

Amplifier

Control box

Headphones

Telescopic pole

Search coil unit

British Mine Dectector

The first mine detectors were actually produced in Poland; the designs were then brought to England and developed. The detector relied on setting up a magnetic field around the head; if this field was disturbed by passing across a piece of metal, a distinctive tone sounded in the earphones and alerted the operator

A mine, part of the British defences along the Gazala Strip, explodes during Rommel's attack in June, 1942

BIG GUNS FOR INFANTRY

The mortar is generally considered to be the infantryman's private close support artillery; as a result it may come as a surprise to many people to find that infantry were often provided with genuine artillery guns and howitzers of their own. This practice was most common in continental armies, much less common among British and US troops. But sooner or later everybody had a try at it, some giving up quicker than others. In the British Army there was a period after the First World War when special light artillery batteries were attached to infantry battalions for close support, but in the late 1920s it was decided that the infantry would do better with their 3-in mortar and the light artillery was turned to other tasks. But during the war, probably spurred by consideration of what the German Army was doing, it was decided to give the infantry their own howitzer, a 95-mm model put together from a variety of bits and pieces in current production for other weapons. The barrel was 95-mm calibre since this was 3·7-in and they could be made on machinery set up for making 3·7-in anti-aircraft gun barrels. The breech mechanism was that of the 25-pounder, the recoil system that of the six-pounder. The whole thing was built up on a steel trail and had a large shield to give protection to the crew; it was towed by a Bren Gun Carrier.

Unfortunately, nobody had thought to ask the infantry if they wanted such a weapon in the first place, and when the idea was produced for their inspection they were distinctly cool about the whole scheme. It was simply a question of manpower. The battalions were already expected to man mortars, anti-tank weapons, machine-guns and even anti-aircraft weapons, and nobody was clear as to where the men were going to come from to man this new addition. In course of time the question resolved itself; the design of the gun ran into severe mechanical problems which were not satisfactorily sorted out until the war was almost over. By the time the answers had been found there was obviously no need for the gun, so it was quietly dropped.

The US Army had a very similar experience. Their idea was to produce a shortened and lightened version of the standard artillery 105-mm howitzer mounted on a 75-mm gun carriage. Infantry Cannon Companies were formed and trained in the use of this weapon, and were sent to North Africa in 1942. Again the manpower problem raised its head, and it became obvious that

Soviet 76·2-mm Gun
In the early months of the invasion of Russia, the Germans captured hundreds of Soviet 76·2-mm guns. These had their chambers reamed out to take a heavier cartridge, and were handed over to the infantry. They became very efficient anti-tank guns, and formed the basis for many post-war Russian designs

there were grave disadvantages to having one's own pocket artillery compared to being allocated a battery of regular artillery. One drawback was that if the cannon company didn't have the firepower to deal with a particular target that was the end of the story, whereas an artillery battery had lines of communication through the artillery set-up which could call in extra firepower to deal with such problems. For these and other reasons, the cannon company idea was abandoned after the North African campaign. But the howitzer was not a total loss, since it was found to be ideal for

A Russian sniper, the holder of three Soviet medals for valour in combat, leads a patrol on the Volkhov front

airborne artillery units, giving a combination of heavy firepower and light weight.

The German Army had the best co-ordination of infantry guns. Each infantry regiment had a gun company with six 75-mm light guns and two 15-cm heavy guns. In some regiments these proportions were varied, and it was common to find four 75-mm and four 15-cm. The 15-cm *Schwere Infanterie Geschutz* was an elderly design but it was quite efficient and stayed in service throughout the war. Firing an 83-lb shell to a range of 5,150 yards, it could provide excellent support and devastating shell-power on to any obstacle which stopped the advance of the forward troops. It was provided with a hollow charge anti-tank shell, and also a very unusual demolition bomb.

This weighed almost 200 lb and was a bomb with a tail rod and a set of fins attached to the body and clear of the rod. The rod was slipped into the gun muzzle so that the fins lay outside the barrel and the bomb's warhead was also outside the muzzle. It was launched by a special blank charge loaded into the gun breech and, after leaving the gun, the tail rod dropped off. The head contained 60 lb of explosive and could go to a range of 1,000 yards, and it was a highly effective device for breaching minefields, removing barbed wire, or demolishing pill-boxes or other strong points. It was not supposed to be an anti-tank weapon, but from some accounts of battles on the Eastern Front it appears that it was so used in emergencies, with an astounding effect on the tanks that were hit.

Absolutely foolproof

The smaller weapons changed their design several times during the war. The original pattern was quite unique mechanically, being the only artillery piece ever to be built on the lines of a shotgun. The 75-mm *Leicht Infanterie Geschutz* 18 did not use the usual type of sliding wedge breech normal in German artillery, but instead held the barrel in a long box, the rear end of which acted as the breech block. When the lever was pulled to open the breech, the rear end of the barrel tipped up to expose the gun chamber, while the 'breech block' stayed still. Once opened, it was held there by a catch which was automatically released as the shell and cartridge were loaded, so that the loaded gun dropped back into position in front of the breech by its own weight. It was a clever idea, rapid to operate and absolutely foolproof. The rest of the weapon was quite normal, a two-wheeled equipment with a tiny shield, weighing 880 lb in action. It could fire its 13-lb shell to 4,150 yards, and as well as high explosive shell it was provided with an anti-tank shell and a blue smoke shell for signalling targets to dive bombers. Mountain units had a special lightweight version using a tubular steel carriage.

After the campaign in France in 1940 the infantry requested a more powerful weapon, but for various reasons they didn't get one until 1944. Then they got two, the IG37 and the IG42. These were identical as far as the gun barrel and breech went, but were on different mountings. The gun had been designed by Krupp in 1942 and used a normal sliding wedge breech with a semi-automatic action, which opened the wedge and threw out the empty cartridge case automatically as the gun recoiled, leaving the block open and held against a spring. As soon as the next round was loaded the block closed automatically.

This was a useful device since it saved one man in the crew; without it, it would have been necessary to have one man doing nothing but open and close the breech. The IG37 model was mounted on the carriage of the obsolete PAK 37 anti-tank gun, while the IG42 was on a carriage which was in production for the 8-cm PAW 600 anti-tank gun. In 1944 the problem confronting the Germans was production, and by adapting the design to items which were already in production they simplified matters and also speeded up the introduction of new weapons.

The new gun fired the same ammunition as the old IG18 but due to having a longer barrel and an extra part to the propelling charge it could reach out to 5,600 yards.

The Germans also obtained one more infantry gun, known as the 7·5-cm *Infanterie Kanone* 290(r). The (r) indicates 'russian' and shows that it was a captured Soviet gun; it was originally the 76·2-mm Regimental Gun Model 1927, which is said to be the first artillery gun designed by the Soviets after the Revolution. Vast numbers of these were captured in the early months of the German invasion in 1941 and, since the German artillery weren't particularly interested, they were given to the infantry

to reinforce their gun strength. It was very popular with its new owners, since, like most Russian weapons, it somehow managed to get more range for its weight than anybody else ever managed. It weighed 1,720 lb, but it reached out to 9,300 yards with a 14-lb shell, which was a very useful performance. It was so well received that the Germans found it worthwhile to produce their own ammunition for it when stocks of captured Soviet ammunition were used up. (In fact, although the Germans called it '7·5-cm' it was actually 76·2-mm in calibre; misleading, to say the least.)

In the Japanese Army the business of artillery support was treated a good deal

differently than in other armies; a lot of this may have been due to the fact that they had, for several years, been engaged in fighting in China where the use of large amounts of artillery was never necessary. Consequently the infantry relied much more on their own infantry guns for support, using them in ones and twos and handling them very boldly, bringing them close to the front where they could fire over open sights at short range as well as deploying them further back to give long range indirect covering fire.

One of the most common was the 70-mm 'Model 92', a tiny gun weighing less than 500 lb. It could fire an 8-lb shell to just over

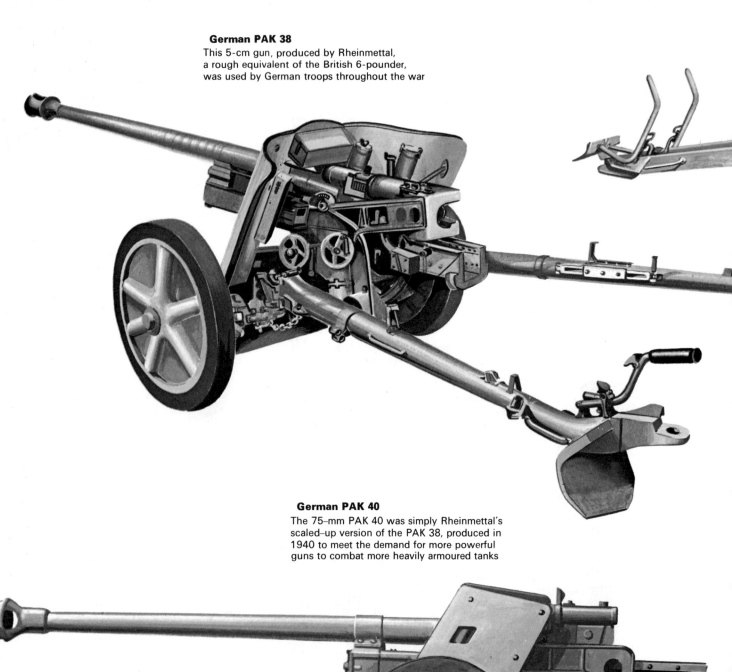

German PAK 38
This 5-cm gun, produced by Rheinmettal, a rough equivalent of the British 6-pounder, was used by German troops throughout the war

German PAK 40
The 75–mm PAK 40 was simply Rheinmettal's scaled–up version of the PAK 38, produced in 1940 to meet the demand for more powerful guns to combat more heavily armoured tanks

3,000 yards however, and was ingeniously designed with wheels mounted on cranked stub axles so that the gun could be set low down for direct shooting – or swung up to give clearance beneath the breech for the gun to recoil when fired at high angles as a mortar. Although some reports claim that it was unreliable and flimsy, the fact remains that it was always in evidence throughout the war and was in use for years afterwards by the Chinese Communist Army, who captured vast stocks of them in Manchuria and used them as late as 1952 in Korea. So it couldn't have been all that ineffective.

The other Japanese infantry gun was a hand-me-down which began life as the

German PAK 41
Krupp's answer to the need for a more powerful anti–tank gun, this was a brilliant design. The barrel tapered from 75–mm at the breech to 55–mm at the muzzle, and its overwhelming performance enabled it to defeat any tank in the world. Unfortunately, the shortage of tungsten, needed for the special shot, limited its life

US 105-mm 'Priest'
The US Howitzer Motor Carriage M7, called Priest because of its pulpit–like machine–gun mount, used the M4 tank chassis as its base. The first models produced went to the British 8th Army in the Western Desert, where it proved highly successful

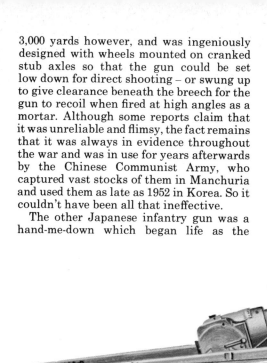

standard field artillery 75-mm gun in 1908. When the field artillery got a better gun they passed the 'Type 41' across to the Mountain Artillery, and in 1935, when they were in turn issued with something more modern, they gave the guns to the infantry. Consequently this weapon was a full-sized artillery piece which weighed almost a ton and fired a 12½ lb shell to almost 12,000 yards. Four of these guns were issued to each infantry regiment, and were kept at regimental headquarters under the Commander's control to be sent as heavy support to points where they were most needed.

The other class of gun which normally fell to the infantry's lot was, of course, the anti-tank gun. For some years it was held that the proper instrument to deal with a tank was another tank, and all the infantry needed was an anti-tank rifle for use in emergency. But as the tank became better understood it was realised that the original vision of grand fleets of tanks battling it out like naval fleets, with the rest of the army standing aside, was unlikely to happen. Tanks instead would be likely to drive through the front and fan out in small parties to do whatever damage they could,

and it looked as if it was up to the people in the front line – in other words, the infantry – to stop them. As a result of this re-thinking, infantry anti-tank guns began to be developed.

It might be said at this point that the British Army, while generally agreeing with the principle, found themselves up against the manpower problem and solved it by issuing their anti-tank guns to specialised artillery regiments and not to infantry. Later in the war infantry regiments did, at last, get anti-tank guns, but not until the artillery regiments had had their full issues.

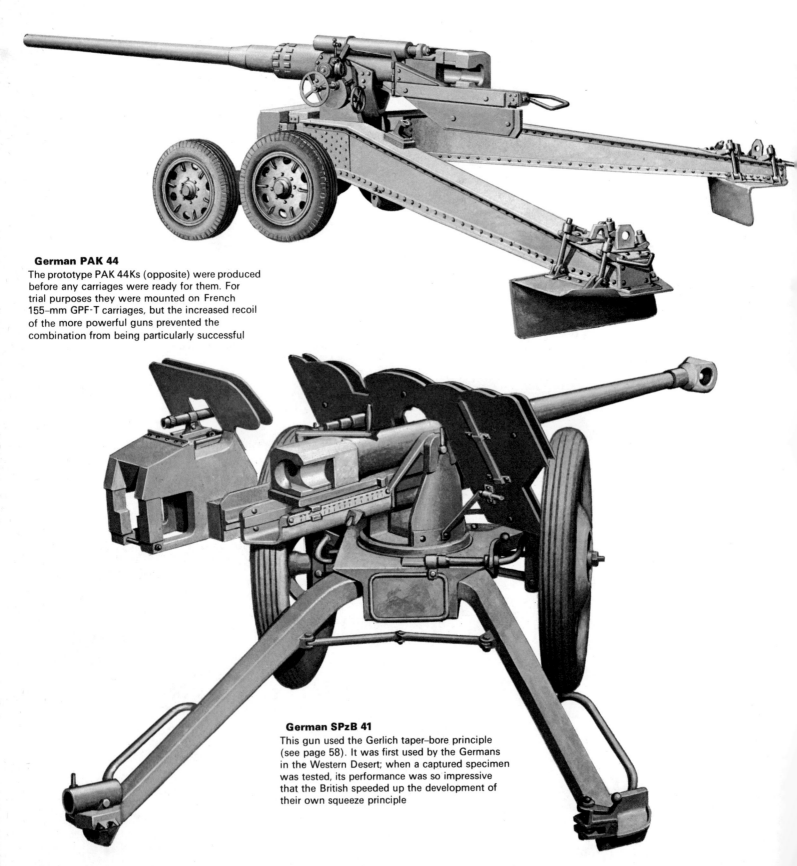

German PAK 44
The prototype PAK 44Ks (opposite) were produced before any carriages were ready for them. For trial purposes they were mounted on French 155–mm GPF-T carriages, but the increased recoil of the more powerful guns prevented the combination from being particularly successful

German SPzB 41
This gun used the Gerlich taper–bore principle (see page 58). It was first used by the Germans in the Western Desert; when a captured specimen was tested, its performance was so impressive that the British speeded up the development of their own squeeze principle

In general, everybody's ideas about a suitable anti-tank gun were similar; something of about 37- to 40-mm calibre, firing a shell of about 2 lb in weight. Rheinmettal of Germany produced a very good design of 37-mm gun which entered the German Army as the PAK (*Panzer Abwehr Kanone* – Tank defence gun) 37, and this design was more or less copied by several other nations. The American equivalent was the 37-mm gun M3, weighing 990 lb and firing a 2-lb shell at 2,900 feet per second to penetrate two inches of armour at 1,000 yards range. Its use was mainly in the Pacific Theatre of

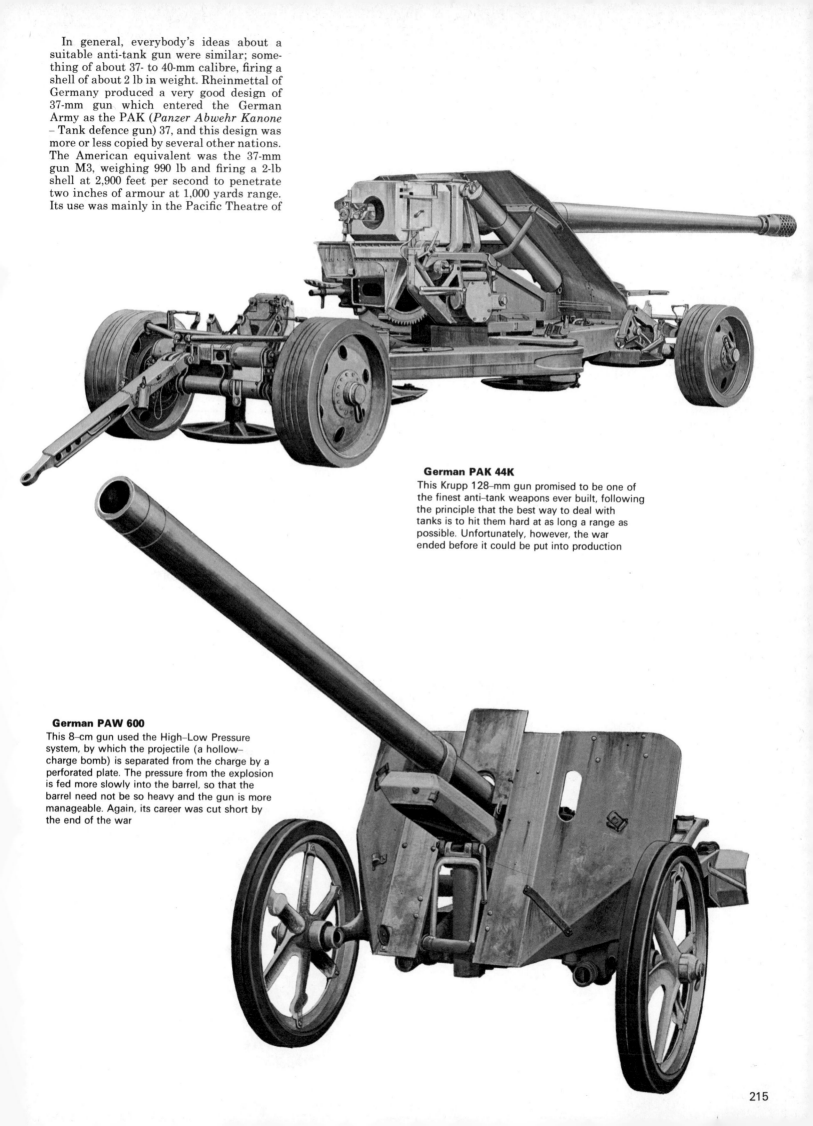

German PAK 44K
This Krupp 128–mm gun promised to be one of the finest anti–tank weapons ever built, following the principle that the best way to deal with tanks is to hit them hard at as long a range as possible. Unfortunately, however, the war ended before it could be put into production

German PAW 600
This 8–cm gun used the High–Low Pressure system, by which the projectile (a hollow–charge bomb) is separated from the charge by a perforated plate. The pressure from the explosion is fed more slowly into the barrel, so that the barrel need not be so heavy and the gun is more manageable. Again, its career was cut short by the end of the war

operations, since by the time the US Army entered the war in Europe and Africa a 37-mm gun was useless against the current German tanks, but it could still cope with the less well-armoured Japanese vehicles.

The Japanese for their part also had a 37-mm gun of very similar performance, but later replaced it with a more powerful 47-mm gun based on a Russian design which they had captured on the Manchurian border in 1939. This weapon fired a 3-lb shell to penetrate 70-mm of plate at 500 yards, but fortunately for the US Army there were never enough of them and for the most part the Japanese had to rely on their older 37-mm which had much less performance and was less of a threat.

Shattering impact

The Russians first equipped their infantry with the German 37-mm PAK 36 which they bought from Germany in 1937. But the Russians were very skilled at tank design, as the Germans found to their cost, and they were of the opinion that a 37-mm gun would not stay superior for very long if a war broke out. They therefore designed a new barrel of 45-mm calibre to fit the German carriage and then redesigned the carriage to produce an entirely new weapon, the Model 42. As well as firing a standard type of piercing shell, this was provided with a special high velocity shot known as an 'Arrowhead' shot, the name coming from its distinctive shape. The body of the shot was of mild steel, but concealed inside was a core of tungsten carbide.

The reason behind this type of shot was a problem which was met by every nation in turn; in fact it was the Germans who solved it first, and the Russian projectile was a copy of a German design. Ordinary anti-tank shell and shot were made of hard steel so as to smash their way through the armour, and the harder they were fired, the more likely they were to penetrate, which was why anti-tank guns had large cartridges and fired at high velocity. But this is only true up to a point. At a point where the shell strikes the armour at about 2,700 feet per second, the force of the impact is such a

shock to the projectile that the point breaks off and the shell simply smashes to pieces without penetrating. It is possible to get some results by placing special protective caps on the tip of the shell, but this only defers shatter to a slightly higher velocity and is not much of a solution.

The only real solution is to use tungsten carbide, a very hard composition which can penetrate without breaking at almost any velocity a gun can reach. But tungsten carbide is three times heavier than steel, is difficult to machine and expensive, so that it would not be practical to make an entire shot out of it. Which all led up to the idea of making a slender core of tungsten and surrounding it in a light steel body of the required calibre. Then, in order to cut down the weight as much as possible, so as to be able to drive the shot at the utmost velocity, the excess steel around the centre of the shot was machined away, leaving two bands at front and rear to support the shot in the gun barrel and a pointed nose to cut through the air; and that in turn leads us to the reason for the shot being called 'Arrowhead' – because of its shape after the cutting away of the centre section.

The Russian 45-mm arrowhead travelled at 3,500 feet per second as it left the muzzle and cut through 54 mm of armour at 500 yards; which brings out another odd point. The ordinary steel shot could penetrate 50 mm at the same range, and the obvious question is "Why go to the trouble for an extra 4 mm of performance?" The answer lies in ballistics and is somewhat involved, but put simply it is that a light projectile (the arrowhead weighed just under 2 lb) 'carries'

less well than a heavy one – one can compare this by throwing a golf ball and a ping pong ball and seeing which goes furthest. So at longer ranges the performance of the arrowhead shot soon fell off. Below 500 yards, (and shooting below 500 yards was preferred anyway since the gunner had more chance of scoring a hit) the arrowhead was superior.

But even with the 45-mm arrowhead it proved impossible to stay ahead of the game, and in 1943 the Soviets produced a 57-mm gun. This was the same calibre as similar weapons in use in the British and US Armies, but had a longer barrel, was more powerful and had better penetration; again, it was

given arrowhead shot. However, the jump from 45-mm to 57-mm was not really enough to get the anti-tank gun well ahead. By this time everybody else was using 75-mm or more, in the hope of producing a gun which would see the next generation of tanks taken care of as well as the current crop. As a result of not thinking big enough, the Russians had to start thinking about a new gun almost as soon as the 57-mm was in service, but the lesson had been learned, and they took a big jump to produce a 100-mm weapon. This was, as might be imagined, exceptionally potent, but very few were produced before the war ended, and eventually it became the standard divisional anti-tank gun for several years after the war.

The Americans had realised the 37-mm was out of date before they ever got into the war, and took an intelligent short cut; they obtained drawings and jigs for the British 6-pounder, converted them as necessary to

Japanese 70-mm Model 92 Howitzer
Maid–of–all–work for the Japanese infantry was this weapon, which could function either as a direct-fire gun or as a mortar, since it had a maximum elevation of 70°

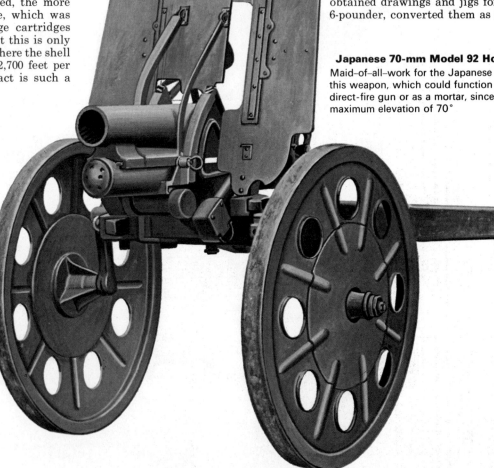

suit American manufacturing methods and standards, and put it into production as the 57-mm Gun M1. As with the British, this was the largest anti-tank gun provided for the infantry, heavier weapons being reserved for special 'Tank Destroyer' battalions. In fact, although the US Army did have some heavier anti-tank guns, there were not very many of them because their policy was to put their anti-tank guns on tracks and go chasing the tanks. This was slightly modified in the latter part of the war, but nevertheless the major US anti-tank strength was in self-propelled weapons.

As befitted a country which had put the tank on the map of modern war, Germany

had the most impressive line-up of anti-tank guns. They had begun by a rational programme of increases in calibre to keep up with the estimated improvement in tanks, but their calculations were upset when the Russian T-34 appeared on the scene and from then on their programme was somewhat patchy, as design after design was rushed into service in order to try and beat the Soviet armour back. They were the first to put tungsten-cored projectiles to use, but they were also the first (and only) country to abandon them, due to the pressure of economic warfare.

Tungsten was not native to either Germany or its possessions, and all supplies had to be imported. Consequently there was a shortage, which in the middle of 1942 became so acute that the decision had to be made whether to use the available supplies for machine tools or for ammunition. Finally, the decision was taken to reserve the scarce material for machine tools, and when the existing stocks of tungsten-cored shot were used up, the German Army had to revert to using ordinary steel shot.

One of the most remarkable German developments in the anti-tank gun field was

Japanese 47-mm Anti-tank Gun
Japanese tanks were poorly armoured and generally inferior. Consequently, their anti–tank weapons, like this one, were somewhat under–powered

the 'taper-bore' gun. In this design the calibre of the gun barrel gradually decreased as it reached the muzzle; thus, since the base area of the shot grew smaller, the pressure behind it increased and the shot was propelled faster. The first weapon to use this was the 'Heavy Anti-tank Rifle 41' in which the barrel tapered from 28-mm to 21-mm, pushing the shot out at over 4,000 feet per second. As might be imagined from the calibre, the shot wasn't very big, but it had good penetrating power at short range. A similar weapon tapering from 42-mm to 30-mm was also produced, just in time for the tungsten shortage to put it out of action.

To replace their 37-mm gun, a 50-mm had been designed but did not appear until 1940. This was quite a useful weapon, and although bigger guns appeared later on it stayed in service throughout the war and was still giving a good account of itself at the end. It was superseded by a 75-mm model which was little more than an enlargement of the 50-mm, so much so that it is hard to tell them apart at a casual glance. But the contract for the 75-mm was put out to two companies, Rheinmettal-Borsig and Krupp; Rheinmettal produced the scaled up model, but Krupp developed a brilliant taper-bore gun, in which the calibre dropped from 75-mm to 50-mm. This had an overwhelming performance and would have been one of the finest anti-tank guns of the war, but, like the other taper-bore weapons, the tungsten

Russian 57-mm Anti-tank Gun
The Russians produced this weapon in 1943, and although the increasing weight of armour carried by tanks had caused the British and US armies to abandon 57-mm guns by then, it continued in service until well after the end of the war

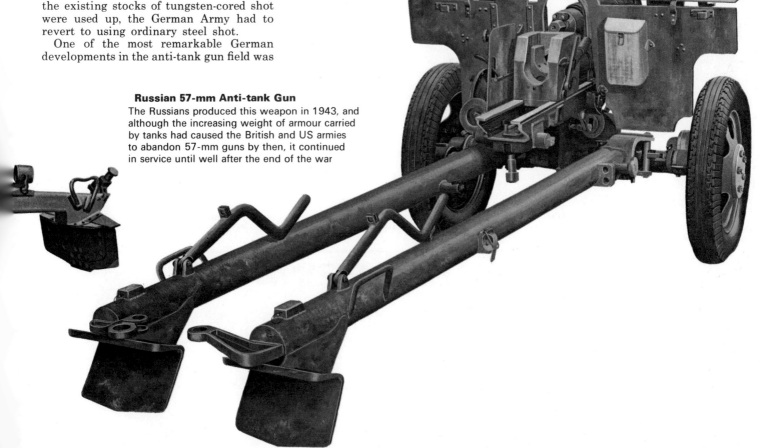

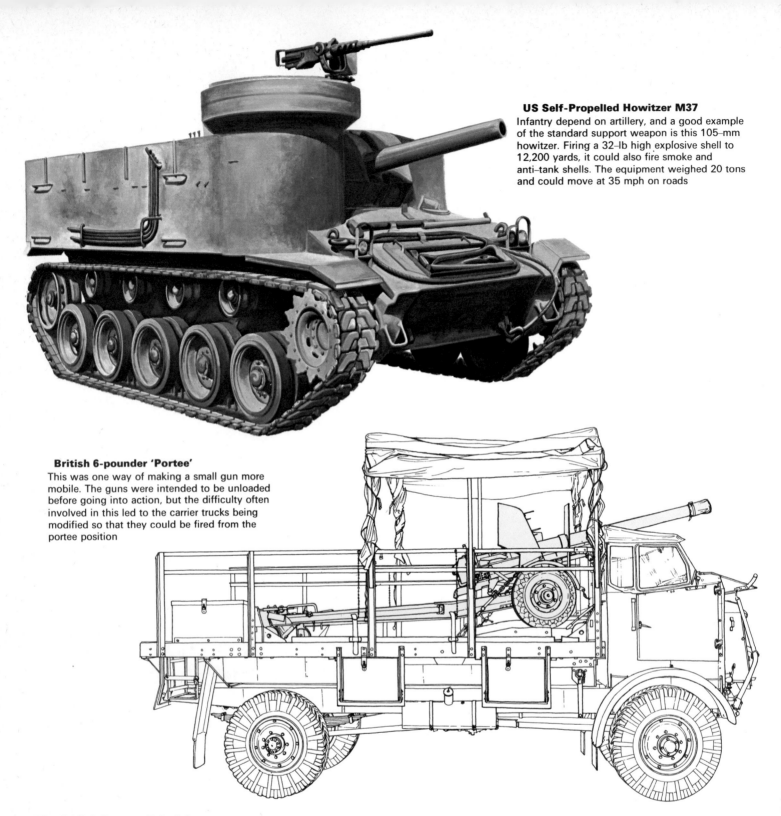

US Self-Propelled Howitzer M37
Infantry depend on artillery, and a good example of the standard support weapon is this 105-mm howitzer. Firing a 32-lb high explosive shell to 12,200 yards, it could also fire smoke and anti-tank shells. The equipment weighed 20 tons and could move at 35 mph on roads

British 6-pounder 'Portee'
This was one way of making a small gun more mobile. The guns were intended to be unloaded before going into action, but the difficulty often involved in this led to the carrier trucks being modified so that they could be fired from the portee position

The Gerlich Squeeze Principle
The squeeze principle involves using a barrel tapering towards the muzzle. As the shell travels along the barrel, the supporting studs are forced into their recesses and the steel sealing band is compressed. Since the shell is smaller at the muzzle than at the breech, the force of the explosion is more concentrated, and the tungsten-cored shot attains a higher velocity

THE SQUEEZE PRINCIPLE

US M8 Self-propelled Howitzer
The M2 75-mm howitzer mounted on a tracked chassis, and given a remarkable 1½ in of armour protection. In general, self-propelled guns did away with such heavy armour; after all, their function was to give guns flexibility and mobility, not to become armoured fighting vehicles

ban finished it. Tungsten was necessary for the ammunition of these taper-bore guns since the tungsten core could be fitted with mild steel skirts which deformed as the barrel tapered; and of course, it was of little use to shoot steel projectiles at velocities over 4,000 feet per second knowing they would shatter on the target.

The most famous German anti-tank gun was, of course, the Eighty-eight. The original '88' was an anti-aircraft gun which was found to have a useful performance against tanks. As a result of this Krupp were asked to produce a purely anti-tank Eighty-eight, which they did in 1943. This was known as the PAK 43 and, instead of being the usual two-wheeled pattern, was carried on a four-wheeled mounting. In action the wheels were removed and the gun sat on a girder platform so that it could swing completely round in a circle and thus fire in any direction. Unfortunately this platform was difficult to make, and since things on the Russian Front were looking bad, the barrel was mounted on to a normal two-wheel pattern put together from stocks of spare parts for other guns. The result looked odd and was heavy and cumbersome to move, but it had an excellent performance. It was claimed by some troops on the Russian Front that they habitually stopped T-34 tanks at ranges up to 3,000 yards; even allowing for exaggeration in telling the tale, it was still a remarkable performance.

Towards the end of the war came the most potent anti-tank gun ever developed, a 5·9-in monster firing a 60-lb shell. Its performance was twice that of the Eighty-eight, but fortunately for the Allies only a handful were made before the war ended.

ANTI-TANK GUNS

NAME	CALIBRE	WEIGHT IN ACTION (lb)	SHELL WEIGHT (lb)	VELOCITY MUZZLE f.p.s.	PENETRATION	REMARKS
Germany						
S Pz B 41	28 mm	500	0.26†	4,595	94 mm @ 100 m	Taper bore
PAK 36	37 mm	950	1.5	2,500	65 mm @ 100 m	
PAK 41	42 mm	1,400	0.75†	4,150	120 mm @ 100 m	Taper bore
PAK 38	50 mm	2,175	1.81†	3,925	120 mm @ 500 m	
PAK 40	75 mm	3,150	7.04†	3,250	154 mm @ 500 m	
PAK 41	75 mm	2,990	5.72†	3,700	209 mm @ 500 m	Taper bore
PAK 36(r)	76.2 mm	3,800	8.9†	3,250	158 mm @ 500 m	Ex-Russian field gun
PAW 600	81 mm	1,325	5.9*	1,700	140 mm	Smoothbore
PAK 43	88 mm	8,150	16.0†	3,700	274 mm @ 500 m	PAK 43/41 similar
PAK 44	128 mm	22,500	6.20	3,200	230 mm @ 1,000 m	Very few made
Great Britain						
Six-pounder	2.244 in	2,525	6.25	2,693	80 mm @ 500 m	
United States						
37 mm M3	37 mm	990	1.92	2,900	60 mm @ 500 m	
57 mm M1	2.244 in	2,810	7.3	2,700	80 mm @500 m	
Italy						
Model 37	47 mm	660	3.25	2,100	43 mm @ 500 m	Czechoslovakian design
Japan						
Model 94	37 mm	815	1.42	2,300	32 mm @ 500 m	
Model 1	47 mm	1,660	3.08	2,700	70 mm @ 500 m	
Soviet Russia						
Model 36	37 mm	896	1.8	2,500	51 mm @ 500 m	Purchased from Germany
Model 42	45 mm	1,257	1.9†	3,500	54 mm @ 500 m	
Model 43	57 mm	2,550	3.8†	4,200	100 mm @ 500 m	
Model 44	100 mm	7,620	20.0†	3,600	181 mm @ 500 m	

General Note: † signifies tunsten-cored ammunition * signifies hollow charge shell

INFANTRY GUNS

NAME	CALIBRE	WEIGHT IN ACTION (lb)	SHELL WEIGHT (lb)	VELOCITY MUZZLE f.p.s.	RANGE MAXIMUM (yards)	REMARKS
Germany						
1e 1G 18	75 mm	880	13.2	690	3,775	
IG 37	75 mm	1,124	12.2	918	5,625	IG 42 had same performance
sIG 33	150 mm	3,750	84.0	790	5,150	
Geb G 36	75 mm	1,654	12.75	1,560	10,000	Mountain gun; 8 mule loads
Geb G 43	75 mm	1,282	12.75	1,575	10,375	Mountain gun; 7 mule loads
Geb H40	105 mm	3,660	32.0	1,870	18,300	Mountain how; 5 mule loads
Inf K 290(r)	76.2 mm	1,720	13.75	1,270	9,350	Ex-Russian
Great Britain						
95 mm inf How	3.7 in	2,100	25.0	1,083	10,000	Never issued
United States						
105 How M3	105 mm	955	33.0	1,020	7,250	
Japan						
Inf How M92	70 mm	468	8.3	650	3,050	
Regtl Type 41	75 mm	2,158	12.5	1,672	11,900	

Although these ever-larger guns gave ever better performance, they also got heavier and heavier, until they got past the stage where the crew could manhandle them in battle. Where the guns were handled by artillery this was less important, for they usually managed to provide more men and have a tractor of some sort handy, but where it was an infantry weapon, operating with the absolute minimum crew, the escalating size became a severe problem. The Germans had, however, produced the answer a long time before, and suddenly everybody began to move in the direction that they had indicated – towards the recoilless gun.

The heaviest part of a normal gun is not usually the actual gun, but the mounting which goes underneath it. This has to be heavy and strong to withstand the shock of recoil, and a lot of the weight comes from the complicated hydraulic system used to check recoil. If recoil could be done away with, it would be possible to dispense entirely with this weighty system and make the rest of the mounting a lot lighter. In fact it had been done in the First World War in a design for an aircraft gun in which the chamber was between two barrels, and firing the cartridge shot a shell out of the front and a balance weight out of the back. Since both articles weighed the same amount and were travelling at the same speed, they recoiled equally and cancelled each other out. But although this weapon worked, it was scarcely practical on the ground.

Surprises in Crete
When the German Air Force began putting their airborne parachute troops together they cast around for some lightweight form of artillery with which to arm them. After some trials, the Rheinmettal company produced a recoilless gun working on new principles: instead of shooting a counterweight out of the back of the gun at the same speed as the shell, they shot a stream of gas out at very much higher speed. The weight and speed of the gas came to the same product as the weight and speed of the shell, so the gun balanced without recoil. To get the stream of gas, a special cartridge was built which had the base made of thick plastic material, and the gun breech had a hole in it leading to a rearward-pointing jet pipe. When the cartridge fired, the pressure built up and started the shell moving and then blew out the plastic disc and allowed gas from the cartridge to blow out through the jet.

Obviously, since some of the charge was being used to produce gas, it meant that the shell would not go so far as it would have done with a conventional gun of the same calibre. Moreover it was necessary to make a larger charge, since about four-fifths of the cartridge actually went back in flame and gas, and only one-fifth was pushing the shell. But the idea worked, and the resulting 'Light Gun' (so called as a security cover) had quite enough performance for the Parachute Infantry. It was designed so that it could be taken to pieces and carried either by individual jumpers or dropped in containers; once on the ground it could be quickly put together on a simple tripod mount with small wheels. The first 75-mm model was so successful that 105-mm models were produced next, and all these guns were revealed when the parachute troops attacked Crete in 1941.

Until then, although the Allies knew about the original counterweight type of

RECOILLESS GUNS
MORE POWER LESS KICK

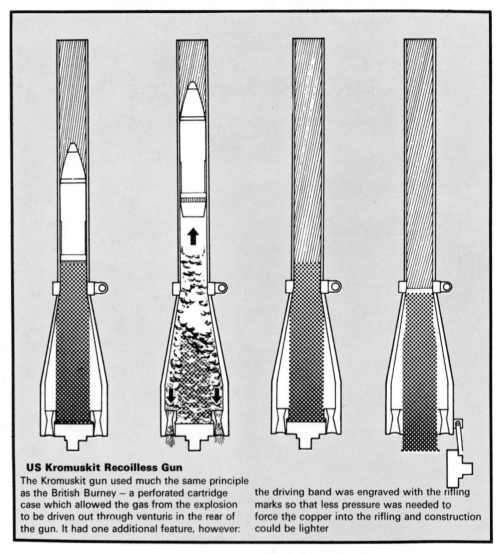

US Kromuskit Recoilless Gun
The Kromuskit gun used much the same principle as the British Burney – a perforated cartridge case which allowed the gas from the explosion to be driven out through venturis in the rear of the gun. It had one additional feature, however: the driving band was engraved with the rifling marks so that less pressure was needed to force the copper into the rifling and construction could be lighter

gun, they had not given much thought to recoilless guns, and the performance of these weapons in Crete gave them a surprise. They therefore began work on their own designs. The British models were generally called 'Burney' guns after their inventor, and they operated on a different system. The breech of the gun was quite normal, but the gun chamber was perforated and surrounded by a second chamber which led back to jets alongside the breech. The cartridge case was pierced by a number of large holes in its side, closed with thin metal sheet. When the charge fired, the metal sheet held up long enough for the shell to start moving, then blew out allowing gas to pass into the outer chamber and then through the rear jets. The end result was the same, though the method of reaching it slightly different.

The British weapons were much longer than the German, but were intended solely for direct shooting at short range, the

principal intention being to use them in the jungles of Burma where such a weapon would allow heavy fire power to be carried through the jungle to otherwise inaccessible places. A 3·45-in gun was first developed; this acquired the nickname of the "twenty-five pounder shoulder gun" as it was the same calibre as the 25-pounder field gun, though firing a much lighter shell. A 3·7-in on a lightweight wheeled carriage was also designed, and a large 95-mm howitzer intended as an artillery weapon for airborne troops. But before any of these could be produced in quantity the war ended, and only a few were issued to selected infantry units so that they could evaluate them and come to some decisions as to the future use of this type of gun.

The American development of recoilless guns was on similar lines to the British and they developed a 57-mm and a 75-mm model, choosing these calibres because they could adapt ammunition already in production for other guns and thus speed things up by having something less to worry about. These also use perforated cartridge cases, and the 57-mm was fired from the shoulder or from a tripod while the 75-mm had a tripod mounting only. Numbers of these were produced in time to be flown to the South Pacific with both high explosive and anti-tank ammunition to be tried out in combat. They were highly successful, and were to remain in service for many years before being replaced by improved designs.

The backlash

But the recoilless gun had two drawbacks: one was the large amount of propellant powder it consumed, and the other was the huge flash and blast which came from the back end when it was fired, and which made an area up to one hundred yards behind the gun very dangerous to be in. No army liked this latter problem, though they have learned to live with it, but it was the former problem which worried the German Army because, by 1944, there was a growing shortage of the chemicals needed for the production of powder, and propellant was practically rationed. It became imperative to try and find some system which would still give the infantry the lightweight gun

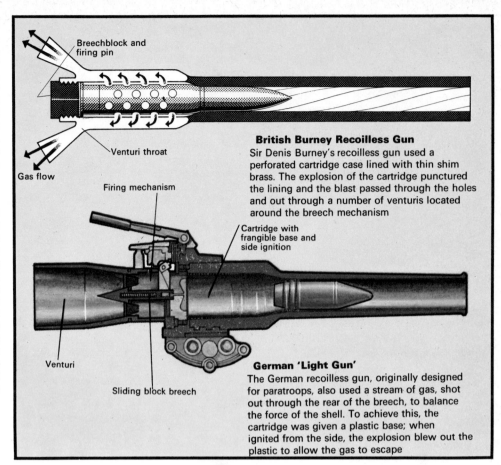

British Burney Recoilless Gun
Sir Denis Burney's recoilless gun used a perforated cartridge case lined with thin shim brass. The explosion of the cartridge punctured the lining and the blast passed through the holes and out through a number of venturis located around the breech mechanism

German 'Light Gun'
The German recoilless gun, originally designed for paratroops, also used a stream of gas, shot out through the rear of the breech, to balance the force of the shell. To achieve this, the cartridge was given a plastic base; when ignited from the side, the explosion blew out the plastic to allow the gas to escape

they demanded, still give a respectable performance against tanks, but do it more economically.

The answer to this seemingly impossible compromise was the PAW (*Panzer Abwehr Werfer* – Tank defence discharger) 600 (see page 55). It was called 'Werfer' rather than 'Kanone' because it had a smooth-bore barrel and fired a fin-stabilised hollow charge bomb of 81-mm calibre. The unusual thing about it was the fitting of a steel plate pierced with jets in front of the cartridge case; when the charge was fired it generated high pressure inside the case which then leaked out into the barrel to propel the bomb. This meant that only the area of the gun chamber had to be strong and heavy, the rest of the barrel could be much thinner as it had less pressure to withstand. This 'High-and-Low Pressure System' was one of the few really new ballistic discoveries to come out of the Second World War, but it has seen relatively little employment since.

As we said in the beginning, the object of the infantry is to defeat the enemy in detail and occupy the ground. To reach this result needs more than just the weapons discussed here, the weapons actually placed in the hands of the infantry for their use. It must be borne in mind that the remaining fighting and supply services – artillery, armour, even air forces and navies – exist solely in order to get the infantryman on to that piece of ground. As a result, every weapon, from the pistol to the dive-bomber, the sub-machine-gun to the battleship is, in the long run, an infantry-supporting weapon. The infantry form the spearhead; the remaining forces of a country are the handle which guides it and gives it power.

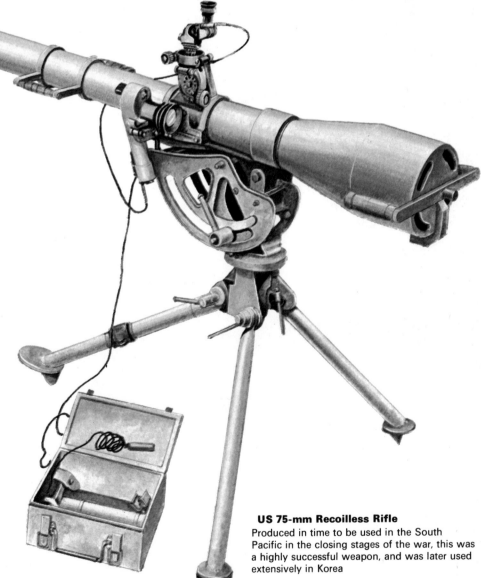

US 75-mm Recoilless Rifle
Produced in time to be used in the South Pacific in the closing stages of the war, this was a highly successful weapon, and was later used extensively in Korea

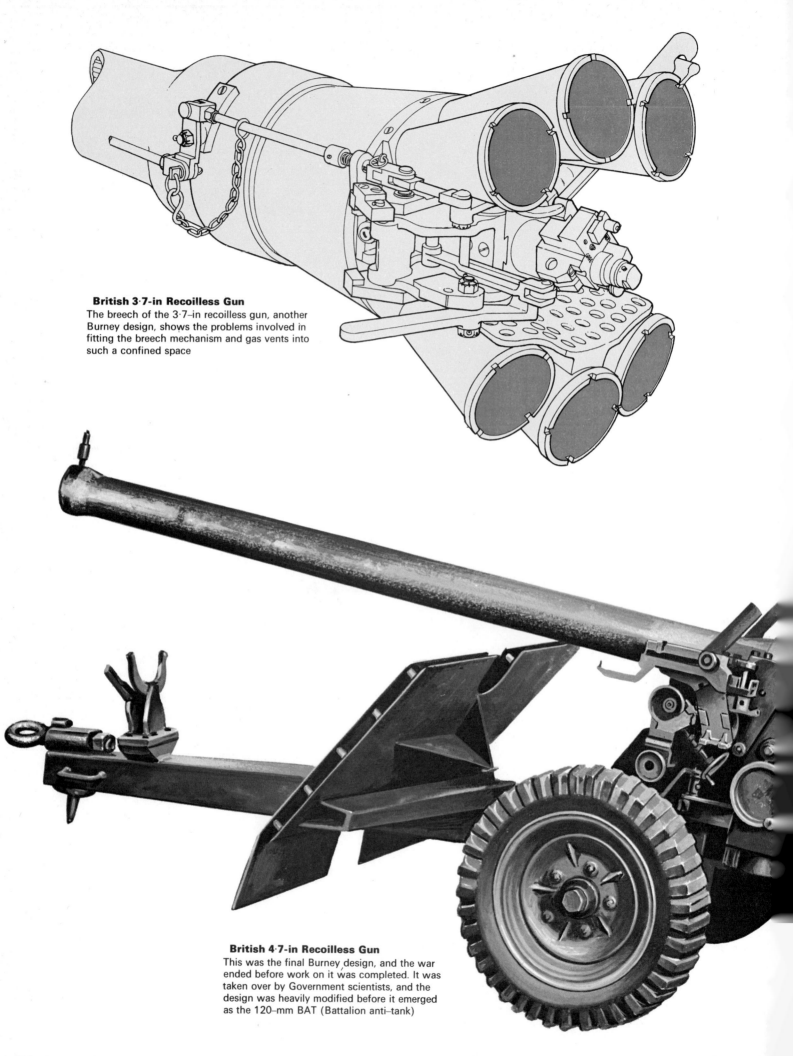

British 3·7-in Recoilless Gun
The breech of the 3·7–in recoilless gun, another Burney design, shows the problems involved in fitting the breech mechanism and gas vents into such a confined space

British 4·7-in Recoilless Gun
This was the final Burney design, and the war ended before work on it was completed. It was taken over by Government scientists, and the design was heavily modified before it emerged as the 120–mm BAT (Battalion anti–tank)

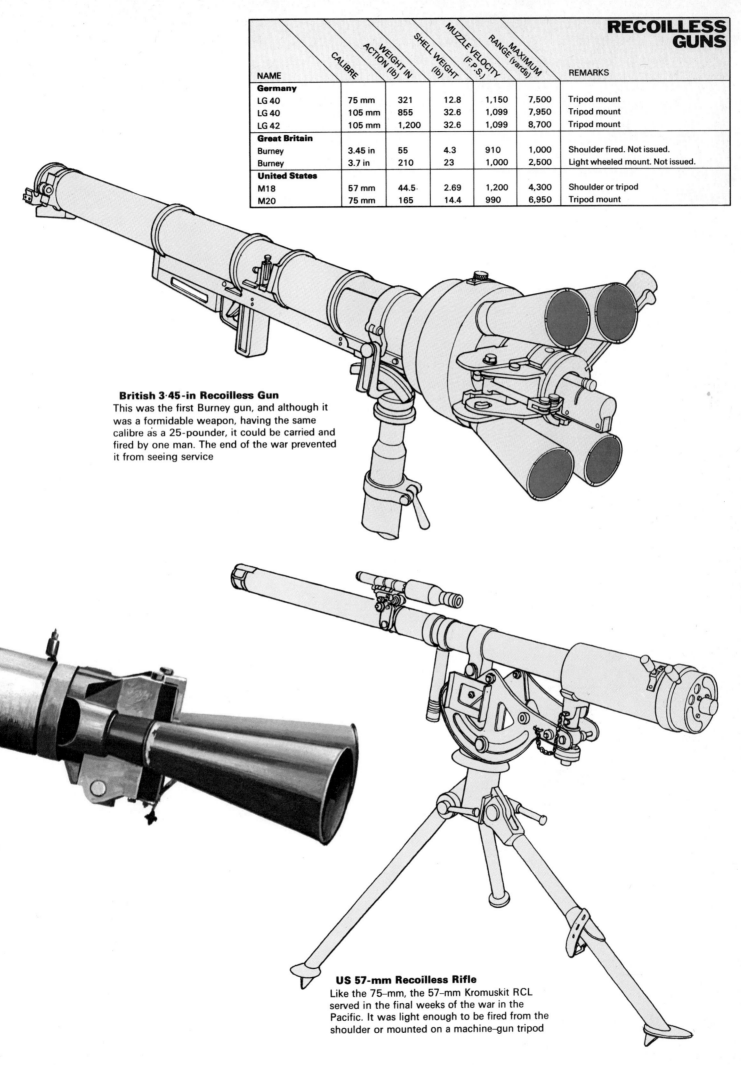

NAME	CALIBRE	WEIGHT IN ACTION (lb)	SHELL WEIGHT (lb)	MUZZLE VELOCITY (F.P.S.)	MAXIMUM RANGE (yards)	REMARKS
Germany						
LG 40	75 mm	321	12.8	1,150	7,500	Tripod mount
LG 40	105 mm	855	32.6	1,099	7,950	Tripod mount
LG 42	105 mm	1,200	32.6	1,099	8,700	Tripod mount
Great Britain						
Burney	3.45 in	55	4.3	910	1,000	Shoulder fired. Not issued.
Burney	3.7 in	210	23	1,000	2,500	Light wheeled mount. Not issued.
United States						
M18	57 mm	44.5	2.69	1,200	4,300	Shoulder or tripod
M20	75 mm	165	14.4	990	6,950	Tripod mount

British 3·45-in Recoilless Gun
This was the first Burney gun, and although it was a formidable weapon, having the same calibre as a 25-pounder, it could be carried and fired by one man. The end of the war prevented it from seeing service

US 57-mm Recoilless Rifle
Like the 75–mm, the 57–mm Kromuskit RCL served in the final weeks of the war in the Pacific. It was light enough to be fired from the shoulder or mounted on a machine–gun tripod

American troops with an M7 Howitzer Motor Carriage, more familiarly known as Priest, which was designed as an infantry support weapon and used in the North African campaign

GERMAN TANKS
Peter Chamberlain and Chris Ellis

PETER CHAMBERLAIN and **CHRIS ELLIS** have collaborated on a large number of publications and are widely recognised as leading experts in the weapon and fighting vehicle field. Peter Chamberlain is a consultant to the Imperial War Museum.

The Germans did not invent the tank, but they learnt the right lessons from the early British and French machines that broke through their trenches in 1918. Although forbidden tanks by the peace treaty, twenty years later a potent and subtle array of fighting machines burst upon an astonished and helpless Europe.

In this section the German tanks, from *Panzerkampfwagen I* to the King Tiger, are described and illustrated and the story of the initial success and eventual downfall of the German panzers is vividly recreated.

CONTENTS

THE FIRST PANZERS
THE LESSONS OF DEFEAT

Tanks were instrumental in the final defeat of the German Imperial Army. Forbidden heavy weapons by the Treaty of Versailles, a group of young German officers were already planning a new kind of warfare. The cardboard chariots of the *Reichswehr* became the hard steel of Nazi Germany's first Panzer Divisions

The story of the German panzers really starts in 1926, well before the rise of the National Socialist party as a major force in Germany, and about ten years before the massive 'Guns before Butter' re-armament of Germany under the Nazi regime gained real momentum in 1936. Back in 1926 the German armed forces were still very much affected by the savage restrictions of the Treaty of Versailles, signed in 1919 after the First World War. Among the conditions of the treaty were a limitation of the army to 100,000 men without tanks of any kind, and an allowance of a small number of armoured cars solely for border patrol work.

General von Seeckt, C-in-C of the *Reichswehr* (German Army) until 1926, was a far-sighted man who had learned the

lessons of the First World War – that mobility and flexibility, as demonstrated by the early British tanks in 1917/18, pointed the way to the warfare of the future. Within the limitations imposed on him, von Seeckt made good use of his available manpower, with a big concentration on hard training and war exercises. The future exponents of armoured warfare like Guderian and Rommel had their formative training during von Seeckt's years, while the victors of the First World War ran down their armies and retired to peacetime soldiering.

The first tanks

National pride had to be restored in those uneasy years in Germany, and the *Reichswehr* played their part by training hard. Von Seeckt's swan song was to initiate a strictly unofficial programme of experimental tank construction. The first tank to appear was made by the firm of Rheinmetall-Borsig, in mild steel and described as a *Grosstraktor* (big tractor) – a thin disguise to circumvent the terms of the Treaty. The vehicle was similar in layout to the contemporary British Medium Mk II tank, weighed around 20 tons and had a 75-mm gun. A year later, in 1927, the firm produced a smaller version of the same thing, called a *Leichtetraktor* (light tractor), this time with a 37-mm gun. A second vehicle of the same type was fitted with a 75-mm gun, and in 1929 there was an improved version, weighing around 10 tons and with the same 37-mm gun and turret as the 1927 model.

These early vehicles gave design experience and were tested secretly in western Russia (with Russian co-operation) at a Soviet tank school, so that strictly speaking they were not contravening the Versailles Treaty terms. Meanwhile tactical experience was gained on *Reichswehr* exercises by the use of dummy tanks, and Major Heinz Guderian, who was a staff tactical instructor at the *Reichswehr's* motor transport school, is credited with the idea. He had dummy tank outlines fabricated from sheet metal and wood, and attached to the BMW Dixi light car which was then a standard type of staff and liaison vehicle. These were supplemented by unpowered 'soapbox' dummies on wheels which were pushed around by men inside them, rather like carnival floats.

Keen disciple

Guderian was a keen disciple of the writings of Captain B H Liddell-Hart, the distinguished British military commentator of the 1920s and 1930s, and of Major-General J F Fuller and other leading British tank warfare theorists who had had experience in the Royal Tank Corps in 1917/18. While in Britain these theorists had only limited success in seeing their ideas realised – for a variety of reasons which included a conservative outlook by the General Staff and a severely limited defence budget – their views were keenly studied by Guderian. In 1931 Guderian, by then a Lieutenant-Colonel, became Chief of Staff to General Lutz, Inspector of Motorised Troops. Lutz and Guderian were convinced that the future tactical development of tank forces should involve the formation of Armoured (Panzer) Divisions. As far as tanks were concerned they postulated the development of two types, a medium tank in the 20-ton class, armed with a 75-mm gun in the turret and two machine-guns, for tank-to-tank battles, and a lighter vehicle intended primarily for reconnaissance, armed with a 50-mm armour-piercing gun and two machine-guns. These conclusions were accepted by the General Staff, except for the gun on the lighter tank which was changed to 37-mm, because this was the size of gun with which the German infantry were already being equipped for the anti-tank role. The Chief of the Ordnance Office and

The first German tank, an A7V of 1918 with its crew. Inset: A dummy tank mounted on a car chassis, used for practising the Blitzkrieg techniques of the future

Peter Chamberlain

the Inspector of Artillery favoured the 37-mm gun and the desirability of standardisation enhanced the argument.

By this time further Rheinmetall prototypes had been built, notably a large multi-turreted vehicle known as NbFz A (*Neubaufahrzeug* [new model] A). This was clearly influenced by the British 'Independent' (a very large one-off vehicle) and the Medium Mk III, the NbFz A being quite close in size and layout to the latter. However, none of the experimental models was considered completely suitable to fill the two roles in the proposed Panzer Divisions. New models were necessary, and these eventually emerged as the PzKpfw III and IV, described later.

Until they were ready, which would take several years, a training tank was needed. In the interests of speedy production and for AFV experience a light tank was the obvious answer. Light tanks were cheap and could be built easily, a fact already realised and one which influenced tank development in most other countries. Accordingly the Germans bought a British Carden-Loyd Mark VI chassis which was sold commercially to several powers in the 1930s. The German purchase was announced as being for use as a carrier for a 20-mm anti-aircraft gun. This modest acquisition created no undue alarm in other countries.

Keen competition

In pursuance of the new tank policy, the German Army Weapons Branch (*Heereswaffenamt*) issued a requirement for a tank of approximately five tons weight with two machine-guns mounted in a turret with all-round traverse and protected by armour immune to attack by small arms fire. Five firms, Rheinmetall-Borsig, Daimler-Benz, MAN, Henschel and Krupp were invited to submit their proposals for a design to meet the specification. Germany was lucky to have so many firms with the necessary engineering experience and design staff. But in these relatively lean economic times there was keen competition.

After close scrutiny, LKA1, a design submitted by Krupp and based on the imported Carden-Loyd Mk VI chassis was selected, and Krupp were made responsible for the development of the chassis, while Daimler-Benz were to construct the turret and the hull. To ensure secrecy and to hide the project from the outside world, the machine was given the totally fictitious code name of *Landwirtschaftlicher Schlepper* (agricultural tractor), abbreviated as La S. The drawings and design were finished in December 1933, by which time Hitler and the Nazi Party had assumed power and were about to overturn openly all the terms of the Versailles Treaty. Henschel were given orders to construct three La S prototypes.

LKA 1 (above left)
Krupp's prototype for a five-ton light tank was designed to meet the needs of the rearming *Reichswehr* in 1933. Accepted for trials, this vehicle was the forerunner of the PzKpfw I

PzKpfw I Ausf A
The Panzer I A was tested during the Spanish Civil War, and saw service during Hitler's early Blitzkriegs
Engine: 60 hp Krupp M105 *Weight:* 5·4 tons
Speed: 25 mph *Crew:* 2 *Armour:* 13 mm max
Armament: 2 × 7·92-mm mg

The first of the vehicles ran in February 1934, an extremely short period even allowing for the fact that the tank was a very simple one with a derived chassis.

Full-scale production began in 1934 with an order for 150 machines given to Henschel under the designation 1A La S Krupp, and this was followed by another version known as 1B La S. About 1800 vehicles in all were built and of these roughly 1500 were the B model, longer and with a more powerful engine.

The La S designation was retained until 1938 when the Germans introduced a new standard code for tank designation. Experimental machines were given an identifying serial number – 700 or 2000 for example. The first number or pair of numbers indicated the weight class of the vehicle, eg 7-ton or 20-ton or 30-ton. The last two numbers were used to indicate the number of the prototype. A prefix VK indicated that the vehicle was fully tracked and where a multiple order had been given to competing firms, the firm's initial letters followed in a bracket after the serial number, eg VK

2001 (H) and VK 2002 (DB) would indicate tanks in the 20-ton class under experimental construction by Henschel and Daimler-Benz respectively.

Designating the panzers
When a tank had been accepted for service it became known by its class name followed by the model number. *Panzerkampfwagen*, abbreviated into PzKpfw or PzKw, was used. For example, PzKpfw I Ausf C indicated Model C of the first tank type. On acceptance into the service a tank also received an Ordnance Dept inventory number, eg *Sonderkraftfahrzeug* (special motor vehicle) 101, abbreviated to SdKfz 101.

The PzKpfw I Ausf A (SdKfz 101) was a straightforward machine with no unexpected characteristics or technical devices. Its 3·5 litre air-cooled Krupp M 305 four-cylinder petrol (gasoline) engine developed 57 hp at 2500 rpm and was housed in the engine compartment at the back of the tank together with a large oil cooler. The drive was taken forward to a five-speed sliding pinion gearbox and thence through cross shafts, carrying on each side a clutch and brake steering system, to the front driving sprockets. Several machines were fitted with the Krupp M 601 diesel engine which developed 45 hp at 2200 rpm, but the experiment proved unsuccessful and the Krupp petrol engine was used in all production models of the PzKpfw I Ausf A.

Refining the suspension

Krupp's original prototype had four suspension wheels each side with a rear idler touching the ground; movement of the wheels was controlled by coil springs and three return rollers were mounted on the hull. The layout was changed in the production models, which had an external girder covering the two rear suspension wheels. The ends of this girder were connected to the axle of the second suspension wheel and to the rear idler wheel axle by forked links carrying elliptical springs whose tips rested on the axles of the third and fourth suspension wheels. Movement of the leading suspension wheel was controlled by coil springs and three return rollers were mounted on the hull. The suspension was satisfactory at low speeds but pitched badly when the tank was moving faster, probably accentuated by the rear idler wheel which remained in contact with the ground. The overall length was, in round figures, 14 ft, the width 6 ft 10 in and the height 5 ft 8 in. This was indeed a small tank. The vehicle, however, was very versatile and with its two 7·92-mm machine-guns could give a good account of itself. The armament consisted of MG 34s, although it is quite possible that tanks of early manufacture were armed with old MG 13s, taking as evidence some pre-war photographs.

Evolved from the Ausf A, the PzKpfw Ausf B (SdKfz 101) prototype appeared in 1935. Superficially the appearance was the same, but there were considerable differences in detail. A more powerful engine, a water-cooled Maybach NL 38 TR, was installed and this required a longer and higher engine compartment. The tank was lengthened to provide the extra room and the sides of the superstructure were raised. The engine developed 100 hp at 3000 rpm and this extra power raised the speed of the tank from 23 to 25 mph

The armament remained the same as in the Ausf A and despite the many disadvantages the two-man crew was retained. Armour thickness remained at 13 mm and the turret showed no change except that an internal mantlet was used, a design feature that appeared on all German tanks until the introduction of the 50-mm gun on the PzKpfw III. In 1940/41 a redesigned transmission incorporating a five-speed gearbox and a better final drive reduction gear was substituted for the type in the Ausf A. The nose of the tank was redesigned to provide room for the final reduction gear which resulted in a complicated design pattern for casting. In both the Ausf A and B no special provision was made for observation by the commander who was, of course, the gunner as well.

To allow for the extra room needed by the bigger engine in the Ausf B the suspension was modified and an extra wheel, making five in all, was added on each side. The rear idler wheel was raised clear of the ground which greatly improved the ride, and the additional suspension wheel meant that the same amount of track as before was in contact with the ground. Four return rollers were used on the hull in place of the three of the earlier model.

The turret was set over on the right hand side of the superstructure; the driver sat on

Imperial War Museum

General Heinz Guderian, one of the early prophets of armoured warfare, and one of the outstanding tank leaders of the Second World War

the left hand side of the hull. This gave a wider hull with the tracks a little further apart and therefore the tank possessed rather more lateral stability than most contemporary light tanks.

VK 1801

1940 prototype for the PzKpfw I *neuer Art* (new model) featured heavier armour as an infantry support vehicle. Thirty of the type were built
Engine: 150 hp Maybach HL 45 *Weight:* 18·5 tons *Speed:* 15 mph *Crew:* 2 *Armour:* 82 mm max *Armament:* 2 × 7·92-mm mg (never fitted)

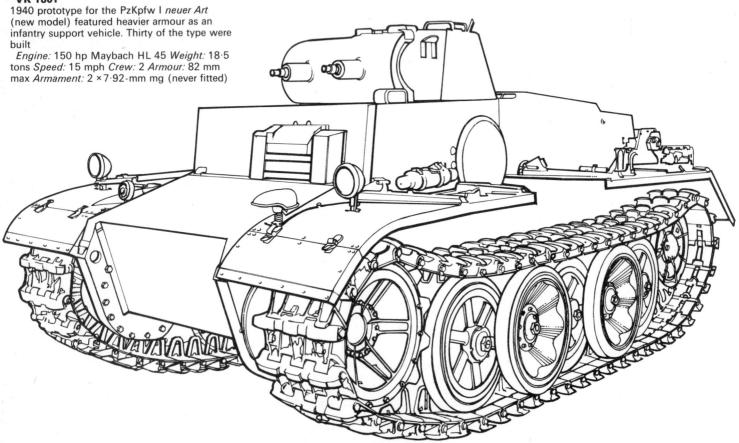

TRIAL BY COMBAT

Columns of tanks looked impressive on the parade-ground, but a propaganda show of strength was a long way from the realities of war. The fledgling weapon needed the chance to test tactics and hardware. The Spanish Civil War presented that opportunity. Above: PzKpfw I Ausf A in Catalonia, 1938

Both models of the Panzer I were 'blooded' in the Spanish Civil War (1936–1939). In this war, the Germans had seen the opportunity to test many of their new weapons. Here were the real battles and along with the Condor Legion, the Panzer I was put to the test. But from observations in Spain it became clear that tanks with far heavier weapons, better armour and longer endurance would be necessary for a future war of mobility.

Guderian was largely responsible for planning the whole complementary series of tanks to equip the new Panzer Division. He postulated a light reconnaissance tank, and a major medium battle tank (ultimately known as the PzKpfw III) which would be fitted with an armour-piercing gun as well as hull and turret machine-guns. The other major type (later to be known as the PzKpfw IV) would be a support vehicle with a 75-mm low velocity gun. Development of the Panzer III and IV proceeded more slowly than forecast. To cover the delay in getting these tanks into the hands of troops it was decided to build a tank in the 10-ton class as a successor to the Panzer I. The new tank was intended to be a training machine stop-gap; it was used in the Spanish Civil War and in the opening stages of the Second World War, being important in the 1939/40 period.

A specification for the new design was issued in July 1934. Three prototypes were submitted, one of them being Krupp's LKA II which looked quite like their prototype LKA I for the PzKpfw I. Prototypes were also built by Henschel and MAN, both resembling the Krupp design except for radical differences in the suspension;

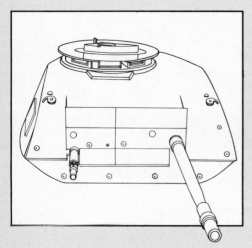

LKA II
Krupp's prototype for the PzKpfw II closely resembled their LKA I. It was rejected in favour of the MAN proposal

PzKpfw II Turret
Retaining the layout of 20-mm cannon and co-axial MG 34 7·92-mm machine-gun, the later model Panzer II turrets were equipped with an all-round periscope cupola

PzKpfw II Ausf A on the Western Front in 1940

with the designation La S 100, the MAN vehicle was selected for production.

Twenty-five tanks were built in 1935 as 1 La S 100 and taken into service as PzKpfw II a1 (SdKfz 121). They weighed 7·2 tons and had a crew of three. They were armed with a 20-mm KwK 30 gun and a 7·92-mm machine-gun mounted coaxially in the turret. The vehicle was powered by a Maybach HL 57 six-cylinder petrol engine developing 130 hp at 2100 rpm. A plate clutch and a six-speed gearbox took the power to a cross shaft; this carried at either end the usual clutch and brake steering mechanism for each track and a driving sprocket. The suspension consisted of six small road wheels in pairs in bogies which were sprung by leaf springs. An outside girder connected the outer ends of the bogie pivot pins, the inboard ends being housed in the hull which

also carried three return rollers. The adjustable rear idler wheel was clear of the ground. The nose plate was a rounded casting, a distinct change from previous German tanks.

These 25 vehicles were followed, also in 1935, by a second batch of 25 PzKpfw II Ausf a2. Externally they were exactly the same as the Ausf a1 but had a better cooling system and more room had been found in the engine compartment. A further batch of 50 machines appeared in 1936 – the PzKpfw II Ausf a3. Further improvements had been effected in the cooling system and the tracks and suspension had been altered for the better in comparison with the earlier machines.

The 2/La S 100, or PzKpfw II Ausf b, appeared in 1936. One hundred machines were built with front armour increased to

30 mm and an all-up weight of 7·9 tons. The armament remained unchanged, but a Maybach HL 62 petrol engine was fitted which developed 140 hp. Externally there was little change. These tanks had a new reduction gear in the cross drive and a new type of driving sprocket which incorporated a geared final drive. These sprockets, together with new pattern track plates that appeared with the machine, were adopted as standard fittings for all subsequent PzKpfw IIs.

In 1937 the third version, 3/La S 100 or PzKpfw II Ausf c, appeared. Slight alterations were made to the turret, which still housed the same 20-mm Solothurn armament, and the driver's front plate extended right across the tank. In other models the superstructure sides tapered slightly towards the front and the driver's plate was narrower than the width of the hull. A

PzKpfw II Ausf B or C

Ausf B and C of the Panzer II were virtually unchanged from the Ausf A, except that the commander was provided with a cupola instead of a periscope. Panzer IIs of all three marks suffered a terrible mauling in Russia in 1941
Engine: 140 hp Maybach HL TR *Weight:* 9·5 tons *Speed:* 25 mph *Crew:* 3 *Armour:* 30 mm max *Armament:* 20-mm cannon; 1 mg

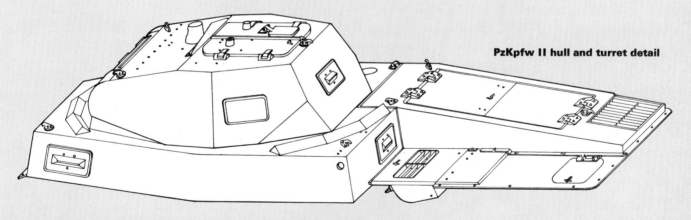

PzKpfw II hull and turret detail

radical change had been effected in the suspension. The outside girder and the small bogies disappeared and were replaced by five medium-sized suspension wheels each individually controlled by quarter elliptic springs. Four return rollers were used on the hull. This suspension was used for all following models of PzKpfw II.

The early versions of the PzKpfw II were tried out under operational conditions in the Spanish Civil War. The performance of the vehicles showed that though they were only intended as training machines they were well made and could play their part in armoured warfare provided that the opposition was not too strong, but even against the primitive anti-tank guns encountered in Spain the Panzer IIs were virtually outclassed. It would appear that the vulnerability of the German tanks during the war

in Spain was somewhat overlooked by the German General Staff who went ahead and approved the continued large-scale production of the PzKpfw II which by 1938/39 was on the way to becoming obsolescent. Even though its armour was increased it was barely proof against the current anti-tank guns in Europe, such as the British 2-pounder. While the armament was adequate for taking on its own kind, the PzKpfw II was too weakly armed and armoured to deal with heavier hostile tanks and could not fire HE (High Explosive) shot. However, despite these shortcomings – which were well known to all German Panzer officers – production continued well into 1942.

The PzKpfw II Ausf A, B and C appeared between 1937 and 1940. There was little difference between these models. The 1937 tanks which marked the start of real mass

production showed little change from PzKpfw II Ausf c. To improve protection the nose plate was changed and became angular and of welded construction instead of being round in shape and cast in construction. The gun mantlets were very slightly changed with flanges at the top and bottom of the internal moving shield, to reduce shot splash. The turret was otherwise unchanged except that provision was now made for a commander's periscope in the Model A and a cupola in Model B and subsequent models.

The German Army had 955 PzKpfw IIs for the attack on France in May 1940, and 1067 when the Russian campaign began in 1941. By April 1942 this had been reduced to 866 largely because of the vulnerability of these small tanks against superior Russian machines.

Notwithstanding the considerable effort made in the late 1930s to equip the Panzer Divisions with effective vehicles, the outbreak of war in September 1939 found the Germans woefully short of their intended AFV (Armoured Fighting Vehicle) establishment. Popular myth – which was to a great extent created by clever German propaganda – has left to this day the legend that the Panzer Divisions which stormed into Poland in 1939, and France and Flanders in 1940, were a vast and unstoppable armoured force. But the truth is that it was only the acquisition of the entire Czechoslovak arms industry after the Munich Agreement of 1938 which gave the German forces sufficient tanks with which to put an adequate Panzer Army into the field.

Armaments firms in Czechoslovakia, prior to the occupation by Germany in March 1939, were concerned with the design, development and production of tanks and other armoured fighting vehicles – both for use by the Czech Army and for commercial sale to foreign armies. The two main tank models were the Skoda LT-35 and the CKD (*Cesko-moravska Kolben Danek*) TNHP, which the Germans took into service as the PzKpfw 35(t) and the PzKpfw 38(t) respectively, the t being an abbreviation of *tscheche*, the German for Czech.

In 1933 the CKD firm of Prague began the design of a new light tank series intended for export. This model received the factory designation LT (Light Tank) L and was subsequently called the TNHB. For export purposes it was often referred to as the LT-34.

During October 1937 the Czech Defence Department formed a tank evaluation committee to conduct thorough testing of all available Czech tank designs. A new tank testing centre was established outside the factory during January 1938. Several factories submitted vehicles for tests apart from CKD, among these being the famous Skoda firm and the lesser known Adamov firm. By this time the CKD LTL-H series had resulted in the LTL-P (TNHS) model with improved armament and armour. Results of the trials showed the TNHS to be the most exceptional model of those submitted and after a gruelling 3000-mile test, some 1000 miles of which were across country, the tank showed virtually no mechanical defects. Throughout its life this tank chassis earned great respect for its reliability and durability. The maintenance and servicing needed were found to be minimal and could be carried out in the field.

Following a report on these tests the

The Czech tanks: a great arms industry falls into Hitler's hands

Czech Defence Department specified that the TNHS should enter production and become the standard tank of the Army. Orders were issued for 150 vehicles. After alteration the new tank was given the designation TNHP.

Export models

Prior to, and during the course of, the tests by the Czech Army, CKD had received orders for most of the developed models from foreign governments. These included Sweden, Switzerland, Peru, Latvia, Yugoslavia and Afghanistan. A total of 196 tanks of this series was exported. One vehicle was purchased by the War Mechanisation Board in Great Britain, who tested it extensively.

The original 8-ton tank mounted a 37·2-mm tank gun (Model Skoda A7) L/47·8 and a coaxial 7·92-mm Besa machine-gun in a turret with all-round traverse. The bulge at the rear of the turret was fitted for ammunition stowage. A further 7·92-mm Besa machine-gun (a British-designed weapon) was ball-mounted at the front of the hull. The elevation gear of the 37-mm gun could be locked for firing and it was intended to fire only when the vehicle was stationary. The coaxial 7·92-mm machine-gun could be used independently when required, by virtue of its ball mounting.

The traversing gear, which was fast and light in action, was operated by a wheel on the left hand side of the gunner. It could be thrown out of action and the turret could then be pushed round by the gunner. The turret ring was 47·5 inches in internal diameter and there was no turntable or turret basket. The cupola, which was fixed, was replaceable with 4 periscopes, each with mirrors and protective glasses. The forward machine-gun could be operated if necessary by the driver, via a Bowden cable attached to one of the steering levers. Construction was riveted with the exception of the top of the superstructure which was bolted. Protection was 25 mm basis at the front, 19 mm on the sides, and 15 mm on the rear.

Four rubber-tyred single wheels were provided on each side, each wheel being mounted on a cranked stub-axle and each pair of wheels being controlled by a semi-elliptic spring freely pivoted. There were two return rollers on each side, mounted well forward. Front sprocket drive was employed, the sprocket being mounted high off the ground. The tracks were engaged by twin sprockets and each sprocket was driven through an internally toothed gear by a pinion attached to the cross shaft. The cross shaft carried two steering units comprising epicyclic and clutch elements giving

PzKpfw 35(t)
Power steering and gear change made this an exceptionally easy tank to steer. It could also achieve remarkably high cross-country speeds. Only a few hundred were used by the Wehrmacht. Right: A 35(t) in France, 1940
Engine: 120 hp ohv *Weight:* 9·9 tons
Speed: 25 mph *Crew:* 4 *Armour:* 35 mm max
Armament: 37·2-mm Skoda L/40; 2 × 7·92-mm mg

two steering ratios and was driven by a bevel gear from an epicyclic five-speed gearbox situated between the driver on the right and the machine-gunner on the left. The forward end of the vehicle was therefore quite congested.

The propeller shaft passed through the centre of the fighting compartment, and a six-cylinder, water-cooled, Praga TNHP ohv petrol engine (as used in commercial trucks) was mounted vertically on the centre-line of the vehicle in the rear compartment. A single dry-plate clutch was installed. The engine had a dry sump and was cooled by a finned cylinder incorporating an Auto-Klean filter. Bosch magneto ignition was employed and all the sparking plugs were screened. A 12-volt dynamo was belt-driven from the crankshaft. Cooling was effected by a centrifugal fan driven through a universal joint from the crankshaft. The air was drawn partly through the bulkhead, but mainly through a mushroom type louvre over the engine compartment and thence through a radiator of the continuous fin and tube type. The air was ejected through an opening in the rear top plate protected by armour-steel slats covered by expanded metal. The fuel tanks were situated on either side of the engine compartment and the total capacity was 49 gallons. The floor plates immediately below the fuel tanks were secured by a few small-diameter bolts, the idea being that in the event of an explosion resulting from damage to the fuel tanks the floor plates would be blown out and so reduce the possibility of damage within the vehicle.

The track was made up of cast steel shoes and the pins were each secured by a clip. Detachable spuds were provided to increase track grip in snow and ice. These were located on the extremities of the pins, which projected beyond the faces of the lugs.

Commandeered by the Wehrmacht
Following the German occupation of Czechoslovakia, from 15 March 1939, all tanks in service with the Czech Army – as well as those in production under export contracts – were taken over by the Wehrmacht. The Germans designated the TNHP the PzKpfw 38(t) (37-mm) and continued its production until early 1942, when Czech tank production was suspended. Production of the vehicle under German guidance was also carried out by the Skoda firm. In 1940 the CKD firm was redesignated BMM (Böhmisch-Mährische Maschinenfabrik AG). The Germans initially requested a monthly production figure of 40 vehicles, although this fluctuated greatly according to the availability of materials and manpower. A total of 1168 tanks of this type was built for the Wehrmacht: 275 in 1940, 698 in 1941 and 195 in 1942. In 1940 228 were in service with VII and VIII Panzer Divisions. By 1 July 1941 there were 763, but this figure dropped to 522 by April 1942.

The PzKpfw 38(t) saw service with the Wehrmacht in Poland, France, Yugoslavia, Greece and Russia, and formed a major part of the tank strength of Rommel's VII Panzer Division during its drive across Northern France in the 1940 campaign. During 1940/41 the PzKpfw 38(t) formed 25% of the total German tank force and its importance was therefore considerable, the vehicle being much superior in hitting power to either the PzKpfw I or II. In 1940 a total of 90 vehicles being built for Sweden were also

PzKpfw 38(t)
The 38(t) equipped two Panzer Divisions during the invasion of France. The proven chassis served in numerous roles throughout the war.
Left: A German-built 38(t)
Engine: 125 hp ohv *Weight:* 8·5 tons
Speed: 20 mph *Crew:* 4 *Armour:* 25 mm max
Armament: 37·2-mm L/47·8; 2 × 7·2-mm mg

taken over, and these were fitted with extra radio equipment to become command tanks – designated *Panzerbefehlswagen* 38(t).

Shortly after gaining control of the Czech production facilities, the Germans ordered the manufacturers to increase the frontal armour to 50 mm and that on the sides to 30 mm. As the result, the turret front had a basic thickness of 25 mm with an additional 25-mm plate. The front vertical hull plate was similarly armoured; the side superstructure armour was 30 mm thick. (In some vehicles the Germans substituted the 37-mm KwK L/45 for the original Czech gun.) The tank's weight correspondingly increased to 11 tons. This new model became the TNHP-S (S meaning *Schwer*, or heavy).

The other Czech tank
The other Czechoslovakian tank to see service with the Panzer Divisions in the early days was what the Germans designated PzKpfw 35(t). The design of this vehicle went back to 1934 when the Skoda firm produced a

prototype 10½-ton tank, Model T-11 which was usually referred to as the LTM-35 (S II a).

Particular care was taken in the design of this vehicle to enable it to travel long distances under its own power. In addition to having a high degree of manoeuvrability considerable emphasis was placed on crew comfort and the durability of the power train. The general design requirements of this vehicle were as follows:

1 Rear sprocket drive so as to have the fighting compartment as free as possible from all power train elements
2 Engine design as short as possible so as to have a large fighting compartment
3 A six-stage transmission with an air-operated gear shift
4 Power steering through the use of compressed air so as to permit long driving hours without excessive driver fatigue
5 The suspension to be of such a design as to obtain equal pressures on all bogie wheels
6 The main accessories were to have double

installations so as to ensure a high degree of reliability and performance

Satisfactory results were achieved with the prototype and the vehicle was put into production during 1935.

The vehicle was armoured with plate up to 35 mm thick. Its armament consisted of a 37-mm gun in a traversing turret – the first Skoda tank to be fitted with one. The gun had a monobloc barrel, was semi-automatic and used a dial sight. The elevation range of the installation was from $-10°$ to $+25°$. Horizontal movement was made by hand traverse of the entire turret, while fine sighting adjustment was secured by traverse through a handwheel. This arrangement proved successful with light tanks since a counterweight at the rear of the turret balanced the gun's weight. The gun could be elevated either by direct action of the gunner's shoulder or through an elevating mechanism. At the moment of firing, however, the gun was arrested hydraulically.

A coaxial machine-gun was used as a secondary weapon. Both weapons could be fired simultaneously or individually and the dial sight of the gun was fitted with reticule scales. In addition each machine-gun had its own sighting telescope. A further machine-gun was mounted in the hull, and fired, in a similar fashion to that on the TNHP tank. The A3 gun was further modified to adapt itself to the narrow turret by shortening the recoil and modifying the elevation wheel so that it was unnecessary for the gunner to release it for firing. The gun was improved by increasing the muzzle velocity to 2620 fps. The general internal layout was otherwise similar to that of the PzKpfw 38(t).

The particular advantage in the design of this tank was the operating efficiency which reduced driver fatigue. The vehicle was very fast and easy to steer, thanks to its 12-speed gearbox and pneumatic-servomechanical steering unit. Trips of 125 miles per day at average speeds of 12–16 mph could be achieved, although the maximum speed of the vehicle was only 25 mph. The durability of the suspension was also remarkable in that track and bogie-wheel life ranged from 4000–8000 km.

This vehicle was adopted by the Wehrmacht as the PzKpfw 35(t) during 1939, and was issued to VI Panzer Division. Originally the Germans had 106 of these tanks in service. During service in the Russian winter it was found that the steering system froze, and consequently a heater was installed. When the 35(t)s were phased out of service they were used for towing purposes and were sometimes employed for tank recovery purposes with a two-man crew. These then – the PzKpfw I, the PzKpfw II, the PzKpfw 38(t) and the PzKpfw 35(t) – formed the bulk of the Panzer Division strength in the key build-up years of 1936/39. The PzKpfw III and IV were also just in service when Germany invaded Poland on 1 September 1939, but these vehicles are more appropriately covered in the next section.

EARLY HEAVY TANKS
GOOD PROPAGANDA

The German pre-war heavy tanks were all built to the same basic specification, but three different models were constructed by different companies. *Grosstraktor* I was built by Daimler-Benz, *Grosstraktor* II was built by Rheinmetall, and *Grosstraktor* III by Krupp. The vehicles were superficially similar in shape and size, although all differed in suspension and details.

For main armament, *Grosstraktor* I had a 105-mm gun and the other two a 75-mm gun. A feature of all the models was an auxiliary machine-gun turret at the rear for enfilading enemy trenches as the vehicles crossed over.

In concept the *Grosstraktor* followed the rhomboid shape of the tanks which appeared at the end of the First World War such as the Anglo-American Mk VIII, with a central traversing turret instead of sponsons containing the main armament. These tanks were only experimental and never went into production.

As a result of these experiments, plans were drawn up by 1934 for a new design of tank very similar in size and layout to the *Grosstraktor* but incorporating new features, such as auxiliary gun turrets fore and aft, in the style of contemporary Russian and British heavy tanks. Designation given

to the new design was *Neubaufahrzeug* (NbFz) or 'new construction vehicle'. Six of these vehicles were built by Rheinmetall and Krupp, and as prototypes they were all of mild steel construction.

The NbFz was powerfully armed with either 75-mm and 37-mm guns coaxially mounted (Model A), or 105-mm and 37-mm guns mounted coaxially in a vertical plane (Model B). It is believed that only one of the latter was actually built. The two auxiliary turrets each mounted twin MG 13 machine-guns. Power was provided by a 360 hp six-cylinder Maybach engine with six speeds with drive to the rear sprocket. The crew numbered seven – commander, driver, two gunners, two machine-gunners and a radio-operator. The maximum thickness of the mild steel plate was 14·5 mm. Maximum speed was about 15 mph.

In the event no order for the NbFz was ever placed, even in later years, for the equipment for the Panzer Divisions evolved into the family of four vehicles – the PzKpfw I, II, III and IV. Of these the last two were considered to be adequately armed and armoured to form the backbone of the armoured strength of the Wehrmacht and there was little need for the NbFz. A new design was thought to be more desirable as

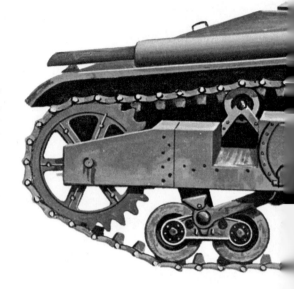

PzKpfw VI (NbFz B)
The second *Neubaufahrzeuge* had as its main armament a 105-mm gun with coaxial 37-mm, with the same auxiliary turret armament as the PzKpfw V (below). Only two were built, and they were scrapped along with the model Vs in 1941

PzKpfw V (NbFz A)
The only difference between this and the PzKpfw VI were the V's main armament of a 75-mm gun with coaxial 37-mm. The 360 hp engine gave it a maximum speed of about 25 mph; maximum armour thickness was 14·5 mm; and the two auxiliary turrets each had two MG 13 machine-guns

a heavier successor to the Panzer IV and this appeared as the DW I 'breakthrough' tank of 1937.

Meanwhile, in 1939, the NbFz was redesignated within the German ordnance classification as a standard design. The Model A (75-mm gun) became the PzKpfw V and the Model B (105-mm gun) became the PzKpfw VI. After the swift demise of these vehicles in 1940/41, the PzKpfw V and VI designations were transferred to the later Panther and Tiger respectively.

If the NbFz models were not to see production, they did achieve fame in 1940 as the visual symbol of German armoured might. In April of that year they were shipped to Oslo at the time of the German invasion of Norway. Five tanks were landed at Oslo docks and another at Putlos. Cleverly posed propaganda photographs of the few vehicles in existence were flashed around the world as 'heavy tanks of the German Army in Oslo', although in fact the photographs actually showed *all* the heavy tanks the Germans then possessed. After the Norwegian campaign the NbFz tanks were shipped back to Germany and disappeared into obscurity. After the war the Allies found documents ordering their scrapping in 1941. So ended the brief career of the pre-war heavy tanks. The hour belonged to two medium designs.

Propaganda photo of NbFz heavy tanks in Oslo during the German occupation of Norway, April 1940

Chris Ellis

BETTER TANKS BETTER TACTICS

The experience gained with the PzKpfw I and II led German planners to bold new concepts of design and tactics. By the outbreak of war, two finely balanced designs formed the back-bone of the Panzer Divisions which secured the Blitzkrieg victories

From 1935 onwards the collective knowledge gained during the design and development of the PzKpfw I and PzKpfw II tanks enabled the German tank-building industry to produce its own design ideas. No longer did it have to rely on foreign developments. Native ideas were, however, sometimes very complicated and did not always take into account the practical difficulties which were later encountered in mass production.

We have already seen that Guderian, the key staff officer, considered two types of armoured fighting vehicles to be required for the full strength Panzer Divisions of the future. The first was to be fitted with an armour-piercing gun as well as bow and turret machine-guns; the other, thought of as a support vehicle, was to have a larger calibre low velocity gun for support purposes. It was planned to equip the three light companies of tank battalions with the first of these two: the PzKpfw III.

Fundamental differences

There were fundamental differences of opinion on the question of arming the vehicle. The Weapons Department and the Artillery Inspectorate considered the 37-mm gun to be sufficient, while the Inspectorate for Motorised Troops demanded a 50-mm gun. The infantry was already equipped with a 37-mm anti-tank gun and, in the interests of standardisation, it was decided to use the same armour-piercing weapon in the PzKpfw III. The installation of the 50-mm gun was rejected at that time. But one provision made was that the PzKpfw III's turret ring would be of a diameter sufficiently large to make possible the fitting of a bigger calibre weapon at a future date.

The question of German road bridge limitations determined that the maximum permissible fighting weight of both the PzKpfw III and IV be 24 tons. A maximum speed of 25 mph was required. The crew was to consist of five men: commander, gun layer and loader in the turret, and driver and wireless operator in the forward compartment. The commander was to have a raised seat, between the aimer's and the loader's places in mid-turret, which would have its own cupola, allowing an all-round view. Throat microphones were to be used for communication among the crew members and also for the radio link from tank to tank, while on the move.

In 1935 The Army Weapons Branch (*Heereswaffenamt*) issued development contracts for the PzKpfw III to MAN, Daimler-Benz, Rheinmetall-Borsig, and Krupp. The Weapons Department's cover name for this vehicle was *Zugkraftwagen* (Platoon Commander's Vehicle) and from 1936 onwards the prototypes were thoroughly tested. The outcome was that Daimler-Benz were made responsible for development and production. In contrast to that of the PzKpfw IV, the PzKpfw III's suspension arrangements showed the influence of the motor car industry in that only torsion rod suspension was used from the fourth model onwards.

The selection of tank building contractors seems to have been made regardless of those manufacturers who had had experience in the mass production of vehicles. The conclusion which can be drawn from this is that, at that particular time, no mass production requirement for these tanks was foreseen. The two largest motor car firms in Germany, Ford and Opel, were deliberately excluded from the tank programme because of their American ownership.

In 1936 the first PzKpfw III tank was produced by Daimler-Benz and ten machines went for troop trials, designated as 1/ZW. Eight of these were fitted with the 37-mm gun. Although the armoured hull, housing and turret did not change substantially throughout its life, the 1/ZW suspension was of an experimental nature and consisted of five large double bogies which were hung on coil springs. The remainder of the suspension was made up of a front driving wheel and a rear idler together with two return rollers. Armour was between 5 and 14·5 mm thick and the overall weight was 15 tons. The engine installed was a development of the Maybach DSO 12-cylinder, high performance 108 TR, which with an approximate capacity of 11 litres, produced a maximum power of 250 hp and a top speed of 20 mph. Transmission was a ZF SFG 75, five gear drive. A total of 150 rounds was carried for the main armament and 4500 rounds for the three machine-guns, two of which were co-axial to the main armament, in the turret. This vehicle was known unofficially as the PzKpfw III Ausf A.

New models

Models B and C appeared during 1937. A new suspension system was tried on these, consisting of eight small bogie wheels on longitudinal leaf springs, and return rollers were increased to three. Armament remained the 37-mm tank gun L/45 in an internal mantlet with two MG 34s, while a third machine-gun, fitted into the front compartment, was worked by the wireless operator. Fifteen each of the Ausf B model (type 2/ZW) and the Ausf C model (type 3a/ZW) were constructed. Armour thickness

remained constant at 14·5 mm all round.

The Ausf D version (type 3b/ZW), which finally went into quantity production appeared at the end of 1938. The same suspension was retained, but the armour was increased to 30 mm all round, thus raising the total weight to about 19 tons. A slightly improved transmission was used. From the Ausf E onwards the more powerful Maybach 12-cylinder HL 120 TR, was fitted which increased the maximum output to 320 hp by enlarging the bore and increasing the cylinder capacity to 11·9 litres. The gearbox in this machine was the Maybach Variorex pre-selector with ten forward and one reverse speeds. This complicated transmission was intended to make gear changing easier, as the change was carried out by a vacuum after the gear had been pre-selected and the release valve was activated by depressing the clutch pedal. The ninth and tenth gear positions were overdrives. Top speed was 25 mph. Fifty-five examples of this version were produced, but the PzKpfw III was still only at 'troop trial' stage when the Blitzkrieg on Poland began on 1 September 1939. All the pre-production machines (Models A–D) were used in the campaign if in only nominal numbers: nevertheless valuable combat experience was gained.

Into production

It was not until 27 September 1939, that the Wehrmacht officially announced that '. . . the *Panzerkampfwagen* III (37-mm) (SdKfz 141), has been adopted for service as a result of its successful troop trials'. Mass production was started just as quickly as a scheme could be set up. With Germany now in a state of war, and with firm official control, a consortium of companies was contracted for the work. These included such famous names as Alkett (one factory on assembly, one fabricating hulls), Daimler-Benz, FAMO, Henschel, MAN, MIAG and others.

The first full production type, Ausf E (or Model) (or 4/ZW), had the final type of hull shape and suspension which changed only in detail from then on. The first vehicles came off the line in late 1939 and were in service by spring 1940. There were now six bogies on each side mounted on torsion bars fitted across the hull. The vehicle weighed about 19·5 tons, had an armour basis of 30 mm and retained the Maybach HL 120 TR engine of the later pre-production machines. The hull alone weighed 13·8 tons. The coupled machine-guns (MG 34s) in the turret which had previously been coaxial with the main armament were replaced with a single

Bundesarchiv

PzKpfw III Torsion Bar Suspension (right)
Used on Panzer IIIs from the Ausf E onwards, this ingenious suspension system relied on the tensile strength of an anchored metal bar to float the armoured vehicle's weight

PzKpfw III Ausf B or C
Only 15 each of these models were built, in 1937, with a new leaf spring suspension system. A third machine-gun was fitted for the wireless operator in the front compartment

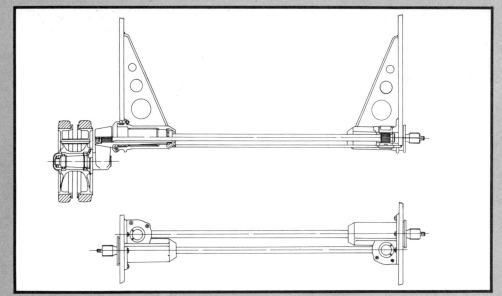

machine-gun, but some early Ausf E machines retained the pre-production type turrets with internal mantlet and two machine-guns.

However, on the production type turret the 37-mm tank gun mantlet was now external. By early 1940 one hundred of these machines had been built and they were rushed into service to provide the main hitting power of tank regiments, which until then had to make do with the PzKpfw I and II and the ex-Czech vehicles. Industry could still only produce tanks in limited numbers, and this low production capacity became more and more of a problem until experience was gained and manufacturing techniques became streamlined. For the big attack on France and Flanders on 10 May 1940, a total of only 349 PzKpfw IIIs of all kinds was available, including the pre-production vehicles which had been brought back from Poland.

Improved models

If the vehicles were few and slow in coming, there was no lack of ideas and foresight at staff level. As early as January 1938 the Weapons Department were asked by *Heereswaffenamt* to improve on the basic model and re-arm it with a 50-mm tank gun. Daimler-Benz built a prototype hull and Krupp designed an improved turret. It was proposed to install the 50-mm KwK L/42 with a muzzle velocity of 2250 fps against the 37-mm gun's 1475 fps. Vehicles equipped with the new 50-mm gun were already coming off the line in May 1940 and a few saw service before the end of the campaign in France.

The improved vehicle was designated PzKpfw III Ausf F and included a slightly up-rated engine. A somewhat lower commander's cupola was a distinctive new feature, as was a prominent equipment box on the back of the turret. From this version on the idler wheels were also distinctively altered to simplify production, in that the new idler was spoked. Some 450 examples of this model were produced. By November 1940, moreover, the production output of PzKpfw III had built up to about 100 vehicles per month.

In October 1940 a new model appeared, the PzKpfw III Ausf G, and in quantitative terms it finally became the real backbone of the tank regiments. For North African service special radiators and air filters were used. These *Fiefel* filters, partly protected by armour, were fitted to the exterior of the engine compartment. Vehicles with this sort of equipment received the designation Tp (Tropical) and were the mainstay of Rommel's famed Afrika Korps. The PzKpfw III was also the main type of German tank used during the fighting in Yugoslavia and Greece in 1941. About 450 Model Gs were constructed and altogether a total of 2143 Panzer IIIs of all types was produced during the years 1940 and 1941.

Hitler himself took a great personal interest in armaments of all kinds, tanks and guns especially, perhaps a legacy of his own days as a ranker in the trenches during the First World War. Some of Hitler's own ideas were practical and sensible and made a worthwhile contribution to development as far as armoured fighting vehicles were concerned. As we will see later, however, he could also stubbornly force through ideas against the better judgment of his technical advisers at *Heereswaffenamt*, often with perverse or wasteful results.

PzKpfw III Ausf E
The first mass-produced version of the Panzer III established the basic torsion bar suspension system and hull configuration for the rest of the series
Engine: Maybach HL 120 TR *Weight:* 19·5 tons *Speed:* 25 mph *Crew:* 5 *Armour:* 30 mm basic *Armament:* 37-mm KwK L/45; 2 mg

PzKpfw III Ausf A
Ten Ausf As were produced by Daimler-Benz in 1936 with an experimental coil spring suspension system. Most other features of the vehicle were standard for the series
Engine: 250 hp Maybach HL 108 TR *Weight:* 15 tons *Speed:* 20 mph *Crew:* 5 *Armour:* 14·5 mm max *Armament:* 37-mm KwK L/45; 3 mg

However, in the case of the Panzer III, Hitler was somewhat ahead of his technical advisers. Aside from sanctioning Guderian's ideas in the pre-war years, he foresaw the need to keep constantly ahead of technical development of arms and armour in enemy countries. For, despite the myth of invincibility, the quality of equipment in the first few Panzer Divisions involved in the Polish and French campaigns was relatively poor. The tanks were mostly the small pre-war types and, apart from the élite divisions (which were the ones most publicised in contemporary photographs), much of the German Army still depended heavily on horses or commandeered commercial trucks for their transport.

What was superior was the tactics. Guderian and Kleist, along with a number of brilliant divisional generals – notably Erwin Rommel, commanding VII Panzer Division – in May 1940 overthrew all the conventions of static warfare for which the Allies on the Western Front were best prepared during the 'Phoney War' period of spring 1940.

A chink in the armour

Here and there, however, there was, almost literally, a 'chink in the armour'. In the right conditions, and properly deployed, the few suitable battle tanks available to the British Expeditionary Force were more than a match for the German AFVs. This was vividly demonstrated when the Matildas of the British 1st Tank Brigade gave elements of Rommel's VII Panzers a particularly bloody nose during a counter-attack at Arras on 21 May 1940. Much of the opposition proved to be light tanks and half-tracks; nevertheless, Rommel recorded the loss of three of his precious Panzer IIIs and six Panzer IVs, while the British Brigade Commander later reported that one Matilda took as many as 14 hits from German 37-mm tank or anti-tank guns with only some gouging out of the armour plate to show for it. The Matilda was admittedly exceptionally well armoured for its day, though lacking in speed and size, but this action, and the Matilda's invincibility in Wavell's Libyan Campaign of late 1940, was clearly a pointer to improvements to be desired in future German tanks.

Britain was by this time known to be pressing on with development of a 6-pounder (57-mm) anti-tank gun, and by the time Rommel's Afrika Korps had become involved in the Libyan fighting there was an instruction from Field Marshal Keitel to the Army High Command, dated 7 July 1941, stating: 'The Führer considers it advantageous to up-armour our new production tanks, by fitting spaced armour plates, additional to the main armour, and thereby to neutralise the increased penetrating power of the British weapons. The increase in weight and the loss of speed must, in the Führer's opinion, be accepted.' Hitler had drawn his own conclusion and this time was right.

Meanwhile the PzKpfw III Ausf H had already appeared in late 1940, with stronger suspension and wider track width (from 360 mm to 400 mm). The hull now weighed 15·8 tons and the fighting weight had risen to 21·6 tons, while the complicated Maybach Variorex drive was replaced by a normal six gear drive with syncromesh gear box and dry plate clutch. The 50-mm L/42 gun was retained, although Hitler had, in fact, suggested when the 50-mm gun was first

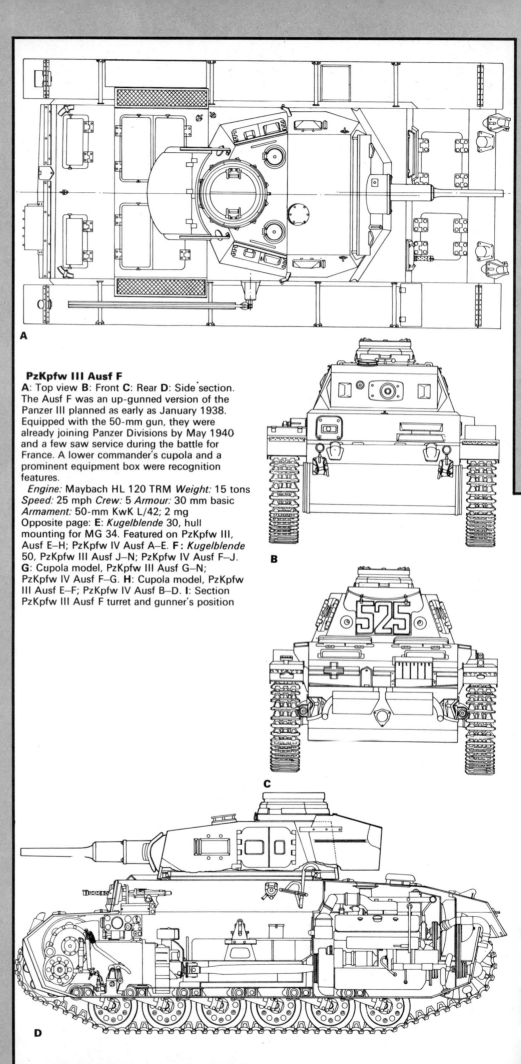

PzKpfw III Ausf F

A: Top view **B**: Front **C**: Rear **D**: Side section. The Ausf F was an up-gunned version of the Panzer III planned as early as January 1938. Equipped with the 50-mm gun, they were already joining Panzer Divisions by May 1940 and a few saw service during the battle for France. A lower commander's cupola and a prominent equipment box were recognition features.

Engine: Maybach HL 120 TRM *Weight:* 15 tons *Speed:* 25 mph *Crew:* 5 *Armour:* 30 mm basic *Armament:* 50-mm KwK L/42; 2 mg Opposite page: **E**: *Kugelblende* 30, hull mounting for MG 34. Featured on PzKpfw III, Ausf E–H; PzKpfw IV Ausf A–E. **F**: *Kugelblende* 50, PzKpfw III Ausf J–N; PzKpfw IV Ausf F–J. **G**: Cupola model, PzKpfw III Ausf G–N; PzKpfw IV Ausf F–G. **H**: Cupola model, PzKpfw III Ausf E–F; PzKpfw IV Ausf B–D. **I**: Section PzKpfw III Ausf F turret and gunner's position

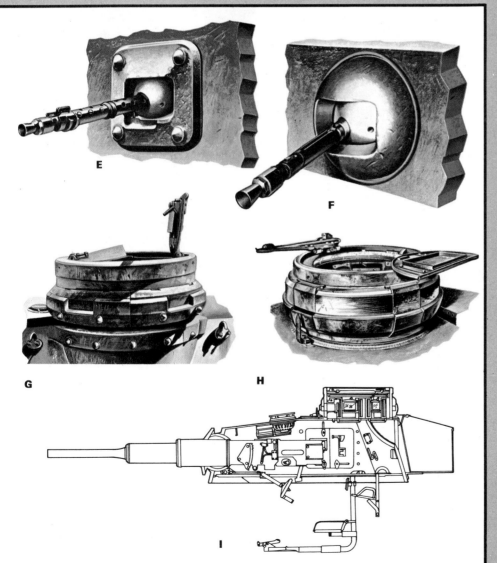

E

F

G

H

I

adopted for the PzKpfw III that the longer high velocity L/60 model was more desirable than the L/42 version.

However, the big inadequacies in German tank armament and armour were only fully realised in late 1941 on the Eastern Front after the appearance of the very superior Russian T-34. In July 1941 (Germany invaded Russia on 22 June 1941), the German staff estimated that 36 Panzer Divisions were needed and these would require 7992 Panzer IIIs. However, by November 1941, after the sensational battle debut of the Russian T-34, there was a complete reversal and entirely new designs of tank were being considered. At this point doubts were already being entertained as to the effectiveness of the Panzer Divisions: Hitler personally described the Panzer III as an unsuccessful design. It must, however, be made clear that, for its time, this vehicle was an extremely advanced design, and if it had been fitted with a 50-mm gun from the outset it would have been the best fighting tank in the world in 1940/41.

Better guns

In the event an order to introduce the improved L/60 50-mm gun (KwK 39) was given in 1941. Using the armour piercing *Panzergranate* 40 shell, this gun had a muzzle velocity of 3875 fps. The production version with the L/60 gun was designated PzKpfw Ausf J (SdKfz 141/1). All earlier Panzer IIIs returned to Germany for general overhaul after April 1941 and were up-gunned with this weapon, though this meant that only 78 rounds could be carried, against the 99 rounds which could be carried for the 50-mm L/42. Certain small technical differences distinguished the Ausf J from its predecessors. The reverse gear change, for instance, was originally pedal operated, but from the Ausf J onwards a hand lever was used. The internal expanding brakes, too, for this and for subsequent versions were

Top: *PzKpfw III Ausf L with long 50-mm gun and spaced frontal armour. Opposite page: PzKpfw IV Ausf D shoots its accompanying infantry into a blazing Russian village*

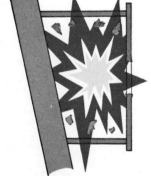

PzKpfw III Ausf L
Further refinement of the basic Panzer III produced the Ausf L, incorporating the long 50-mm KwK 39 L/60 main armament of the Ausf J. Ammunition supply was increased, and 20-mm spaced armour plate protection was provided for the driver and on the turret front

Spaced Armour Principle
Hollow charge ammunition exploded against armour plate, blasting a stream of gas and molten metal through it. Spaced armour aimed to explode the projectile against the outer layer, dissipating the blast harmlessly against the inner

concentric rather than eccentric. There were several other detail changes, and the total weight became 21·5 tons. In 1942 production figures of 150–190 vehicles a month were achieved.

The PzKpfw III Ausf L was a further improvement introduced at the end of 1941 and had increased front turret armour and additional plates 20 mm thick in front of the turret shield and the driver's plate. Increasing the front and the turret front armour to 50 mm + 20 mm (70 mm) upped the combat weight to 22·3 tons, and the machine-gun ammunition supply was increased from 2000 to 4950 rounds.

At this point we must consider the story of the last major tank to be produced under the original Guderian plans of the 1930s. This was the vehicle originally intended for the support role – with a low velocity 75-mm gun to fire HE and smoke shells, and big enough to act as a commander's vehicle if required.

In spring 1935, Krupp, Rheinmetall, and MAN all sent in designs to fit the specification drawn up by *Heereswaffenamt*. This vehicle, in the 20-ton class, was the VK2001, known under the code or 'cover' designation of BW (*Battaillonsführer Wagen*), and the Krupp design was chosen for production. The prototype trials took place at Ulm and Kummersdorf in 1937.

Small beginnings

As with the Panzer III, some pre-production models were built in small numbers for 'troop trials'. Three models, Ausf A, B and C, had been built by 1939, and the few available vehicles took part in the Polish campaign. There was much less variety in detail of these, and the relative unimportance of the PzKpfw IV as originally conceived and ordered is demonstrated by the fact that only one contractor was involved as against eight for the PzKpfw III. Also, in the 'Blitzkrieg era' of 1939/41, there was little change in the PzKpfw IV, for in service it was fulfilling the role Guderian had envisaged for it. In the event, however, the PzKpfw IV was destined to supplant the PzKpfw III as the mainstay of the Panzer Divisions for its larger size allowed it to be more effectively up-gunned and up-armoured when the urgent need arose for a more effective answer to the new Soviet and American tanks of 1942/43. The PzKpfw IV, indeed, had the distinction of remaining in production throughout the war both as a battle tank and as a major basis (with all the

other standard types) for the dozens of self-propelled guns and tank destroyers which the Germans produced.

With the outbreak of war in 1939, the design was 'frozen' and large scale production was ordered as the PzKpfw IV Ausf D. Against the PzKpfw III, the PzKpfw IV's production was modest, as will be evident from the following numbers of PzKpfw IV on Army strength during the first three years of the war: end of 1939 – 174; end of 1940 – 386; end of 1941 – 769. In fact, the total Panzer IV production during 1941 amounted to only 480, despite an order in July 1941 which requested production of 2160 to equip the planned 36 armoured divisions. A monthly production goal of 40 per month was set for 1941, while in January 1942 a monthly output of 57 units was anticipated. In the event this target was exceeded and 964 urgently needed vehicles were produced during 1942. Originally the main assembly was by Krupp of Gruson, with hulls and turrets supplied by Krupp of Essen and Eisen of Bochum.

This picture changed considerably during 1942 due to Allied air raids. The relocation of key war industry to areas not readily accessible to the bombers was begun in 1940 and established several new tank factories. One of these was *Nibelungenwerke* at St Valentin, Austria, managed by Steyr-Daimler-Puch. Initially intended for the production of a replacement vehicle for PzKpfw IV – the Porsche *Leopard* (Porsche Type 100) – it became operational just in time to take on the expanded Panzer IV production. From 1943, the Panzer IV was assembled almost exclusively at this factory and remained in production there until the end of the war. Its proximity to the Hermann Göring steel mills at Linz established a good source of material for hulls and turrets.

The raw material used in one PzKpfw IV (without weapons, optical instruments or radio equipment) comprised 86,000 lb of steel, 2·6 lb tin, 430 lb copper, 525 lb aluminium, 140 lb lead, 146 lb zinc, $\frac{1}{3}$ lb magnesium and 256 lb rubber. These totals illustrate the enormous strain placed on German industry by tank production and go far to explain its limitations, even in the early days of the war, compared with the achievements of Allied industry in this field.

The Panzer IV hull was a comparatively simple design. All joints were austenitic steel welds and the plates were high-quality chromium-molybdenum steel made by the electric furnace process. Two bulkheads separated the hull into three compartments – driving, fighting and engine. The front driving compartment housed the transmission and final drive assemblies as well as seats for the driver and radio operator/hull gunner. Three petrol (gasoline) tanks with a capacity of approximately 105 gallons were located beneath the floor of the centre fighting compartment.

Overhanging superstructure

A most noticeable and characteristic feature of the vehicle was the superstructure, of welded construction, bolted to the top flange of the hull. To accommodate the rather large turret ring, it projected well beyond each side wall of the hull. One bolted and two hinged maintenance hatches were provided in the front glacis plate, while access hatches for driver and radio operator were provided in the roof plate, though there were many detail changes incorporated in later models.

The welded turret provided seats for three crew members – commander, gunner and loader. The sides were sloped so that the overall width was appreciably greater than

PzKpfw Ausf D
A: Longitudinal section B: Layout of driving, fighting and engine compartment C: Section through engine compartment D: Section through fighting compartment E: Section through driving compartment

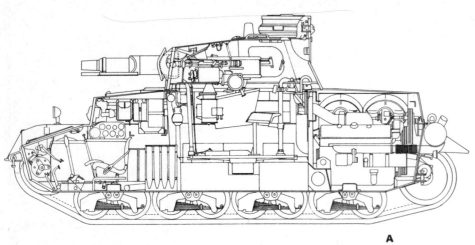

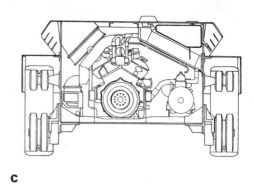

A

C

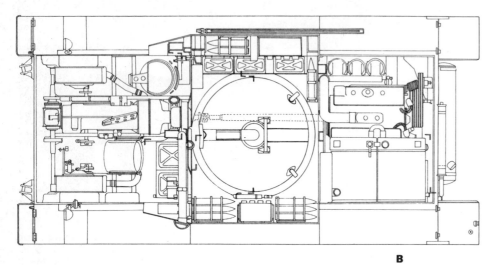

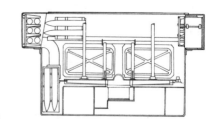

D

B

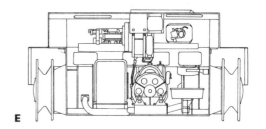

E

PzKpfw IV Ausf D
First produced in 1938, the Ausf D was similar to the A, B and C models that preceded it. It had an improved commander's cupola, better bow machine-gun mount and various detail changes

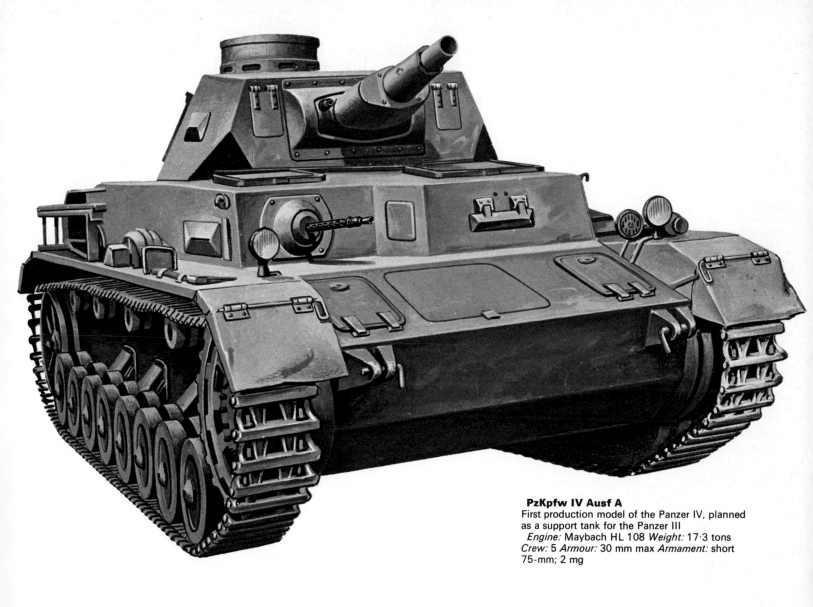

PzKpfw IV Ausf A
First production model of the Panzer IV, planned
as a support tank for the Panzer III
 Engine: Maybach HL 108 *Weight:* 17·3 tons
Crew: 5 *Armour:* 30 mm max *Armament:* short
75-mm; 2 mg

the internal diameter of the turret ring. The
75-mm gun was mounted on a trunnion axis.
The forward end of the recoil mechanism
projected through the mantlet to afford
additional protection. The commander's
cupola, set well back on the turret roof, had
five observation ports equally spaced around
it with the front port pointing directly forward in line with the gun. It was closed by a
pair of semi-circular hatch covers. An
observation port was provided in each side
wall of the turret, in front of the side access
hatches. Additional observation ports were
fitted at either side of the gun mantlet
officially, though these were not found on
later vehicles. A signal port was fitted on
the turret roof, similar to those mounted on
both driving compartment crew access
hatches. There were also two revolver and
carbine ports at the rear of the turret,
while the fighting compartment was venti-
lated by a roof-mounted extractor fan.

The main power plant was the standard
medium tank engine, the Maybach HL 120
TRM, a 12-cylinder, 11,867 cc liquid-cooled

petrol engine. Normally developing an
output of 300 bhp at 3000 rpm, the engine
was in most instances restricted to 2600 rpm,
giving a rating of 265 bhp. It used only low
grade 74 octane petrol. Cooling air entered
through louvres on the left hand side of the
engine compartment, was drawn through
two radiators and over the engine by two
ten-bladed fans. An exceptionally large filter
provided clean air for the power plant.

Transmission details
Engine output was transmitted by a propeller shaft and a three-plate dry clutch to
the synchro-mesh six-speed gearbox. Small
multi-disc synchronising clutches were used
for 2nd, 3rd, 4th, 5th and 6th gears. A Krupp-
Wilson 'Clutch-Brake' final drive and steering mechanism was used. In this, the input
gear drove the annulus of an epicyclic train.
The sunwheel was coupled to a steering
brake drum, which was held stationary by
an external band and compression spring
while the vehicle was in motion. The drive
from the epicyclic annulus was transmitted

through the planet carrier to the spur
reduction gears, which drove the track
sprockets. The six-speed gearbox and the
final drive units had one common oil circula-
tion system.

Each track consisted of 98 links, each 400
mm wide with 120 mm pitch. Manganese
steel was used for this 'skeleton' type of
track which weighed approximately 1400 lb.
Track tension was adjusted by a large
diameter idler wheel mounted on an eccentric axle at the rear of the vehicle. The
suspension system consisted of four bogie
units per side, each of which was fitted with
two 18·5-in diameter rubber-tyred wheels.
Quarter elliptic springs were mounted on
the underside of the leading axle arm of
each bogie. The other end of the spring
rested on a shackle pin and roller, carried
on an extension of the trailing axle arm.
Four support rollers per side completed the
suspension.

Thus it can be seen that the basic Panzer
IV was a simpler vehicle than the Panzer
III, and influenced the later IIIs.

LEADING FROM THE FRONT

Well prior to the outbreak of war, the need for command vehicles to keep up with advanced echelons in an armoured assault was foreseen. Versions of both the six-wheel and eight-wheel armoured cars, comprehensively equipped with radio and map tables, were produced, as were half-track command vehicles. But with the large-scale production of tanks and an expansion to 36 armoured divisions envisaged, command versions of tanks were felt to be necessary.

Once again the ubiquitous ZW design – the Panzer III – was used, and the earliest command vehicles of this type served in the France and Flanders campaigns of May 1940.

This first type was produced by conversion of the pre-production PzKpfw III Ausf D. The official designation was *Panzerbefehlswagen* (PzBefWg) III Ausf D. (*Panzerbefehlswagen* = armoured command vehicle). Like the armoured command car, these vehicles carried a distinctive frame aerial above the engine compartment, though this was replaced by pole-type aerials from 1943. For the defence of the five-man crew there was an MG 34. A dummy gun replaced the main armament since absence of a gun would have made its function apparent to the enemy. The major series of PzBefWeg IIIs used the Ausf E model as a basis and appeared during 1940 as the PzBefWg III Ausf E and later there was an Ausf H version with the usual additional spaced armour plates fitted in front of the driver's plate, and extra armour on the nose. At the start of the French campaign in 1940, there were 39 *Panzerbefehlswagen* with the tank divisions. In November 1940 a production contract was put out for new vehicles of this type. It was planned to turn out a total of 10 machines per month during the first six months of 1941; the final 14 of the contract were delivered in January 1942. They saw wide service on all fronts.

Spacious interior

The absence of main armament gave plenty of interior space for a command staff but made the vehicle of only limited use on active service, and it was also necessary to make special parts. There were, however, urgent demands for command vehicles which could be produced from AFVs in use with the troops in the field. In January 1941 a development contract covering this requirement was given to Daimler-Benz, ordering a new development of the armoured command vehicle. It specified that the new design was to be equipped with a 50-mm tank gun L/42 or L/60 in a fully traversing turret (the turret was fixed in the earlier

August 1942: the commander of a PzKpfw III leads his formation into Russia. Lavish wireless equipment and efficient command vehicles gave an immediate tactical edge over the Soviet opposition

Panzerbefehlswagen I
Modified superstructure on the PzKpfw I Ausf B chassis formed the first German armoured command vehicle. The revolving turret was abandoned to give space for a radio and map table, while nose and turret face armour were increased by 17 mm (turret) and 10 mm (nose)

models). Prototypes were troop tested after August 1941. The type 7/ZW hull was selected as a basis. The official designation of the new model was *Panzerbefehlswagen III Ausf K*. With a five man crew (the commander was also the wireless operator), the vehicle had a combat weight of about 23 tons. Quantity production ran from August 1942 to August 1943. The total of large and small command vehicles available on 1 July 1941 was 331 and on 1 April 1942 the total was 273.

The PzBefWg III was not the first full-tracked armoured command vehicle, however. The first German example was based on the little PzKpfw I Ausf B chassis and was built in 1939.

The superstructure was considerably modified and the sides built up to form a rectangular housing which carried a 7·92-mm MG 34 in a ball mount in the front plate for defensive purposes only. An additional 17 mm of armour plate was added to the turret face and the nose plate was also reinforced by an additional 10 mm.

Two hundred of the PzKpfw II Ausf B chassis were modified and three types were produced. Differences between them were slight, but one of them incorporated a rotating turret which was abandoned because the interior was too cramped. The crew of these tanks was increased to three men. Provision was made for a small table and for the display of maps, and two wireless sets, an Fu 2 and an Fu 6, were fitted. Additional dynamo capacity was provided.

They were first used in the Polish campaign and 96 of them were available for use in the operations in the West in 1940. Some remained in service throughout the Second World War, long after the Panzer I had disappeared from use as a fighting tank, and although the PzKpfw I and II were replaced as swiftly as PzKpfw III and IV production would allow, there were a few further developments.

PzBefWg III Ausf K had a 50-mm gun in a traversing turret and prominent aerials

Bundesarchiv

AMPHIBIOUS TANKS

Bundesarchiv

PzKpfw III Ausf E equipped with snorkel and sealing devices is tested in preparation for Operation Sea Lion, the invasion of Great Britain

Some interesting other developments took place during the classic 'Blitzkrieg era'. In September and October 1940, volunteers from II Panzer Regiment at Putlos were formed into Panzer-Battalion 'A' and trained for 'Operation Sea Lion' (*Seelöwe*), the planned invasion of Great Britain. Two other special formations, Panzer Battalions 'B' and 'C', were being raised at the same time and the same place. These units later formed the XVIII Panzer Regiment of XVIII Panzer Division and converted their PzKpfw II and PzKpfw IV tanks into submersibles. All openings, vision slits, flaps, and so on, were made water tight with sealing compounds and cable tar.

The turret entry ports were bolted from the inside and the air intake opening for the engine was completely closed. A rubber cover sheet was fixed over the circular port of the tank gun, the commander's cupola and the wireless operator's machine gun. An ignition wire blew off the covering sheet upon surfacing and left the vehicle ready for action. Between the hull and the turret was a rubber sealing ring which, when inflated, prevented the water from entering.

The fresh air supply was maintained by a wire-bound rubber tube with a diameter of about 20 cm and 18 metres long. A buoy with an attached radio antenna was fitted to one end of this tube. The exhaust pipes were fitted with high pressure, non-return relief valves. When travelling submerged, sea water was used to cool the engine and seepage was removed by a bilge pump.

Maximum diving depth was 15 metres, with three metres of the air tube's 18-metre length available as a safety measure.

The 'submersible tanks' were to be launched from ferries which were hastily made from converted river barges or lighters, though river ferries were also used in the tests. The tanks slid into the water via an elongated ramp made of iron rails. Direction was kept by radio orders from a command vessel to the submerged machines. Underwater navigation was carried out by means of gyro compass and the crew was equipped with submarine-type escape apparatus. The submerged machines were relatively easy to steer as buoyancy partly raised them. When Operation Sea Lion was abandoned, these vehicles were used during the Russian campaign, in 1941, for the crossing of the River Bug.

LAST OF THEIR KIND

The PzKpfw II Ausf D and E were built by Daimler-Benz and were intended to be faster versions of the standard tanks. As far as the turret, superstructure, engine and transmission were concerned they showed no difference from the other Panzer IIs. However, the suspension was completely changed and used four large suspension wheels in Christie tank fashion but with their movement controlled by torsion bars. They could reach 35 mph but their performance across country was considerably slower than that of the standard PzKpfw II.

Because the performance of Models D and E did not come up to expectation they were taken out of service and 95 were converted to a flame-throwing role with the designation *Flammpanzer* II (SdKfz 122). Each was fitted with two projectors covering an arc of 180° each side with a flame range of about 40 yards. This was about the maximum that could be obtained with a pump-fed gun. To obtain greater range it was necessary to use a gas pressure system which introduced problems over stowage. Sufficient fuel was stowed in internal tanks for about 80 shots, each of 2–3 seconds duration.

The hollow charge anti-tank missile had by 1940 become a menace to be reckoned with and then it was decided that all future AFVs were to be up-armoured by the addition of spaced plates to reduce the effect of the new missiles. Thus, when the last of the Panzer II series, the PzKpfw II Ausf F, appeared in late 1940 it carried this 'spaced armour' feature.

The vehicle weighed 9½ tons with 35 mm of armour on the front and 20 mm on the sides. Otherwise it was exactly the same as the PzKpfw II Ausf C in appearance with engine, transmission, armament and suspension remaining unchanged. The top speed was considerably reduced but this was a penalty that had been foreseen when the order authorising additional armour was issued. Crew losses had been heavy enough to justify the reduction in performance in the hope of saving lives. Production of the Model F was to have been at the rate of 45 per month but rarely reached this target.

In 1941 a new specification was issued which called for a ten-ton vehicle with more armour than the Model F and a higher speed. To meet this requirement MAN delivered a chassis in September which had a Maybach HL-P engine developing 200 hp and was capable of a top speed of 40 mph. The tank was to have had 30 mm of armour, a three-man crew, and was to have been armed with a 20-mm type 38 gun of higher velocity than that used in the other models, together with a 7·92-mm machine-gun. Production was scheduled for July 1942 but by then the PzKpfw II was so obviously obsolete in its tank role that the order was cancelled.

Prototype for the Lynx

The prototype of the PzKpfw II Ausf L (Sdkfz 123) *Luchs* (Lynx) (VK 1303) appeared in mild steel in 1942, but its development story goes back to 1938 when Daimler-Benz were given instructions to produce a new version of PzKpfw II 'with principal emphasis on increased speed' under the development number VK 901. A Maybach HL 45 six-cylinder petrol engine was used which gave 145 hp and a top speed for the tank of 32 mph. The specified speed was 37·5 mph but that was difficult to achieve because no engine of the necessary power (200 hp) was available at the time. VK 901 had 30 mm of front armour and weighed 9·2 tons. It was armed with a 20-mm KwK 38 tank gun and a 7·92-mm machine-gun which was mounted coaxially; both guns were installed in a stabilised mounting.

The manufacture of 75 pre-production machines began in 1940. The third prototype VK 903 had a turret from VK 1303 which became the Lynx and which had been equipped with a range finder and locating instruments. This substitution of turrets gives some idea of the complexity of the German tank programme, for at that time the Lynx chassis had been built by MAN. The interplay of one model on another is difficult to disentangle, but development was on a most extensive scale possibly because of Hitler's continued interest in all sorts of new projects.

To complicate matters even further, Daimler-Benz and MAN together received another contract in December 1939 for a very different type of machine. VK 1601 was to carry 'the thickest possible armour' with a crew of three men. The Maybach III 45 engine of 200 hp was used for the project giving a top speed of 20 mph with an all-up weight of 16½ tons. Frontal armour was 80 mm thick and the side armour was 50 mm. Armament was the 20-mm KwK 38 gun and a machine-gun in a stabilised mount.

Both VK 901 and VK 1601 used a new type of suspension with five large overlapping suspension wheels and no return rollers. Torsion bar springing was used. This type of suspension, which had already been tried out on VK 601 (the PzKpfw I prototype of the 6-ton tank) ultimately led to the Panther and Tiger suspension where overlapping became interleaving, a necessary step to reduce ground pressure but one

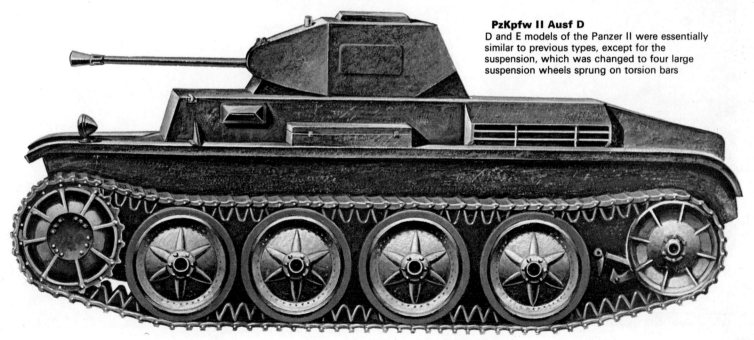

PzKpfw II Ausf D
D and E models of the Panzer II were essentially similar to previous types, except for the suspension, which was changed to four large suspension wheels sprung on torsion bars

which brought its own problems, mainly track jamming.

Out of these two models VK 1301, *Luchs*, was born. VK 901 was considered too light for its proposed role and VK 1601 was much too heavy. The prototype VK 1301 in mild steel ran in April 1942, looking very like VK 901. Various alterations were made to this first prototype and VK 1303, the third prototype, was accepted for production at a weight of 11·8 tons, a reduction of a little over a ton on the first prototype VK 1301.

Different designations
Intended primarily for reconnaissance the Lynx was also given the designation *Panzerspahwagen* II (20-mm KwK 38) *Luchs* with the same Ordnance number, SdKfz 123. It weighed 11·8 tons, had a crew of four men, and was fitted with a Maybach III 66P six-cylinder engine which developed 180 hp. The drive was taken to the front sprocket through a six-speed syncromesh gear box and controlled differential steering on the

cross shafts. The maximum speed was 38 mph. The Lynx used the same five overlapping suspension wheel suspension that had been developed on VK901 and VK 1601 with torsion bar springing. Frontal armour was 30 mm and the side plates 20 mm. MAN built the chassis and Daimler-Benz the hulls and turrets. One hundred of these tanks were fitted with 20-mm guns and a further 31 were fitted with 50-mm KwK 39 L/60 guns.

In 1941 the Army Weapons Branch called for a vehicle capable of undertaking battle reconnaissance, in contrast to the Lynx which was intended for general reconnaissance and was not intended to take part in the main battle. The new contract given to MAN and Daimler-Benz was for VK 1602 (inspired by VK 1601) which has already been mentioned as one of the forerunners of the Lynx.

The new tank, Leopard, was to have 80 mm of armour on the turret and the front and 60 mm on the sides. It was to have a 550

hp engine to give it a top speed of 37 mph and it was to be armed with a 50-mm type 39 L/60 gun and a machine-gun coaxially mounted. It was to have a crew of four men, but the tank never went into production although the turrets were used for the famous eight-wheeled armoured car, the Puma.

The last light tanks
Aside from several minor experimental versions of the Panzer I (including a proposed heavily armoured Ausf C), the Lynx and the Leopard represented the final German efforts at light tank production. Like the other combatants in the Second World War, the Germans discovered only too quickly that the light tank's recce function could be fulfilled just as easily by armoured cars – but in the Wehrmacht at least the Panzer I and II hulls soldiered on until 1945 as the basic chassis for numerous self-propelled guns, as driver trainers, or as munitions carriers.

A PzKpfw II Ausf B passes an abandoned Russian BT-7A. Soviet tank losses in the opening phases of Barbarossa were immense, and by December 1941 over 15,000 machines had been destroyed

PzKpfw II Ausf F

Basically similar to Ausf C, with spaced armour plates added in an attempt to reduce crew losses. An extra 35 mm of armour on the front and 20 mm on sides increased its weight to 9·5 tons, with a consequent loss of performance

Panzer II Suspension Development

Top row: Two examples of sprung girder types derived from British Carden-Loyd types and used on pre-production machines. Bottom left: Leaf Spring (Ausf A, B and C). Right: Christie type (Ausf D and E)

Bundesarchiv

257

PzKpfw II Luchs (Lynx)

The ultimate in German light tank development, the Lynx never supplanted armoured cars in its intended reconnaissance role

Engine: 180 hp Maybach HL 66P *Weight:* 11·8 tons *Speed:* 20 mph *Crew:* 4 *Armour:* 30 mm max *Armament:* 20-mm KwK 38

VK 1601

A 1939 prototype for a heavily armoured reconnaissance tank, VK 1601 proved much too heavy, though its suspension was later used in the Lynx light tank

Engine: 200 hp Maybach HL 45 *Weight:* 16·5 tons *Speed:* 20 mph *Crew:* 3 *Armour:* 80 mm max *Armament:* 20-mm KwK 38; 1 mg

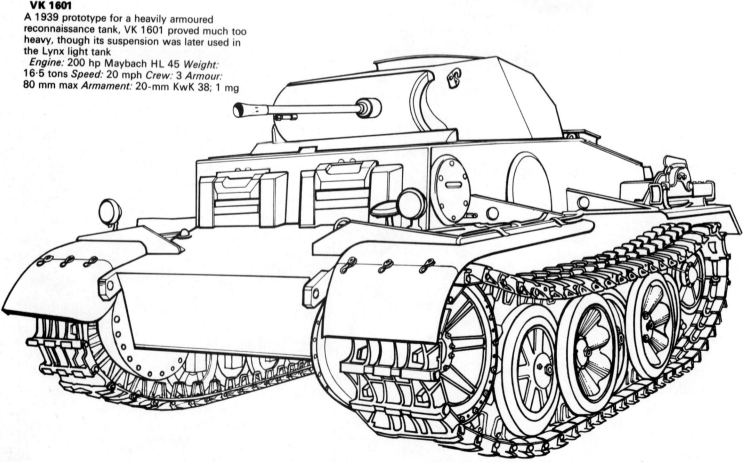

THE DEFENSIVE PERIOD
THE TIDE TURNS

German tanks had overrun mainland Europe and half of Western Russia – but a bitter lesson waited in the snows of Russia and the response of Allied designers to the German challenge. Before the second generation of tanks appeared, the older models would have to stem the tide . . .

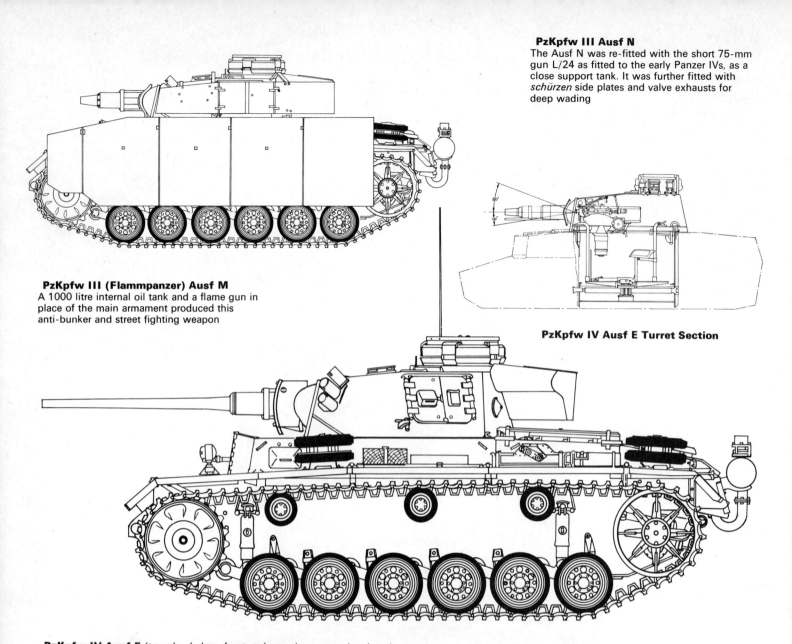

PzKpfw III Ausf N
The Ausf N was re-fitted with the short 75-mm gun L/24 as fitted to the early Panzer IVs, as a close support tank. It was further fitted with *schürzen* side plates and valve exhausts for deep wading

PzKpfw III (Flammpanzer) Ausf M
A 1000 litre internal oil tank and a flame gun in place of the main armament produced this anti-bunker and street fighting weapon

PzKpfw IV Ausf E Turret Section

PzKpfw IV Ausf E (top plan below, front and rear views opposite above)
Ausf E showed several modifications over the Ausf D – the commander's cupola was moved forward, armour improved and several details changed. Production began in 1939, making it available for the invasion of France

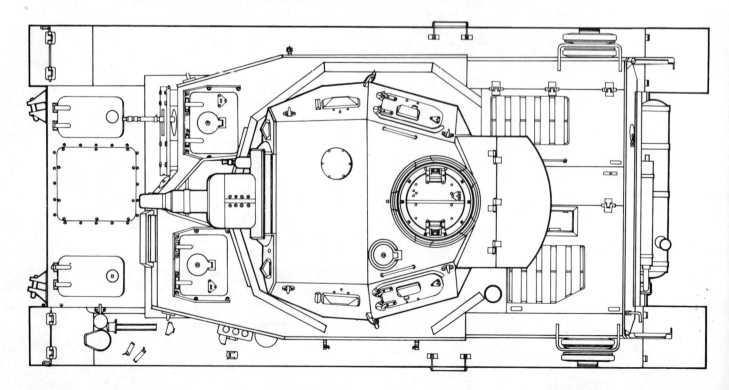

From late 1941 onwards, there was a great change in German fortune as far as tank warfare was concerned. First of all, the Soviet T-34 tank appeared on the Russian Front and caused a fundamental re-appraisal of tank production and a concentration on new designs – though in truth there was no lack of designs in the pipeline at this time. Secondly, in the early part of 1942 the new American-built medium tanks appeared in the Western Desert. First came the M3 medium tanks (Grant and Lee) at the Battle of Gazala in May 1942, then the much superior M4 medium tank (Sherman) reached service for the Battle of Alamein in October 1942.

Both these American vehicles gave Allied tank men, for the first time, a tank with a 75-mm gun which matched the contemporary German weapons, could fire HE and AP ammunition, and was suitable for mass-production. The Sherman was big enough for future development, too, and

by 1944 had been up-gunned to take a 17-pounder high velocity gun (the British Sherman Firefly) or the 76-mm gun (the American 'Easy Eight' model). There was an associated programme of American tank destroyers, too, with 90-mm guns, and overwhelming Allied air superiority on all fronts – not to mention the big Russian Stalin tanks and the American Pershings which were in service in the last year of the war.

On the defensive

All these factors combined to put German AFVs and tactics into a defensive rather than an offensive situation. The Panzer III was soon outmoded, for the 50-mm gun was the biggest it could carry. The Ausf M was a refined version of the Ausf L, with the side escape doors eliminated to simplify production. Last of the line was the Ausf N which, ironically, took over the Panzer IV's intended role as a 'support' tank. It

was fitted with the short low velocity 75-mm KwK L/24 from the early marks of PzKpfw IV, and production of the PzKpfw III as a gun tank ceased completely in August 1943. However, the vehicle soldiered on in service in less important areas until the war's end, was used for training, and was converted to special purpose use. In addition the Panzer III chassis was used for a whole range of assault gun and tank destroyer conversions which played an important part in the Panzer Divisions until the end of the war.

The Panzer IV, being a bigger vehicle, took over the Panzer III's envisaged role of principal battle tank and was truthfully described as the 'workhorse of the Panzer Divisions'. Though inferior in shape and equipment to the T-34 and the post-1942 German designs, the Panzer IV had the virtues of the Sherman – it was reliable and relatively simple to maintain. The PzKpfw IV Ausf E was a major production type with

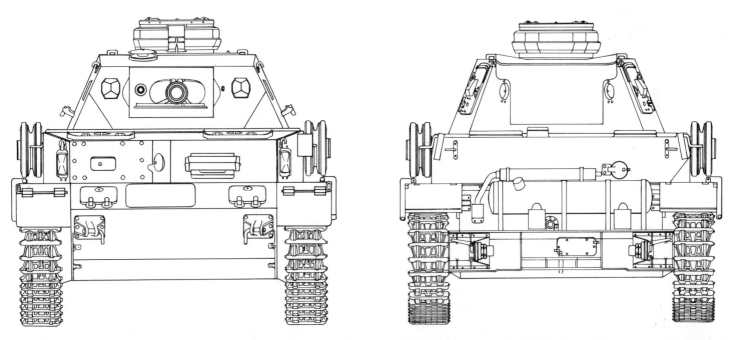

PzKpfw IV Ausf E in a column of armour, clearly showing the additional armour plates

Bundesarchiv

a simplified one-piece front to the super-structure, and retained the 75-mm gun. The Ausf F (later re-designated Ausf F1) was an up-armoured model with basic armour of 50 mm at the front and a ball-mounted hull machine-gun. Like the later models of the Panzer III it had a simplified idler wheel and widened tracks.

The major development, however, was the PzKpfw IV Ausf F2, the original F re-armed with a long high velocity 75-mm gun, and produced specifically to restore the balance of fire-power to the Afrika Korps in 1942 when the American-built tanks with 75-mm guns appeared. This vehicle – known to the British troops as the

'Mk IV Special' – was exceptionally effective, but Rommel could never get enough of them to restore the Panzer Divisions of the Afrika Korps to their original dominating position on the battlefield. The PzKpfw IV Ausf G was a similar vehicle, but built from the start with the high velocity gun, and with an improved up-armoured turret and detail changes.

Improved armament

The new gun, KwK 40 L/43, of the PzKpfw IV Ausf F2 was easily distinguished by its longer barrel and muzzle brake. While the first production model was fitted with a single-baffle globular muzzle brake, later

vehicles had a double brake. The gun itself was capable of penetrating homogenous armour of 77 mm thickness at 2000 yards using PzGr 39 at normal impact. It could fire at least six different kinds of ammunition and 87 rounds were carried, plus 2250 rounds of 7·92-mm ammunition for the two MG 34 machine-guns. One of these was mounted co-axially on the right side of the gun; the other was ball-mounted on the right side of the front vertical plate and worked by the radio-operator. Turret traverse was effected by either hand or electric power supplied from a generator, driven by a DKW two-cylinder two-cycle 10 hp 500 cc petrol engine.

PzKpfw IV Ausf F2 Turret and Hull Front
The long barrelled 75-mm KwK L/43. The vision ports on the turret sides, and the loader's on the turret front, were omitted

PzKpfw IV Ausf F
The Ausf F (or F1) appeared in 1940, and had profited from the experience of the Polish campaign. Armour was increased, track widened and visibility improved, though its armament was still not equal to the T-34's
Engine: 265 hp Maybach HL 120 *Weight:* 22 tons *Speed:* 25 mph
Crew: 5 *Armour:* 50 mm max *Armament:* short 75-mm; 2 mg

By mid-1943, the vehicle had been further refined with the appearance of the PzKpfw IV Ausf H. Similar to the G model, it had a still more powerful 75-mm gun, the L/48, which was about 15 in longer than the L/43. A new cupola with 100-mm armour was fitted, and some vehicles had 30-mm plates of extra armour welded or bolted on the nose. Later vehicles were built new with 85 mm thick frontal armour. Simplified suspension components were used to reduce production costs. For protection from hollow charge anti-tank projectiles of the bazooka type, mild steel skirt armour plates were suspended from rails attached to the superstructure, since hollow charge

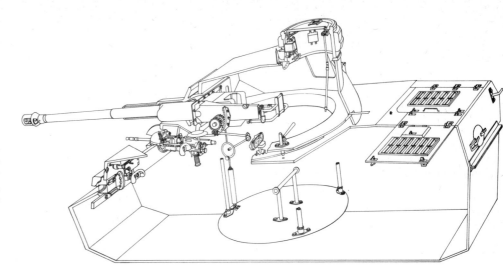

PzKpfw IV Ausf F2 Interior Detail
Showing turret and fighting compartment, driver's and machine-gunner/radio-operator's hull positions. The long 75-mm gun still left plenty of room in the turret cage

weapons were now in Allied service on an increasing scale. Zimmerit anti-magnetic compound also made its appearance at this time to prevent magnetic charges being placed on the vehicle.

Last of the PzKpfw IV line was the Ausf J, with further changes to simplify production: the generator which provided power traverse for the turret was removed and replaced by extra fuel tanks. Heavy gauge mesh wire replaced the steel skirt armour, and most late vehicles had spaced armour plates right round the turret. Appearing in mid-1944, the Ausf J remained in production until the end of the war.

In 1942, an early attempt at rationalisation was the PzKpfw III/IV, combining parts from both tanks to make a standard battle tank using a Famo-type suspension similar to that fitted to the German half-tracks. However, the obvious limitations of the vehicle's development potential led to the hybrid III/IV model being dropped as a project, and work concentrated on the big Tiger tank.

The PzKpfw IV Ausf F1 retained the short 75-mm gun, and was quickly replaced by the Ausf F2

Bundesarchiv

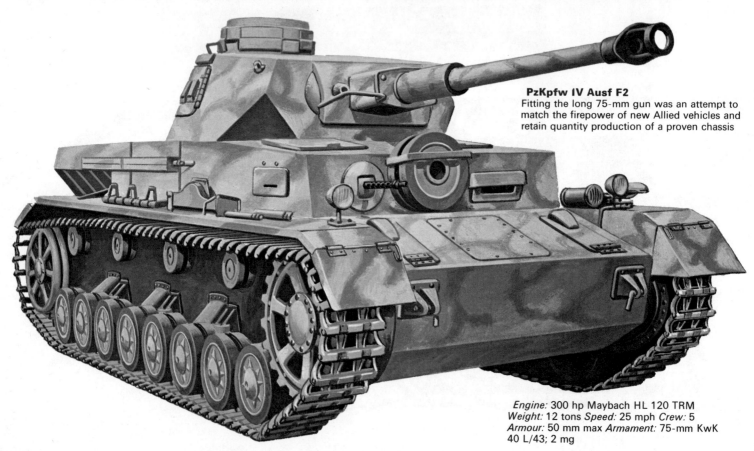

PzKpfw IV Ausf F2
Fitting the long 75-mm gun was an attempt to match the firepower of new Allied vehicles and retain quantity production of a proven chassis

Engine: 300 hp Maybach HL 120 TRM
Weight: 12 tons *Speed:* 25 mph *Crew:* 5
Armour: 50 mm max *Armament:* 75-mm KwK 40 L/43; 2 mg

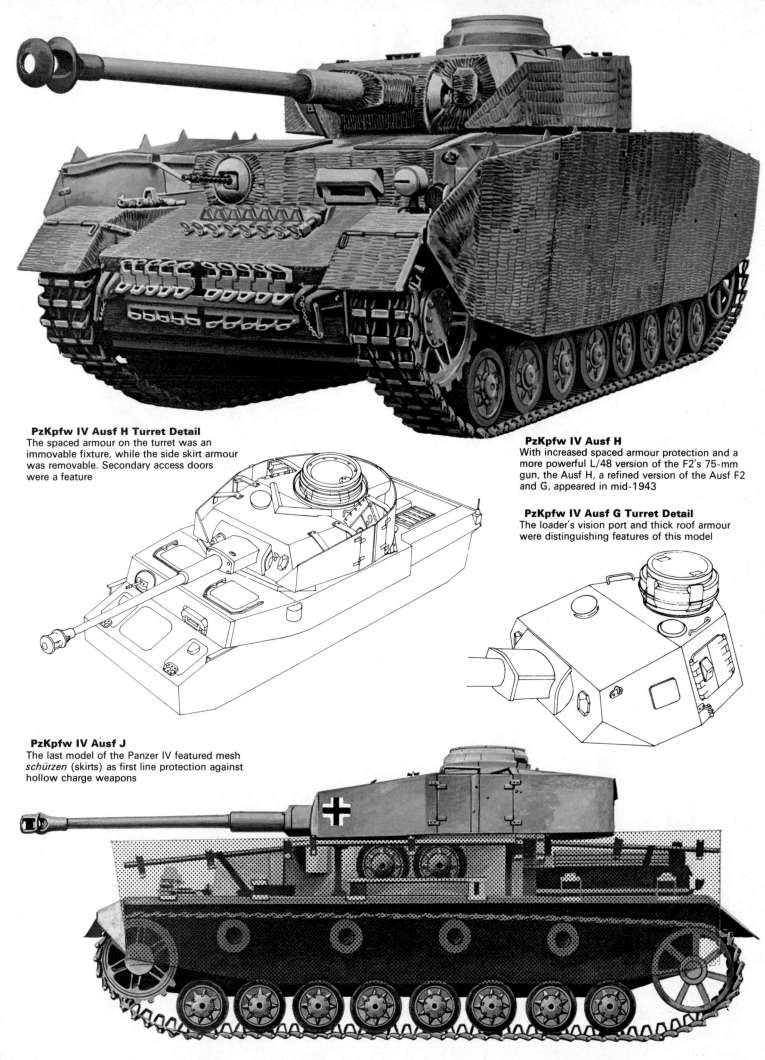

PzKpfw IV Ausf H Turret Detail
The spaced armour on the turret was an immovable fixture, while the side skirt armour was removable. Secondary access doors were a feature

PzKpfw IV Ausf H
With increased spaced armour protection and a more powerful L/48 version of the F2's 75-mm gun, the Ausf H, a refined version of the Ausf F2 and G, appeared in mid-1943

PzKpfw IV Ausf G Turret Detail
The loader's vision port and thick roof armour were distinguishing features of this model

PzKpfw IV Ausf J
The last model of the Panzer IV featured mesh *schürzen* (skirts) as first line protection against hollow charge weapons

THE TIGER
A NEW BREED
OF PANZER

The Tiger tank originated from developments started in 1937. Henschel were instructed to design and construct a 30- to 33-ton tank prototype as a possible successor to the Panzer IV. The new vehicle was known as the DW I, DW being an abbreviation of *Durchbrüchswagen* (breakthrough vehicle). However, after one chassis with interleaved road wheel suspension had been built and testing had commenced, the trials were suspended in 1938 to allow work to be carried out on a further design for a 65-ton tank, the VK 6501. The VK 6501 was itself a further development of the original Panzer VI (the NbFz Model B).

Two prototypes of the VK 6501 were built and were undergoing trials when the project was cancelled and development resumed on the DW I. By 1940 Henschel had so improved the original design that it was re-designated DW 2. In this form it weighed 32 tons and accommodated a crew of five. The planned armament was the short 75-mm gun with two MG 34 machine-guns. Trials were carried out with a prototype chassis until 1941, by which time Henschel had received an order for a new design in the same class and weight as the DW 2. The development designation for the new vehicle was VK 3001, and Henschel's competitors, Porsche, MAN and Daimler-Benz, were also invited to submit designs.

The Henschel version, VK 3001 (H), was a development of the DW 2 and four prototypes were built, two in March 1941 and two the following October, differing only in detail from each other. The superstructure of the VK 3001 (H) resembled that of the Panzer IV, and the suspension consisted of seven interleaved road wheels and three return rollers per side. It was planned to mount the 75-mm L/48 gun in this vehicle, but the appearance that year of the Russian T-34 with its 76-mm gun, rendered the vehicle obsolete and development was discontinued. Two of the VK 3001 (H) chassis were converted to self-propelled guns, by lengthening and mounting a 128-mm K 40 gun. These two vehicles were used in Russia in 1942.

Prototypes from Porsche
The Porsche version, VK 3001 (P), was also known to its designers as the Leopard or Type 100. This turretless prototype incorporated several new design features such as petrol-electric drive and longitudinal torsion bar suspension. MAN and Daimler-Benz also constructed prototypes to this design but like the Henschel project they had become obsolete.

With the order for the VK 3001 an additional order had also been placed for a 36-ton tank under the designation VK 3601. This specification had been personally proposed by Hitler, and included a powerful, high velocity gun, heavy armour and a maximum speed of at least 25 mph. A prototype of this project was built by Henschel in March 1942, but experimental work on both the VK 3001 and VK 3601 was stopped when a further order for a 45-ton tank was received in May 1941.

Designated VK 4501, the intended vehicle was to mount a tank version of the 88-mm gun. With the order came a stipulation that the prototype was to be ready in time for Hitler's birthday on 20 April 1942, when a full demonstration of its capabilities was to be staged. As design time was limited Henschel decided to incorporate the best features of their VK 3001 (H) and VK 3601

(H) projects into a vehicle of the weight and class required. They planned to build two models, the type H 1 mounting an 88-mm KwK 36 L/56 gun, and the Type H 2 with a 75-mm KwK L/70, although the H 2 existed only as a wooden mock-up at that time.

Porsche also received an order for a prototype to the VK 4501 specification and like Henschel they decided to incorporate as many as possible of the design features from their previous model, the VK 3001 (P), which had performed well on trials.

The demonstration of the two competing prototypes, the VK 4501 (H) and VK 4501 (P) duly took place before Hitler at Rastenburg, when the Henschel design was considered to be superior. An order for production to commence in August 1942, was given and the vehicle was designated *Panzerkampfwagen VI Tiger* (Tiger) Ausf E. The Ordnance Dept. number was SdKfz 181.

Production history
The Tiger was subsequently in production for two years, from August 1942 until August 1944, and in this period a total of 1350 vehicles were delivered out of 1376 ordered. Maximum monthly production was achieved in April 1944, when 104 Tigers were built. It is interesting to note that the specified weight of 45 tons was exceeded in production by about 11 tons.

The Tiger was technically the most sophisticated and best engineered vehicle of its time. The hull was divided into four compartments: the forward two housed the driver and hull gunner/radio operator, the centre was the fighting compartment, and the engine compartment was at the rear. The driver sat on the left and steered by means of a wheel which acted hydraulically on the Tiger's controlled differential steering unit. Emergency steering was provided for by two steering levers one either side of the driver operating disc brakes. These brakes were also used for vehicle parking and were connected to a foot pedal and parking brake lever. A visor was provided for the driver and this was opened and closed by a sliding shutter worked from a hand wheel on the front vertical plate.

Fixed episcopes were provided in both the driver's and the wireless operator's escape hatches. A standard German gyro direction indicator and instrument panel were situated to the left and right of the driver's seat respectively. The gearbox separated the two forward crew members' compartments. The machine-gunner/radio-operator seated on the right manned a standard 7.92-mm MG 34 in a ball mounting in the front vertical plate; this was fired by a hand trigger and sighted by a KZF cranked telescope. The radio sets were mounted on a shelf to the operator's left.

The centre fighting compartment was separated from the front compartments by an arched cross member and from the engine compartment in the rear by a solid bulkhead. The floor of the fighting compartment was suspended from the turret by three steel tubes and rotated with the

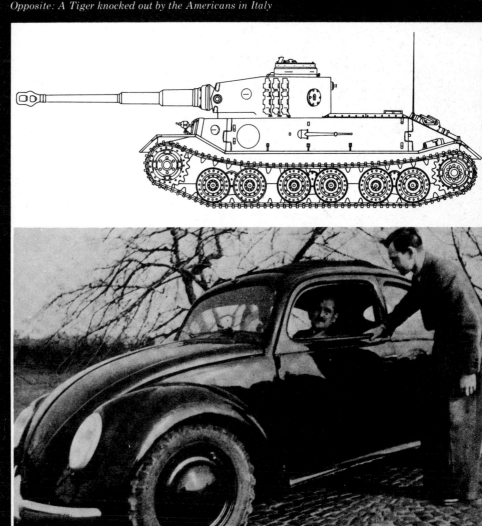

VK 4501 (P), the Porsche prototype for the Tiger, and Dr Ferdinand Porsche, its designer (in car). Opposite: A Tiger knocked out by the Americans in Italy

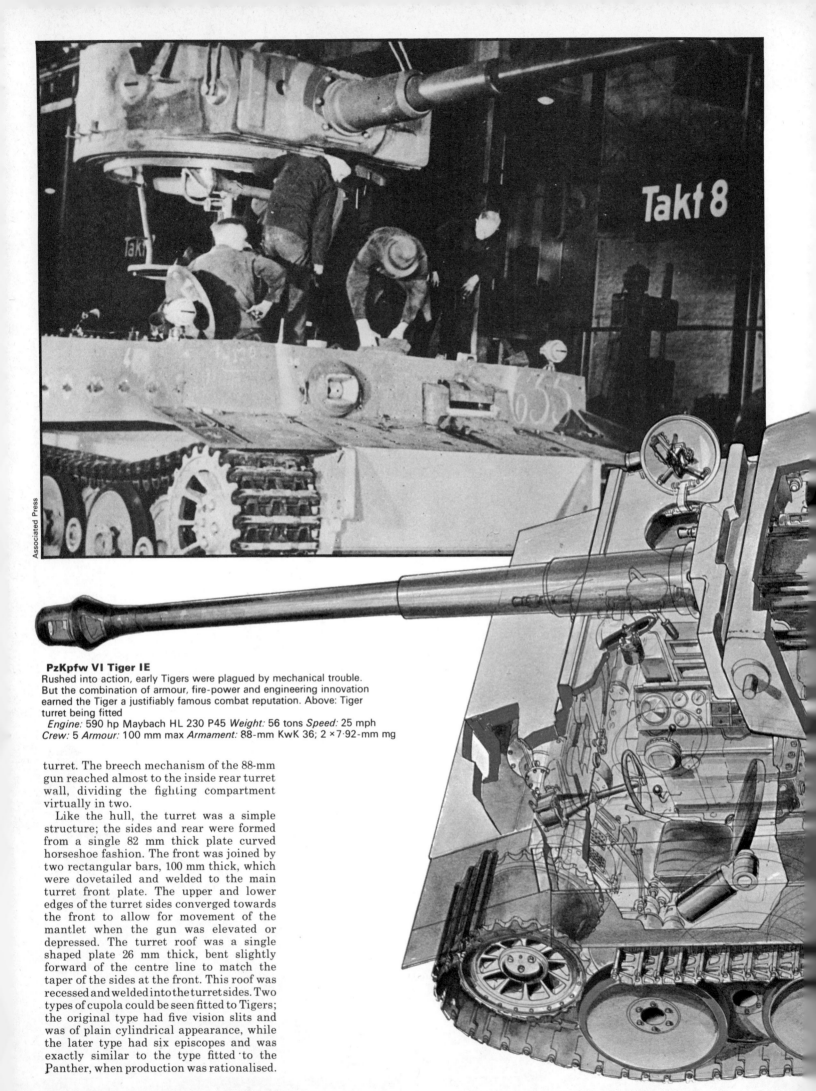

Associated Press

PzKpfw VI Tiger IE
Rushed into action, early Tigers were plagued by mechanical trouble.
But the combination of armour, fire-power and engineering innovation
earned the Tiger a justifiably famous combat reputation. Above: Tiger
turret being fitted
  Engine: 590 hp Maybach HL 230 P45 *Weight:* 56 tons *Speed:* 25 mph
Crew: 5 *Armour:* 100 mm max *Armament:* 88-mm KwK 36; 2 ×7·92-mm mg

turret. The breech mechanism of the 88-mm
gun reached almost to the inside rear turret
wall, dividing the fighting compartment
virtually in two.

Like the hull, the turret was a simple
structure; the sides and rear were formed
from a single 82 mm thick plate curved
horseshoe fashion. The front was joined by
two rectangular bars, 100 mm thick, which
were dovetailed and welded to the main
turret front plate. The upper and lower
edges of the turret sides converged towards
the front to allow for movement of the
mantlet when the gun was elevated or
depressed. The turret roof was a single
shaped plate 26 mm thick, bent slightly
forward of the centre line to match the
taper of the sides at the front. This roof was
recessed and welded into the turret sides. Two
types of cupola could be seen fitted to Tigers;
the original type had five vision slits and
was of plain cylindrical appearance, while
the later type had six episcopes and was
exactly similar to the type fitted to the
Panther, when production was rationalised.

Other external turret fittings were three NbK 39 90-mm smoke generators on either side towards the front, two stowage bins either side towards the front and two stowage bins either side of the centre line at the rear. The bins were used to stow the bedding, rations, packs and other personal effects of the crew.

Tiger I was the first German combat tank to be fitted with overlapping road wheel suspension, arranged with triple overlapping and interleaved wheels of a steel disc type with solid rubber tyres. The overlapping wheel system was adopted for optimum weight distribution. There were eight independently sprung torsion bar axles on each side. In order to carry all the axles inside the hull it was necessary to stagger them on the floor so that the right hand axles trailed aft and the left hand axles led forward. It was thus possible to incorporate the maximum number within the vehicle's length, and this resulted in an

extremely soft and stable ride for a tank of this weight and size.

Two types of track were used: one 28·5 in wide for combat and a narrower one 20·5 in wide for travel and transportation. When the narrow tracks were fitted the outer wheels were removed from each suspension unit. Though this type of suspension gave a superior ride, it also had its drawbacks, one being that the interleaved road wheels were liable to become packed with mud and snow during winter fighting, and if ignored until frozen this could jam the wheels. The Russians discovered this and took advantage of the situation by timing their attacks for dawn, when the vehicles were likely to have become immobilised during the night's frosts.

Very late production Tigers had steel disc type wheels with resilient internal rubber spring rims of the type fitted to the King Tigers and late model Panthers. In Tigers so fitted, the outside run of wheels

was omitted. This obviated to some extent the icing-up problem and also reduced overheating of the axle bearings.

The Tiger was originally fitted with a Maybach V-12 petrol engine, the HL 210 P45 of 21 litres capacity, but it was soon realised that the vehicle was underpowered and an uprated engine, the HL 230 P45 of 24 litres was substituted. The Tigers used in North Africa, and (after early experience of dust) in Russia in summer, were fitted with the Feifel air cleaner system. This was attached to the rear of the hull and linked to the engine over the engine cover plate. These tropical Tigers were known as the Tiger (Tp), but the Feifel air system was soon discontinued to simplify production.

While all earlier designs of German tank had the simple clutch-and-brake type of steering, the Tiger's greatly increased weight necessitated a more refined system. Henschel therefore developed and adopted a special steering unit, similar to the British

Merritt-Brown type, which was fully regenerative and continuous. It had the added feature of a twin radius turn in each gear. The gearbox, which was based on earlier Maybach types, gave no less than eight forward gear ratios and, with its preselector, made the Tiger very light and easy to handle for a vehicle of its size. The Tiger's mechanical layout followed that of previous operational German designs in that the transmission shaft led forward beneath the turret cage to the gearbox set alongside the driver.

The steering unit was mounted transversely in the nose of the tank, a bevel drive leading to a final reduction gear in each front sprocket. Power take-off for the hydraulic turret traverse unit, mounted in the turret floor, was taken from the rear of the gearbox, and it is typical of the Tiger's well-thought out design that the hydraulic unit could be disconnected from the power drive shaft by releasing a dog-clutch, thus allowing the turret to be lifted from the vehicle without the complications of disconnecting any other joints or pipes.

The first production Tigers were elaborately equipped for totally submerged wading to a depth of 13 ft with Snorkel breathing, but this proved an expensive luxury and with the need to simplify production this was discarded. Subsequent Tigers had a wading capability to a maximum depth of 4 ft.

One of the Tiger's biggest advances over any previous design was in its method of construction. In order to simplify assembly as much as possible and allow the use of heavy armour plate, flat sections were used throughout the hull. Hull and superstructure were welded, in contrast to previous German tanks where a bolted joint was used between hull and superstructure. The Tiger front and rear superstructure was in one unit and interlocking stepped joints, secured by welding, were used in the construction of both the lower hull and the superstructure. A pannier was, in effect, formed over each track by extending the superstructure sideways to full width and the complete length of the vehicle was so shaped from front vertical plate to tail plate. The top front plate of the hull covered the full width of the vehicle and it was this extreme width which permitted a turret ring of 6 ft 1 in internal diameter to be fitted which was of ample size to accommodate the breech and mounting of the 88-mm gun. The belly was also in one piece, being a plate 1 in thick and 15 ft 10¼ in long by 5 ft 11 in wide.

High velocity gun

The 88-mm KwK 36 gun which formed the Tiger's main armament had ballistic characteristics similar to those of the famous Flak 18 and Flak 36 88-mm high velocity AA guns from which it was derived. The principal modifications were the addition of a muzzle brake and electric firing by a trigger-operated primer on the elevating handwheel. A 7·92-mm MG 34 was coaxially mounted in the left side of the mantlet and was fired by mechanical linkage from a foot pedal operated by the gunner. The 88-mm had a breech of the semi-automatic falling wedge type scaled up from the conventional type used on smaller German tank guns. The weight of the barrel was balanced by a large coil spring housed in a cylinder on the left hand front of the turret. Elevation and hand traverse were controlled by handwheels to the right and left of the gunner respectively

and an additional traverse handwheel was provided for the commander's use in an emergency.

The hydraulic power traverse was controlled by a rocking footplate operated by the gunner's right foot. Because of the turret's weight, traverse was necessarily low-geared both in hand and power. It took 720 turns of the gunner's handwheel, for instance, to move the turret through 360° and power traverse through any large arc demanded a good deal of footwork (and concentration) by the gunner. Allied tanks – always more lightly armoured – were often able to take advantage of this limitation to get in the first shot when surprising a Tiger from the side or rear. In fact, this became almost standard procedure for engaging a Tiger tank, one tank attracting its attention from the front while one or two others attempted to work round to the flanks or rear to get in a shot at the more vulnerable areas of the vehicle.

For sighting purposes the gunner was provided with a binocular telescope, a clinometer for use in HE shoots, and a turret position indicator dial. Ammunition for the 88-mm gun was stowed partly in bins each side of the fighting compartment and partly alongside the driver and under the turret floor.

Early production Tigers were fitted with 'S' mine dischargers on top of the superstructure, a total of five being mounted in various positions on the front, sides and rear. These devices were installed for protection against infantry attacking with such anti-tank weapons as magnetic mines or pole charges. The 'S' mine was an anti-personnel bomb shaped like a jam jar and was about 5 in deep by 4 in wide. It was shot some 3 to 5 ft into the air, where it was set to explode and scatter its contents – 360 ⅜-in diameter steel balls.

However, when the turret design was amended in late 1943 to incorporate a periscope of the type fitted to the Panther, a standard *Nahverteidigungswaffe* (close-in defence weapon) was fitted in the turret roof in place of the extractor fan, which was itself moved to the centre. The *Nahverteidgungswaffe* had all-round traverse and was internally loaded, rendering the somewhat awkward S mine dischargers superfluous.

There were two Tiger tank variants: Tiger Command Tank (*Panzerbefehlswagen*), was designated PzBefWg *Tiger* Ausf E, SdKfz 267 or 258. The difference between these two sub-variants was solely in the wireless equipment fitted, the SkKfz 267 carrying combinations of the Fu 5 and Fu 8 radio and

Tiger Stowage
Below: Fighting compartment and turret, side view (right) rear view (left). Centre: Driver's compartment (left), hull gunner's compartment (right). Bottom: Tiger I side section

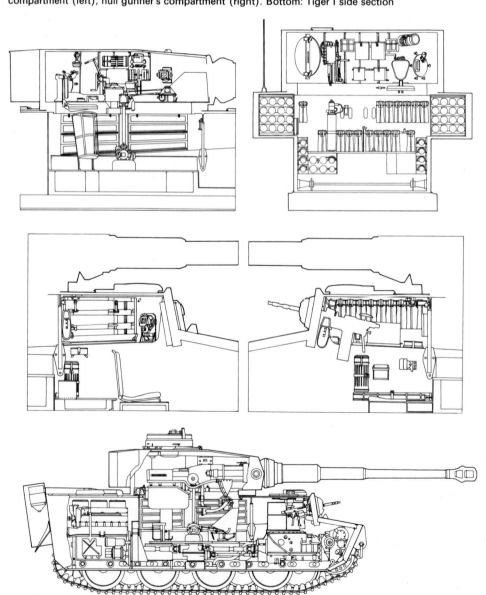

An early production model Tiger Ausf E, a disturbing sight in a French farmyard in 1944

the SdKfz 268 being fitted with combinations of the Fu 5 and Fu 7.

In addition there was the Tiger Armoured Recovery Vehicle, designated *Bergepanzer Tiger* I Ausf E. This was no more than a towing vehicle for assisting crippled or otherwise malfunctioning Tigers back to an area where repairs could be effected. The adaptation involved the removal of the main armament, sealing of the mantlet, fixing the turret in the traversed position and fitting a winch to the turret rear with a wire rope guide on the front. No lifting gear was provided.

At the time of its debut in service in late 1942, the PzKpfw VI Tiger I was an outstanding design among its contemporaries by virtue of its powerful gun and armour protection of up to 100 mm thick. These factors made the 56-ton Tiger the most formidable fighting vehicle then in service. It was, however, relatively costly to produce in terms of man-hours and difficult to adapt for mass production. In January 1944, the heavier and generally superior Tiger II Ausf B went into production with the result that successively fewer E models were produced until they were finally phased out of production completely in August 1944.

It was intended to use the Tiger as a heavy infantry or assault gun, and Tiger battalions were organised as independent units under GHQ troops. Armoured divisions engaged in a major operation would receive an allotment of Tigers to spearhead an attack, but owing to the Tiger's basic lack of manoeuvrability, it was always considered necessary to employ lighter tanks in supporting platoons on the flanks. Normally Panzer IIIs or IVs fulfilled this function.

It was later decided to include Tigers in the basic organisation of armoured divisions, but due to attrition which depleted the number of Tigers serviceable at any one time, it was never possible to put this plan into operation except in Waffen-SS armoured divisions. These divisions were among the first to receive Tigers, which went into service with such famous formations as the I SS Panzer Division *Leibstandarte* SS 'Adolf Hitler', and the II SS Panzer Division

'Das Reich'. The fact that there were never sufficient Tigers to go round was probably the greatest comfort that opposing forces could take from their appearance.

However, the Tiger's debut was not very sensational. Hitler was impatient to make use of the formidable new weapon as soon as possible in September 1942, against the opinion of senior staff officers who favoured building up Tiger strength during the winter, and perfecting tactics and training for a mass Tiger offensive in Russia in the Spring of 1943.

Therefore, in the earliest Tiger action of the war, on 23 September 1942, near Leningrad, the employment of this sinister-looking vehicle was restricted to such limited numbers that resolute action by anti-tank gunners taking full advantage of the situation was more than enough to counter their impact. This attack took place on terrain unsuitable for any successful tank action, and, restricted to single-file progress on forest tracks through swamps, the Tigers proved to be an easy target for the Soviet gunners posted to cover the tracks.

The British first encountered the Tiger in February 1943, near Pont du Fahs in Tunisia. Having received advance warning of the impending attack, the British anti-tank gunners were concealed with their 6-pounders with instructions to hold their fire until signalled. Two Tigers, flanked by nine Panzer IIIs and Panzer IVs, advanced with artillery support and were not engaged until the range had closed to 500 yards on each flank. Fire from the 6-pounders knocked out both Tigers.

Defensive role

Although the Tiger had been conceived as a powerful assault weapon, the changing tide of war in 1943/44 meant that Tigers were more and more used in a defensive role, in which they were very successful. The Tiger's bulk, limited mobility and susceptibility to mechanical failure were severe disadvantages in a war of movement. The limitations of very large, heavily armoured tanks were well demonstrated by the Tiger, but the conclusions were too late to influence the new generations of heavy AFVs which were to appear in 1944/45.

The story of Henschel's Tiger is inextricably tied up with the close rival designs of

Porsche, which were produced to meet the same *Heereswaffenamt* specification. The first Porsche design was to meet the same VK 3001 requirement as the first Henschel prototype. Strictly speaking this nominal 30-tonner was a replacement design for the Panzer IV and not a 'heavy tank' in the later sense of the term.

The Porsche design, known to the Army as the VK 3001 (P) was designated the Type 100 Leopard by the factory, and two prototypes were built in 1940. Porsche used a hull similar in shape to that of the existing Panzer IV, but employed petrol-electric drive with two air-cooled motors powering electric motors driving the front sprockets. Steering and gear changing was also electric, and there was torsion bar suspension. On test the vehicle put up a good running performance.

In May 1941, Porsche received the same development contract for the VK 4501 requirement, calling for a 45-ton vehicle mounting an 88-mm gun and to be ready for production within one year. The prototype was to be ready (with Henschel's prototype) for demonstration before Hitler in April 1942. With time running out, Porsche utilised the same sort of drive arrangement that had been developed for the now-abandoned VK 3001 (P). The new design was designated Type 101 by Porsche and featured petrol-electric drive as before. The petrol engines were air-cooled and proved a constant source of trouble in the design. Porsche contemplated hydraulic transmission as an alternative to electric transmission.

As a result of the competitive trials, however, the conventionally driven Henschel design was preferred to Porsche's, and the main production contract for what eventually became known as the Tiger went to Henschel. Only a few pilot models of the Porsche design were actually completed as gun tanks, and these were designated PzKpfw VI VK 4501 (P), Tiger (P), to distinguish them from the Henschel Tiger.

The only portion of the Porsche Tiger to enter production was the turret. This was designed by Krupp with the adapted L/56 gun. Henschel had contemplated a turret with a tapered-bore gun firing hard tungsten-steel armour-piercing shells. The scarcity of tungsten and the difficulties of machining the gun barrel led to the abandonment of this idea at an early stage, and the simple turret with conventional 88-mm gun intended for the Porsche was put into production for the Henschel Tiger.

Ninety Porsche Tigers had been ordered, however, partly as a safeguard against delays or failure of the Henschel Tiger and partly to appease Porsche and, in turn, Hitler, who admired Porsche's technical brilliance. But the mechanically unreliable Porsche Tigers were not wanted by the Army, and there remained the problem of how to utilise the redundant chassis. It was decided eventually to convert them to *Panzerjager* (heavy tank destroyers) and in late 1942, 85 of the completed chassis were transferred from the Steyr-Daimler factory at Nibelungen (the builders of Porsche designs) to Alkett for completion in their new form. Dr Porsche supervised the design of the new vehicle, which was subsequently designated *Jagdpanzer* Ferdinand (after Porsche's first name) and later *Panzerjager Tiger* (P) *Elefant* (SdKfz 184). These vehicles made their battle debut – unsuccessfully as it turned out – in the great Kursk tank battle.

THE PANTHER
THE EQUALISER

The unexpected appearance of the revolutionary T-34 tank in Soviet hands rendered all existing German front line tanks obsolete almost literally overnight and there was no tank of comparable size or performance available to the Germans, who did not until then suspect the Russians of having anything of such advanced design. This complacency had been caused entirely by the excellence and versatility of the Panzer IV, even though work on a planned successor had started in 1937. By 1941 the prototypes by Henschel, VK 3001 (H), and Porsche, VK 3001 (P), had been completed but, just prior to the invasion of Russia, when the T-34 was met, requirements were changed yet again in favour of a larger design with an 88-mm gun in the 45-ton class, the VK 4501. This eventually became the Tiger heavy tank but because the VK 4501 design was needed urgently, it incorporated many features from the earlier development prototypes and the Tiger thus owed nothing to the T-34 design. The 88-mm gun and the heavy (100-mm) armour specified for the VK 4501 design were, however, influenced by the T-34's appearance, for it was considered essential to have a tank with these features in production as a safeguard against any eventual Soviet development of an up-gunned and up-armoured version of the T-34.

Meanwhile General Guderian, by now the commander of *Panzergruppe* II, in whose sector the T-34 was first encountered in large numbers in November 1941, sent a report to his Army Group commander suggesting that the Armaments Ministry should appoint a commission most urgently to investigate what sort of new tank design – and anti-tank gun – would be needed to counter the T-34 threat and restore tank superiority to the Germans. The commission, Guderian suggested, should include representatives of the Army Ordnance Department, the main tank manufacturers, and the tank design section. The Armaments Ministry acted swiftly, and appointed just such a commission which was sent to Guderian's front for an 'on the spot' investigation on 20 November 1941, to assess the key features of the T-34 design.

The three main characteristics of this vehicle which rendered all existing German tanks technically obsolete were: (1) the sloped armour which gave optimum shot deflection all round; (2) the large road wheels which gave a stable and steady ride; and (3) the overhanging gun, a feature previously avoided by the Germans as impracticable. Of these the first was the

most revolutionary. Having received the commission's report on 25 November 1941, the Armaments Ministry promptly contracted with two principal armament firms, Daimler-Benz and MAN to produce designs for a new medium tank in the 30- to 35-ton class, under the ordnance designation VK 3002. To be ready the following spring, the specifications called for a vehicle with 60-mm frontal armour and 40-mm side armour, the front and sides to be sloped as in the T-34. A maximum speed of 34 mph was required.

In April 1942 the two designs, VK 3002 (DB) (Daimler-Benz) and VK 3002 (MAN), were submitted to a committee of *Waffenprufamt* 6, the section of the Army Ordnance Department responsible for AFV design and procurement.

Contrasting designs

The designs afforded an interesting contrast. The Daimler-Benz proposal was an almost direct copy of the T-34 in layout, with the addition of a few excellent refinements. It had a hull shape similar to that of the T-34 with turret mounted well forward – so far forward in fact that the driver sat within the turret cage – with remote control hydraulic steering. An MB507 diesel engine was fitted with transmission to the rear sprockets, again exactly duplicating the T-34 layout. Paired steel bogies (without rubber tyres) were suspended by leaf springs, and other features included escape hatches in the hull sides and jettisonable fuel tanks on the hull rear in the T-34 fashion.

The VK 3002 (DB) was in fact a remarkably 'clean' design with much potential. Leaf springs, for example, were cheaper and easier to produce than torsion bars, and the use of all-steel wheels recognised the problem of rubber shortage from the start. The compact engine and transmission at the rear left the fighting compartment unencumbered for future up-gunning or structural change, while the diesel engine itself would have been an advantage in later years when petrol supply became acutely restricted.

By comparison, the VK 3002 (MAN) displayed original German (rather than Russian) thinking: it was sophisticated rather than simple. It had a higher, wider hull than either the VK 3002 (DB) or the T-34, with a large turret placed well back to offset as much as possible the overhang of the long 75-mm gun which was called for as the main armament. Torsion bar suspension was used with interleaved road wheels, while a Maybach HL 210 petrol (gasoline)

V-12 engine was proposed, with drive to the front sprockets. The internal layout followed conventional German practice with stations for the driver and hull gunner/radio-operator in the front compartment.

When the respective Daimler-Benz and MAN designs were submitted by the *Waffenprufamt* 6 committee in April 1942, Hitler was most impressed with the Daimler-Benz T-34 type proposal, though he suggested that the gun be changed from the 75-mm L/48 model to the longer and more powerful L/70 weapon. Hitler's intervention in the proceedings at this stage led to an order for 200 VK 3002 (DB) vehicles being placed, and prototypes actually went into production.

The 'Panther Committee'

However, the committee set up by *Waffenprufamt* 6 – which was already being called unofficially the 'Panther Committee' – preferred the VK 3002 (MAN) design, because it was far more conventional by existing German engineering standards. MAN's proposal was accepted in May 1942 and they were asked to go ahead and produce a mild steel prototype as fast as possible. Later in

Bundesarchiv

1942, the order for the 200 Daimler-Benz vehicles was quietly and discreetly rescinded.

Meanwhile Dipl Ing Kniepkampf, chief engineer and designer of *Waffenprufamt* 6, took personal charge of detail design work on the MAN vehicle. This reflected the priority given to the Panther project. Kniepkampf was a key figure in German AFV design at this time, having been with *Waffenprufamt* 6 since 1936 and remaining as chief engineer almost until the war's end in 1945. Among other things he was principally responsible for German half-track development and introduced many of the characteristic features like interleaved road wheels, torsion bar suspension, and the Maybach-Olvar gearbox to German tanks.

In September 1942 the first pilot model of the VK 3002 (MAN) was completed and tested in the MAN factory grounds at Nuremberg. This was closely followed by the second pilot model which was transported to the *Heereswaffenamt* test ground at Kummersdorf for official army trials. By this time, the Tiger tank had just started in production, but its shortcomings – includ-

Sloped armour, a 75-mm high velocity gun, wider tracks and improved suspension gave the Panther a chance against the Russian T-34

ing excessive weight, low speed, and poor ballistic shape – were already recognised. The new vehicle was ordered into immediate production as the PzKpfw V Panther, with the ordnance designation SdKfz 171, and it got absolute top priority rating.

The first vehicle was turned out by MAN in November 1942, only two months after completion of the prototype. It was planned

to build at a rate of 250 vehicles a month as soon as possible, but at the end of 1942 this target was increased to 600 a month. To reach such an ambitious target it was necessary to form a large Panther production group. Daimler-Benz were quickly switched from work on their now-discarded Panther design (prototypes of which had by then been almost completed) and in November 1942 they too began tooling up to build

273

Panthers, the first vehicles coming from Daimler early in 1943.

Also in January 1943, Maschinenfabrik Niedersachsen of Hanover and Henschel, began tooling up to build MAN Panthers – production started in February/March – and scores of sub-contractors were soon involved in what became one of the most concentrated German armaments programmes of the war. Even aircraft production was cut back, partly to conserve fuel for use in tanks but partly, also, to free manufacturing facilities for the urgently needed Panther engines and components.

The monthly target of 600 vehicles was never achieved, however. By May 1943 output had reached a total of 324 completed vehicles and the monthly production average over the year was 154. In 1944 a monthly production average of 330 vehicles was achieved. By February 1945, when production tailed off, 4814 Panthers had been built. Panthers were first used in action in the great Kursk Offensive of 5 July 1943, but the haste with which the design had been evolved, and the speed with which it had been put into production, led to many teething troubles. In particular the complicated track and suspension gave trouble, with frequent breakages, while the engine presented cooling problems and this led to frequent engine fires. In the early months of service, indeed, more Panthers were put out of service by mechanical faults than by Soviet anti-tank guns.

Conventional layout

The Panther conformed to the usual layout of German tanks. It had the driving and transmission compartment forward, the fighting compartment and turret in the centre, and the engine compartment at the rear. The driver sat on the left-hand side forward with a vision port in front of him in the glacis plate. This was fitted with a laminated glass screen and had an armoured hinged flap on the outside which was closed under combat conditions. Forward vision was then given by two fixed episcopes in the compartment roof, one facing directly forward while the other faced half left in the '10–30' position. This restricted vision considerably and in the later Ausf G a rotating periscope was fitted in place of the fixed forward episcope, and the half left episcope and the vision port were completely dispensed with.

The Ausf G was thus easily recognised from the front since it had an unpierced glacis plate. The wireless operator, who was also the hull machine-gunner, sat on the right side forward. In the early Ausf D models, he was provided with a vertical opening flap in the glacis plate – rather similar to a vertical letterbox flap – through which he fired a standard MG 34 machine-gun in action. In the Ausf A and G, however, this arrangement was replaced by an integral ball-mount which took the MG 34 in the standard type of tank mounting. The radio equipment was fitted to the radio-operator/gunner's right and was located in the sponson which overhung the tracks. Episcopes were fitted, duplicating the driver's side.

Between the driver and wireless operator was the gearbox, with final drive which led each side to the front sprockets. The gearbox was specially evolved for the Panther as this vehicle was bulkier and heavier than previous designs and developed considerably more power. Known as the AK 7-200,

the gearbox was an all synchromesh unit with seven speeds. Argus hydraulic disc brakes were used for steering in the conventional manner by braking the tracks. However, the epicyclic gears could also be used to assist steering by driving one or other of the sprockets against the main drive, so retarding the track on that side and allowing sharper radius turns.

In the turret the gunner sat on the left hand side of the gun and was originally provided with an articulated binocular sight; this was later changed to a monocular sight. He fired the gun electrically by a trigger fitted on the elevating handwheel. The coaxial machine-gun, fitted in the mantlet, was fired by the gunner from a foot switch. Traverse was by hydraulic power or hand, the same handwheel being used for either method.

The vehicle commander's station was at

the left rear of the turret, the offset location being necessitated by the length of the breech which virtually divided the turret into two. A prominent cupola was provided which was of the 'dustbin' type with six vision slits in the Ausf D. In the Panther Ausf A and G, however, an improved cupola was fitted which had seven equally-spaced periscopes. This had a hatch which lifted and opened horizontally. Above the cupola was fitted a ring mount for an MG 34 which could be used for air defence, though this mount was sometimes removed.

The remaining crew member was the loader who occupied the right side of the turret. The turret itself had sloped walls and a rounded front covered by a curved cast mantlet. The cage had a full floor which rotated with the turret. Drive for the hydraulic traverse was taken through the centre of the floor to a gearbox, and thence

Armaments Minister Albert Speer tests a Kettenrad tracked motorcycle. His programme of rationalisation was vital to the quantity production of Panthers and Tigers

Panther Armour Disposition

to an oil motor. Turret openings were kept to a minimum and included a large circular hatch on the rear face which was an access/escape hatch for the loader and was also used for loading ammunition. On the left side beneath the cupola was a circular hatch for ejecting expended cartridge cases, but this was eliminated in the Ausf A and G. Similarly eliminated were three small pistol ports, one in each face, which were normally plugged by a steel bung and chain.

The engine, housed in the rear compartment, was a Maybach HL 230 P30, a V-12 23 litre unit of 700 hp at 3000 rpm. This was a bored out version of the HL 210 engine originally planned. The earliest production vehicles had the HL 210 unit, but like most AFV designs, the Panther had increased in weight considerably during the development stage with a heavier gun and heavier armour (among other things) bringing its

weight up from the 35 tons originally envisaged to about 43 (metric) tons. The easy way to increase the power to compensate for the added weight was to enlarge the engine. Access to the engine for maintenance was via a large inspection hatch in the centre of the rear decking. Cooling grilles and fans occupied most of the remainder of the rear decking. Exhaust was taken away through manifolds on the squared off hull rear. Most Panthers had stowage boxes flanking the rear exhaust pipes, but these were not always fitted.

The actual hull and superstructure was a single built-up unit of machinable quality homogenous armour plate of welded construction but with all main edges strengthened by mortised interlocking as had been pioneered in the Tiger. The heaviest armour (80 mm) was on the glacis plate which was sloped at 33° to the horizontal, an angle

specifically selected to deflect shells striking the glacis upwards clear of the mantlet.

The suspension consisted of eight double interleaved bogie wheels on each side, the wheels being dished discs with solid rubber tyres. Some very late production vehicles, however, had all-steel resiliently sprung wheels of the type subsequently fitted to the late production Tiger and the Tiger II (Royal Tiger). The first, third, fifth and seventh wheels from the front were double while the intervening axles carried spaced wheels overlapping the others on the inside and outside. Each bogie axle was joined by a radius arm to a torsion bar coupled in series to a second bar lying parallel to it.

PzKpfw V Panther Ausf D
The first production model suffered from a high breakdown rate, a result of hasty development. But the Panther design was a fine balance of protection, speed and hitting power
Engine: 700 hp Maybach HL 230 P30
Weight: 43 tons *Speed:* 25 mph *Crew:* 5
Armour: 120 mm max *Armament:* 75-mm KwK 42 L/70

The torsion bars were carried across the floor and the bogie wheels on the right hand side of the vehicle were set behind their respective torsion bars while those on the left were set in front. Thus, as in the Tiger, the wheel layout was not symmetrical. Though this suspension was technically advanced and gave the vehicle superb flotation, maintenance was complicated by the size of the wheels and consequent inaccessibility of the axles and torsion bars. Moreover, wheel replacement was a heavy and lengthy task.

The 75-mm L/70 gun mounted in the Panther was developed by Rheinmetall who had been asked July 1941 to design a high velocity version of the 75-mm weapon which could penetrate 140 mm of armour plate at 1000 metres. This high velocity gun requirement, initially envisaged for a field carriage, was among the weapons considered by the Panzer Commission in their deliberations of November 1941 which gave rise to the VK 3002 (Panther) specification. As a result, Rheinmetall were asked to design the turret and mount to hold this gun for installation in the VK 3002 design. The prototype gun was ready in early 1942, a weapon 60 calibres long. Test firing

indicated that performance was a little below the requested minimum, so the barrel was lengthened to 70 calibres, the improved prototype being ready for tests in June 1942. In this lengthened form the gun went into production. Initially it had a single baffle muzzle brake – and was so used on the earliest Panthers – but later a double baffle muzzle brake was adopted.

Panther production

The first Panther models which came off the MAN line from November 1942 were designated in standard German fashion as PzKpfw V Ausf A. The designation PzKpfw V Ausf B was earmarked for a proposed version of the vehicle which was to have the Maybach-Olvar gearbox in place of the specially developed AK-7-200 unit. However, the Maybach gearbox was considered unsuitable for installation in the Panther and the Ausf B never materialised. The first 20 Panthers which originally had the Ausf A designation were 'pre-production' vehicles. They had the 60 mm thick front armour as originally called for, the Maybach HL 210 engine, also as originally specified, a ZF 7 gearbox with clutch and brake steering, the earliest form of the

L/70 gun, and a cupola bulge in the side of the turret. From January 1943, however, Panthers appeared with all the design improvements suggested from trials with the pilot model. The glacis plate thickness was increased to 80 mm, the bored out HL 230 engine was fitted together with the new AK 7-200 gearbox which allowed single radius turns (ie, a definite fixed radius of turn depending on the gear engaged) and also made a neutral turn possible with the vehicle stationary. To simplify turret production, the cupola was shifted slightly to the right, thus eliminating the bulged housing.

Confused classification

This first full production type was designated PzKpfw V Ausf D. No record has been unearthed of an Ausf C model, but it seems almost certain that this was a 'paper project', like the Ausf B with some other proposed mechanical change. Much confusion has always existed over the designations of these early Panthers mainly because the Germans themselves later classed the early Ausf A vehicles with the full production Ausf Ds for record purposes. Early in 1943 they confused the record

PzKpfw V Panther Ausf A
The main Panther model encountered by the Allies in Normandy, the Ausf A featured a new cupola with armoured periscopes, a ball-mounted hull machine-gun, side-skirts and a *Zimmerit* finish

Panther Suspension
The torsion bar suspension proven on the Panzer III was re-worked in an ingenious way for the new weight problems posed by the Panther and Tiger. Interleaved road wheels (top) gave weight distribution in a compact space. Staggered torsion bars running the width of the hull (below) gave excellent flotation

further by identifying the original Ausf A as the Ausf D1 and the Ausf D as the Ausf D2.

Characteristics of the Ausf D were the 'dustbin' cupola, the vision port and machine-gun port on the glacis, smoke dischargers on the turret sides, and a straight edge to the lower sponson sides with separate stowage compartments fabricated beneath the rear ends. On later Ausf Ds the improved type of cupola was fitted and the smoke dischargers were dropped in favour of a bomb thrower installed in the turret roof and operated by the loader. Later Ausf Ds also had the skirt armour, which was adopted as standard to protect the top run of the tracks from bazooka hits. Zimmerit anti-magnetic paste covering to prevent the attachment of mines was another retrospective feature. All except the earliest vehicles had the L/70 gun with double baffle.

Next production model of the Panther was designated Ausf A, an anomaly which has not been fully explained, but may conceivably have resulted from an early administrative, phonetic, or clerical error, since the logical designation was Ausf E. Be that as it may, the Ausf A appeared in the latter half of 1943 and featured several detail improvements. Chief among these was the adoption of the new cupola with armoured periscopes, and the provision of a proper ball-mount for the hull machine-gun. Side skirts of 5-mm armour and a Zimmerit finish were standard. The side skirts were only loosely fixed by bolts and they were frequently removed, either by the crew or accidentally in combat conditions. The gunner's binocular sight was replaced by a monocular one, though this was not noticeable externally. To further simplify turret production, however, the pistol ports and the small loading hatch featured in the Ausf D were eliminated completely, leaving just the big loading/escape hatch in the turret rear. The Panther Ausf A was the main type encountered by the Allies in the Normandy fighting.

Action in Normandy

The final production model of the Panther in its original form, the Ausf G, was also in action in Normandy in June 1944. By this time the designation PzKpfw V had been dropped following a personal directive from Hitler on 27 February 1944, and the vehicle was simply known as the Panther Ausf G.

(At this same time, the Tiger was similarly re-designated as Tiger Ausf E.) Considerable modifications were featured in G model. The superstructure sides were altered, mainly to simplify production, so that the rear stowage compartments were now integral with the hull instead of separate additions. This gave a sloping lower edge to the sponsons. The hull sides were at the same time increased in thickness from 40 mm to 50 mm with the angle of slope altered from 30° to 40°. The driver's vision port was eliminated from the glacis plate and his vision was greatly improved by provision of a rotating periscope in place of the episcopes. New hinged hatches with spring-assisted opening replaced the original hatches provided in the hull roof for the driver and wireless operator. The earlier models had pivoted hatches which were found to jam easily. Internally, armoured ammunition bins were fitted inside each sponson with sliding armoured doors to reduce fire risk. The 75-mm ammunition stowage was also slightly increased in this model from 79 to 82 rounds.

Some amendments were made to external stowage, including the provision of a stronger method of attaching the skirt armour.

In very late production vehicles the cylindrical stowage box for the gun pull-through and cleaning gear was removed from the left side of the hull and mounted across the hull at the rear of the engine compartment. With Tiger and Panther production under way, a new generation of tanks was planned in 1943 which was to incorporate the lessons from existing designs. In particular, attention was to be given to simplifying production, economising on materials, reducing maintenance, and standardising components as far as possible. By this time economic conditions were extremely grim in Germany, with shortages of fuel, raw materials and manpower, and disruption of all aspects of life by continual Allied bombing – not to mention the demands of supplying several hard-pressed fighting fronts at the same time.

In February 1943 *Waffenprufamt* 6 asked MAN and Henschel to produce improved designs for the Panther and Tiger respectively, ensuring maximum interchangeability of parts. Henschel produced the Tiger II which went into production at the end of 1943 since a replacement for the somewhat unsatisfactory Tiger was urgently needed. The improved Panther, the Panther II, officially designated Panther Ausf F, was to have a hull similar to the existing Panther but with the same form of interleaved all-steel resilient wheels as the Tiger II. Other changes were to be the adoption of an improved gearbox and transmission, the AK 7-400, and mechanical parts such as brakes identical to those in the Tiger II.

Smaller turret

The armour on the hull top was to be doubled to 25 mm and the ball-mount was to be altered to take the MG 42. The major change, however, was to be a new design of turret, known as the *Panzerturm Schmal* (small), which as its name implies was much smaller than the original Panther turret. The object was to reduce weight, simplify production, reduce frontal area, eliminate shot traps beneath the mantlet (a weakness in the original Panther turret) and enable a larger gun to be fitted. It was to have a built-in stereoscopic rangefinder, and a gyrostabiliser for both the sight and the gun based on that fitted in American tanks. As part of the experimental work for this a standard Panther was fitted with a captured American gyrostabiliser for firing trials and proved to have its accuracy and effectiveness doubled.

A major tactical requirement was that any Panther II should be capable of instant conversion in the field to the command tank role by providing the necessary brackets and a second aerial in every turret. This meant that the commander's ultra short wave radio set could be 'plugged in' to any vehicle so that command and staff officers could quickly change to other tanks if theirs were damaged or immobilised.

The new small turret was developed as a separate project by Daimler-Benz under the direction of Dr Wunderlich, assisted by Col Henrici, a gunnery expert from *Waffenprufamt* 6. Kniepkampf was in over-all charge of both the Tiger II and Panther II projects. The new turret proved a most successful design. It had the same ring diameter as the old turret, but took 30% less time to make and had 30% more armour plate all within the same weight limit. It

could take the L/70 gun and was also designed to accommodate a proposed lengthened L/100 version of the same weapon. It could take the same 88-mm gun as the Tiger II as yet another alternative. The wide mantlet, difficult to manufacture, which characterised the old turret, was replaced by a relatively simple *Saukopf* (pig's head) mantlet, of conical shape as its description implies. To fit this conical shape the design of the L/70 gun was modified so that the recuperator and buffer cylinders were situated under the gun. The compressor for the barrel blow-out apparatus was eliminated, and compressed air was obtained instead from an additional cylinder round the recuperator which was kept charged by pressure from the gun recoil.

The result of these changes was that the wide welded cradle in the original Panther turret could be dispensed with completely. It also proved possible to eliminate the muzzle brake with the new small turret, even though the recoil forces were massively increased from 12 to 18 tons.

The *schmal* turret was ready before the Panther II, but though running prototypes of this vehicle were produced in 1944, the rapidly deteriorating conditions of the war with facilities curtailed and the need for continued supply of types already proven in service meant that the Panther II, or Ausf F, never went into production and there was thus no chance for this fine design, virtually a perfected version of the original Panther, to prove its mettle. It would have undoubtedly been a much more useful and potent weapon than the very heavy and bulky Tiger II.

Final production models of the Panther Ausf G did, in fact, incorporate one feature intended for the Ausf F. This was the all-steel resilient road wheel which replaced the rubber-tyred type and became standard for late-production Tigers as well as the Tiger II. It is apposite also to mention here the engine improvements which were gradually introduced for the Maybach HL 230 motor. Overheating had been a problem in the early days, as previously mentioned. This was overcome by fitting a second cooling pump, modifying the coolant distribution, and improving the bearings and cylinder head seal. Later Panthers, therefore, proved very much more reliable than the vehicles involved in the Kursk debacle. To improve the power of the HL 230 for the Tiger II and Panther II it was proposed to increase the compression ratio and incorporate fuel injection and, later, superchargers. Though modified prototypes were built and tested, the war had ended before the up-rated engine could go into production.

Special conversions

There were several special purpose conversions of the Panther, two of these for the command role. For unit commanders the *Befehlspanzer* Panther was produced. These were simply versions of either the Ausf D, A or G fitted with extra radio equipment and the associated aerials. A second radio receiver and transmitter were fitted to the inside right wall of the turret and the loader acted as operator. There were two externally similar models, differing only in the radio installation. The SdKfz 267 had Fu 5 and Fu 8 equipment, while the SdKfz 268 had Fu 5 and Fu 7. (Fu 5 was the standard German tank wire-

Below: Schmalturm on a test rig. Note the tube-frame turret cage. The projected turret for the second generation Panther, in spite of being lighter and smaller, was designed to take a larger gun. The saukopf mantlet and sloped armour gave very good defensive qualities. Top right: A Panther is given an engine change by a half-track mobile gantry. Below right: Panther Ausf G, with side track skirts and a generous cover of Zimmerit anti-magnetic mine paste coating

Peter Chamberlain

less for short-range communication within tank regiments and battalions on RT or MCW transmission. Fu 7 was the standard air co-operation set and Fu 8 was the set used for main divisional nets. [Fu = *Funk* = Radio].) In each case ammunition stowage was reduced to 64 75-mm rounds. *Befehlspanzer* Panthers were used by regimental and battalion command and staff officers and could only be distinguished externally by the extra aerials (or the call sign number when this was visible).

The *Beobachtungspanzer* Panther (SdKfz 172) was an old Ausf D converted as an OP vehicle for observation officers, commanders, and staff officers of SP artillery regiments. The gun was replaced by a short wooden dummy, the turret was fixed in place, an extra radio was fitted, and a map table was added inside the turret. A ball-mounted MG 34 in the turret front was the only armament.

Finally there was the *Bergepanzer* Panther (also known as the *Bergepanther*), designated SdKfz 179, which was a recovery vehicle specially for work with tanks in the 45-ton class. The *Bergepanther* replaced the 18-ton half-track in the heavy recovery role, since it took up to three of these latter vehicles to move heavy tanks like the Tiger or Panther. The *Bergepanther* was an old Ausf D model converted by the removal of the turret and the fighting equipment. A movable winch and winch motor were installed in the fighting compartment. A limited superstructure was provided round the former turret opening consisting of heavy wood cladding over mild steel framing. A canvas tilt could cover the complete compartment in bad weather. An 'A' frame was fitted over the rear decking and this supported a towing eye and towing rollers. A heavy earth spade was hinged on the hull rear and was raised and lowered from the vehicle's winch. There was a light demountable jib which could be erected either side for lifting work and there was an MG 34

or a 20-mm cannon for air defence, mounted as required.

Had the war dragged on (and had Germany been able to maintain its planned production programme unhindered), the Panther and Panther II would have become the backbone of the German panzer divisions (together with the Tiger II and *Jagdtiger* in lesser numbers), and from late 1944 a rationalisation programme was introduced (*Richtwert-Programm* IV) which terminated production of all earlier types in favour of the 'new generation' vehicles. The other type to be included in the new programme was a family of *Waffentragers* and SP types developed on a light chassis adapted from the Czech-built PzKpfw 38(t) (and its German-developed derivative the 38(d)).

However, the defeat of Germany in May 1945 brought Panther development to an end with much of the potential of the design still unrealised. A few Panthers served on

for a number of post-war years in the French Army, which equipped some units with captured vehicles. The other victorious nations each took a few Panthers for trials. The British actually built at least one Panther in 1946, using spare and cannibalised parts to assemble a 'new' vehicle which was used in comparative trials with the Black Prince and Centurion.

An interesting clandestine use of the Panther took place during the last desperate German offensive in the West, the so-called 'Battle of the Bulge'. Here at least ten, probably more, Panthers were effectively disguised and marked to resemble US Army M10 tank destroyers. The cupola was removed, together with the external stowage boxes on the hull. The turret and nose were then disguised with thin sheet metal to resemble the shape of the M10, including the distinctive rear overhang of that vehicle's turret counterweight. Despite being finished in very convincing US markings,

the phoney M10s enjoyed little success largely because the subtlety of the idea was nullified by the general confusion prevailing – on both sides – during the frantic days of the Ardennes offensive.

Japan purchased a Panther with a view to producing licence-built versions for use in the Pacific. Colonel Ishide, an AFV specialist with the Japanese Military Mission to Germany, witnessed a demonstration of both a Tiger I and a Panther at the Henschel works on 30 July 1943. As a result the Japanese bought one sample of each vehicle with the intention of shipping them to Japan for further trials and probable production of their own versions. This transaction took place in November 1944 when a Tiger and Panther were formally handed over to the Japanese Mission in Germany. However, by this time there was no means of getting these vehicles safely to Japan and as far as is known they never actually left Germany.

TIGER II
THE KING PANZER

The original Tiger design was finalised before the Soviet T-34 was encountered so it lacked the excellent ballistic shape which was a feature of the Panther. In late 1942 when the first Tigers had entered service, *Waffenamt* asked Henschel if they could produce a modified design with sloping side plates as in the new Panther. By February 1943 the mechanical failings of the Tiger had been revealed and a development specification for a replacement vehicle was issued under the designation VK 4502.

Porsche offered a design, the VK 4502 (P), which had alternative layouts with the turret either well forward or at the back of the hull. These were designated Types 180 and 181 by Porsche. The new vehicle was to mount the longer L/71 KwK 43 gun as in the *Elefant* tank destroyer. A turret design was put in hand by Wegmann of Kassel. Porsche once again offered petrol-electric drive for the Type 180, and once again this was rejected by *Waffenamt* as unreliable and too sophisticated for service conditions. In addition a shortage of copper ruled out electric transmission.

Henschel also put forward a design, the VK 4503 (H), to meet the requirement, and this was powered conventionally like their Tiger. This design was accepted and the project was put in hand as a top priority

effort. The design was not finalised until October 1943, however, three months later than scheduled. This happened because the Panther II had since been designed, and under the new rationalisation policy it was decided that as many parts of the Panther II as possible had to be incorporated, and design features were to be standardised between the two vehicles.

Turret changes

The prototype of the new tank, now designated PzKpfw Tiger Ausf B, was ready in November 1943 and the first production models appeared in February 1944 on the Henschel production line, parallel to that which was still building the Tiger Ausf E. Ironically enough, history repeated itself when the first 50 of the new vehicles mounted the Porsche turret, now surplus from the cancelled Type 180. All subsequent Tiger Ausf Bs had a Henschel-designed turret, which was of more simple construction than the Porsche model, and it avoided the Porsche turret's built-in shot-trap under the mantlet.

Henschel remained the sole builders of the Tiger Ausf B during its whole production life. By September 1944 Tiger Ausf E production ceased completely in favour of the new vehicle. At this particular time a

production rate of 100, increasing to 145, vehicles a month had been fixed. In practice, however, disruption by enemy bombing and shortage of materials meant the best ever monthly output, in August 1944, was only 84. By March 1945, this total had fallen to 25. Final total of Tiger Ausf Bs produced was 484.

Although the official designation was PzKpfw Tiger Ausf B (SdKfz 182) it was variously known as the *Königstiger* (King Tiger), or Tiger II to the Germans, and as the Royal Tiger to the Allies. The King Tiger went into action for the first time in Normandy in June 1944, and the first was knocked out by the Allies in the August of that year.

The King Tiger was derived from the Tiger Ausf E and both tanks had many features in common. At the same time it bore a much closer resemblance to the late model Panther and the projected Panther Ausf F (Panther II), which was due for production in 1945. Many fittings were standard to both the King Tiger and the late model Panther and Tiger. These included cupolas, engines, engine covers and road wheels, to mention just a few. Compared with the other vehicles, the King Tiger had thicker armour (maximum 150 mm) and was dimensionally larger. There

was a small, conical *Saukopf* (pig's head) mantlet, and a well-sloped turret and sloped morticed armour plates making up the hull. Main armament was the 88-mm KwK 43 (L/61) as fitted to the *Elefant*.

Internally the vehicle followed the usual German layout with front sprocket drive and crew positions as for the Panther. The big turret had several interesting features – it lacked the usual basket and was built out very wide over an immense 73-in diameter

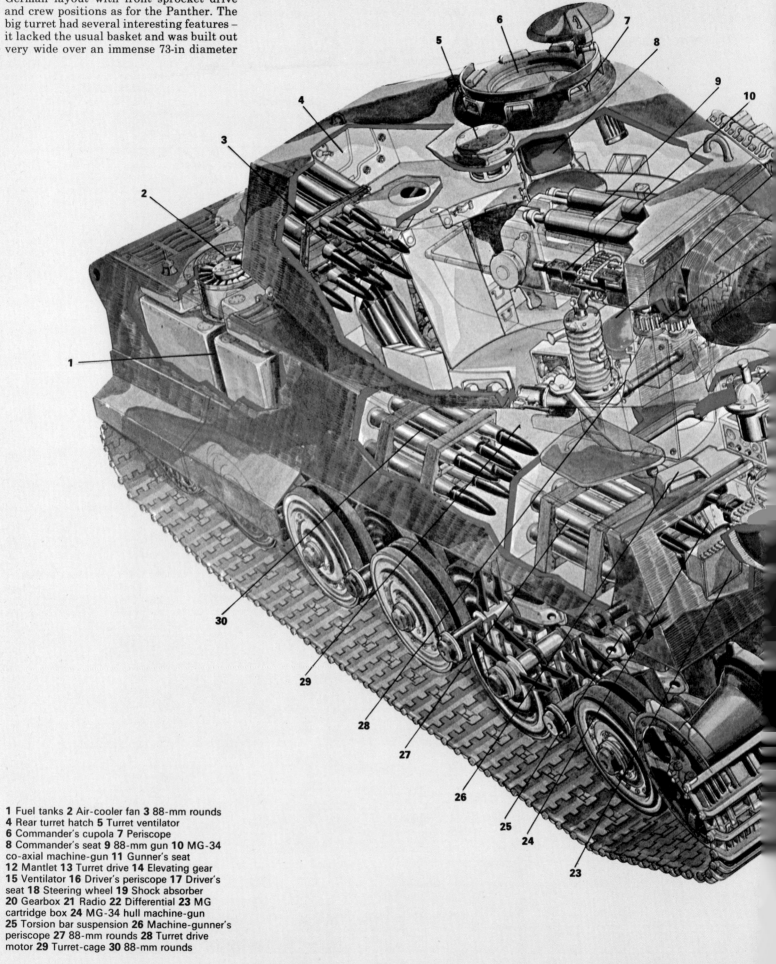

1 Fuel tanks 2 Air-cooler fan 3 88-mm rounds
4 Rear turret hatch 5 Turret ventilator
6 Commander's cupola 7 Periscope
8 Commander's seat 9 88-mm gun 10 MG-34 co-axial machine-gun 11 Gunner's seat
12 Mantlet 13 Turret drive 14 Elevating gear
15 Ventilator 16 Driver's periscope 17 Driver's seat 18 Steering wheel 19 Shock absorber
20 Gearbox 21 Radio 22 Differential 23 MG cartridge box 24 MG-34 hull machine-gun
25 Torsion bar suspension 26 Machine-gunner's periscope 27 88-mm rounds 28 Turret drive motor 29 Turret-cage 30 88-mm rounds

turter ring. To assist in loading the big ammunition rounds carried, 22 ready-for-use rounds were mounted, 11 per side, in the rear turret bulge, thus giving the loader a minimum handling movement. Power traverse was as for the Panther and Tiger.

Suspension was by torsion bars and it followed the same type of arrangement as in the Tiger Ausf E. But, the wheels were overlapped rather than interleaved as on the Tiger E. This change was adapted to simplify the maintenance problems which had been inherent with interleaved road wheels. Similarly, the tendency for the wheels to freeze solid with packed snow was obviated to some extent. Steel-tyred resiliently sprung wheels (which featured a layer of rubber between two steel tyres) were standard on the King Tiger as on the late model Tiger Es and Panthers. A commander's version of the King Tiger, differing only in internal radio stowage, was also produced and it carried the designation of *Befehlspanzer* Tiger B.

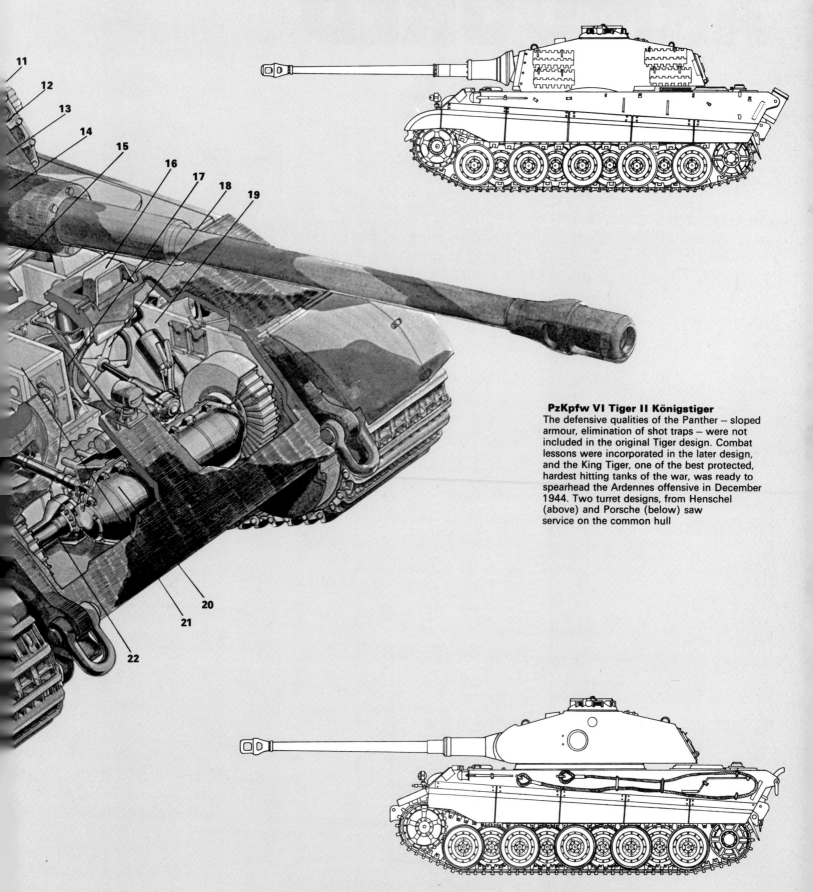

PzKpfw VI Tiger II Königstiger
The defensive qualities of the Panther – sloped armour, elimination of shot traps – were not included in the original Tiger design. Combat lessons were incorporated in the later design, and the King Tiger, one of the best protected, hardest hitting tanks of the war, was ready to spearhead the Ardennes offensive in December 1944. Two turret designs, from Henschel (above) and Porsche (below) saw service on the common hull

THE ARMOURED GIANTS

The structure of the German tank industry was anything but super-efficient. Partisans of rival schemes fluttered around Hitler's court to win approval for grandiose new projects. Thus the 'mad inventors' had their day and the super-heavy tanks were born . . .

The obsession with size culminated in a series of super-heavy tanks of gigantic proportion. Porsche was the driving force behind the first of these, the *Maus*, while the second type to be built, the E 100, was sponsored by the *Heereswaffenamt* as a competitive design.

First consideration of a super-heavy tank was made in late 1941, when Dr Ferdinand Porsche was at his most influential with Hitler. Porsche drew up a project which he called *Maus* (Mouse) and in August 1943, he got approval, again directly from Hitler, to go ahead and build a prototype. At this time Porsche's personal stock was low with the *Waffenamt*; he was removed as head of the Panzer Commission and none of his designs had been selected for production.

Hitler possibly thought that the *Maus* project would recompense Porsche for the past failures, or at least keep him away from other projects. Also, at this period, Hitler was still enthusiastic about the super-heavy tank, and his armaments minister, Albert Speer, had endorsed the idea.

Tank inventor's dream

The *Maus* was a tank inventor's dream, a wild extravagance of design, in short the ultimate example of the divergence of ideas between the designer of little knowledge of tactical requirements and the men with practical experience. The resultant vehicle weighed 180 tons, was over 30 ft long and had 240 mm thickness of frontal armour. A special engine of 1200 hp was designed by Daimler and was tested in both its diesel and petrol forms in two prototypes.

True to Porsche's previous designs, the *Maus* had electric transmission, and the internal layout, with central engine, was similar to that of the *Elefant*, which had been produced from the Tiger (P). The armour was made up of flat rolled plates and the main plates were mortised and welded together. The turret had a rounded front made from a single bent plate, 93 mm thick. The rear plate of the turret was slightly sloped.

Special equipment

Maximum road speed, via electric generators under the turret floor, was $12\frac{1}{2}$ mph. There were 48 partly interleaved wheels in four sets of four wheel bogies providing the suspension, and over-all track height was low in relation to the size of the vehicle. The axles of each set of bogies were sprung, and each bogie was set on a longitudinal torsion bar. The vehicle was gas proof and it could wade to a depth of 26 ft using a snorkel tube to provide air. The snorkel installation was to allow the *Maus* to cross rivers, for few bridges could take the weight of the monster.

The designed turret weighed 50 tons (the weight of a complete Centurion tank) and the armament was a 128-mm gun (as fitted to the *Jagdtiger* tank destroyer) plus a 75-mm gun mounted co-axially. The first turret was not completed, in fact, until the middle of 1944, and the two prototypes completed trials with weighted simulated turrets.

Initially the *Maus* project was known under the code name *Mammut* (Mammoth) and Krupp were contracted to build it. The first prototype was completed in November 1943 and the dummy turret was added the following month. Trials were held

An early prototype of Maus, Dr Ferdinand Porsche's idea of a super heavy tank. Pressures on the German war industry led to work on it being halted in 1944, and the prototypes were blown up in the last weeks of the war in Europe

in the winter, lasting until May 1944. The following month the second prototype was delivered. Trials took place at Krupp's test area in Meppen.

Development abandoned

Two more hulls were under construction during the closing months of the war and a total of six vehicles were known to have been ordered. But, in the last year of war, work on the *Maus* virtually stopped. In April 1944, Hitler personally ordered that all work on giant tank projects was to cease in favour of devoting all resources to building established proven tanks like the Panther and King Tiger. The *Maus* prototypes were blown up in the last weeks of the war as the Russians closed in on Meppen. Guns, turrets, hulls and test firing rigs were found by Allied Intelligence officers abandoned and partially destroyed.

Porsche and the tank's builders, Krupp, had not stopped at producing the *Maus*, however, for design studies were found at Krupp for tanks of 110, 130, 150 and 170 tons and all carried the designation 'Krupp-Maus'. Also discovered was a project study for a version of the *Maus* carrying a 305-mm breech-loading mortar. The project was named 'Bear'. The most fantastic of all, however, was a preliminary layout for a giant 1500-ton vehicle with an 800-mm gun as main armament and two 15-cm guns in auxiliary turrets on the rear quarters. Frontal armour was to be 250 mm at 45° and it was planned to power this monster with four submarine diesel engines.

The E 100 project was virtually the *Heereswaffenamt* rival to the *Maus*. There was considerable opposition to Porsche and his unconventional mechanical ideas within the Ordnance Department, and no

opportunity was lost to play down his projects.

Under Heydekampf at the Panzer Commission a long-term plan to produce a rationalised series of 'new generation' tanks was drawn up, the so-called *Entwicklungtypen* or E series. This range of tanks was to use standardised parts and was to be built in classes of varying sizes to replace existing vehicles. Types were as follows:

Designation	Weight class
E 10	10–15 tons
E 25	25–30 tons
E 50	50 tons (Panther replacement)

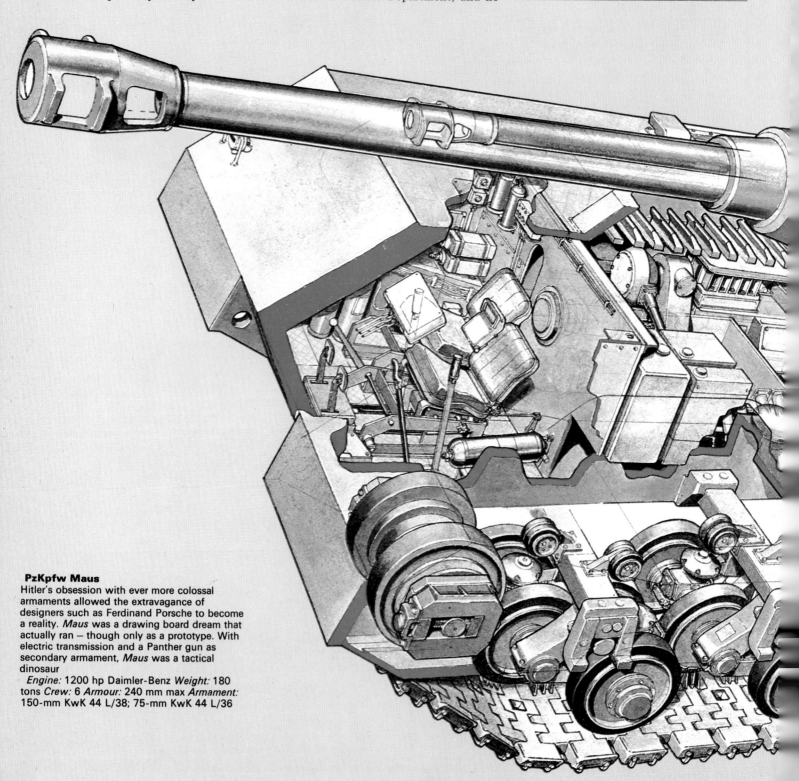

PzKpfw Maus
Hitler's obsession with ever more colossal armaments allowed the extravagance of designers such as Ferdinand Porsche to become a reality. *Maus* was a drawing board dream that actually ran – though only as a prototype. With electric transmission and a Panther gun as secondary armament, *Maus* was a tactical dinosaur
Engine: 1200 hp Daimler-Benz *Weight:* 180 tons *Crew:* 6 *Armour:* 240 mm max *Armament:* 150-mm KwK 44 L/38; 75-mm KwK 44 L/36

E 75	75 tons (Tiger/King Tiger replacement)
E 100	130–140 tons (super heavy type)

Of these only the E 100 project was actually started. This appears to have come about purely as an attempt to rival Porsche's work. When Porsche started work on the *Maus*, the first action of *Heereswaffenamt* was to place an instant contract with Henschel, builders of the King Tiger, for a much enlarged, super-heavy version of the King Tiger. This project was known unofficially as the *Tiger-Maus* and officially as the *Löwe* (Lion) or VK 7001. The armament was to be the same 128-mm gun as the *Jagdtiger*.

With the *Entwicklungtypen* programme drawn up, opportunity was taken to cancel the VK 7001 in favour of the E 100. The E 100 was to be a 140-ton tank with the same turret and gun layout as the *Maus*, except that a 150-mm gun was envisaged as the main weapon. Tiger engines and transmission were specified for the prototype, but an enlarged Maybach engine, the HL 234, was to be developed for production vehicles.

E 100 construction
Road wheels, sprockets and idlers were to be similar to those used on the King Tiger, as was the basic hull shape. Armoured covers were proposed for the tracks, which were one metre wide. To keep the vehicle narrow enough to fit inside the railway loading gauge, the track covers and outer idlers and sprockets were removable. Also, narrow travelling tracks were to be fitted

as on the Tiger. Removable folding jibs on the turret sides were to be used for removing the track covers as required.

After Hitler's order that work on super-heavy projects was to cease, construction of the E 100 prototype proceeded slowly at Henschel's test plant at Haustenbeck, Paderborn. Parts were delayed by the confused industrial situation in Germany during the last year of the war, and at the cessation of hostilities in May 1945 only the bare hull and suspension was partially completed. This was captured by the British, and the E 100 project ceased prematurely. Thus the last of the super-heavy tanks was still-born, and the remainder of the projected 'E' series designs never got beyond the initial drawing board stage. Like National Socialist Germany the development of armoured vehicles had reached the end of the road.

An *SdKfz 233* of the Afrika Korps

GERMAN FIGHTING VEHICLES

Peter Chamberlain and *Chris Ellis*

The German army was unique at the start of the Second World War, both in its range of specialised armoured fighting vehicles and in the tactical thinking behind them. As the war progressed, and new problems were encountered, the types of vehicle proliferated, while a variety of equipments were fitted to the half-tracks and to obsolete or captured tank chassis. And with armies fighting in North Africa and Russia, far from the factories and workshops of Germany, the fine balance was lost. Fighting vehicle production became more and more a matter of improvisation, as the struggle against time grew ever more desperate.

This chapter examines Germany's armoured fighting vehicles, and the ideas that have influenced AFV development ever since.

CONTENTS

ARMOURED CARS
REICHSWEHR RUNABOUTS

The early German victories of the Second World War were achieved largely through the bold use of fighting vehicles, closely supported by aerial bombardment, to bypass enemy fortifications, isolate his forces and attack his nerve centres. But the finely honed Panzer Divisions that shattered Poland, France and – for a while – Russia, had their roots in the *Reichswehr*'s early experiments with armoured cars – the 'paper panzers' of the 1920s and early 1930s

Armoured cars played a more important part in the development of armoured doctrine in Germany in the years between the two world wars than in any other major military power. This was largely due to the severe restrictions imposed by the Versailles Treaty of 1919, which prohibited tanks in the small new post-war German army (the *Reichswehr*) but permitted a number of armoured cars for police and patrol work.

In the First World War the Germans trailed way behind the British in the development of armoured cars and by the time of the Armistice in 1918 had only a token number of types in service. Erhardt were the main builders, and 20 new Erhardt vehicles were, in fact, the first armoured cars to see service with the *Reichswehr*. Some 30 armoured lorries were also pressed into service to help the new Republic's police and army to deal with the smouldering civil unrest of 1919.

A further stopgap type was a conversion of the First World War Daimler KD I four-wheel drive artillery tractor to make quite an effective armoured car. This was thus one of the earliest of all four-wheel drive armoured vehicles, known as the *Panzerkraftwagen* Daimler DZVR (*Daimler-Zugmaschine mit Vier Radantrieb*, or four-wheel drive Daimler tractor). Less than 50 of them were converted to armoured cars with a simple armoured box body with side doors.

The Versailles Treaty conditions were imposed from January 1920 under the aegis of the Allied Control Commission. By late 1920 the *Reichswehr* had organised seven motor transport battalions, each of which was allowed by the Commission to have 15 armoured personnel carriers – 105 armoured vehicles in all.

Dummy armoured cars used for training by the Reichswehr, on parade in 1933. Inset: An earlier type, with the body mounted on a tricycle

Bundesarchiv

The resulting vehicle was again based on the Krupp-Daimler KD I, but the chassis was improved to allow all the wheels to be of the same diameter (the DZVR had rear wheels bigger than the front as in the original tractor). No armament or cupola was permitted on the new vehicle – it merely had a box-like armoured body with loopholes, and provided seating for 12 men in addition to the three-man crew. Known as the *Gepanzerter Mannschaftstransportwagen*, SdKfz 3 (armoured personnel carrier, special vehicle 3) it was the forerunner of the many armoured cars and troop carriers that were to see service with the German forces up to the end of the war.

The SdKfz 3 had four-wheel drive and solid tyres, weighed a little under 11 tons and was 20 ft long; top speed was 31 mph and it had 11 mm of armour. Though of limited tactical value – the cross-country performance was poor – these vehicles remained in service well into the 1930s, some becoming radio-fitted command vehicles when more modern armoured cars were available in large numbers.

The other type of armoured car built in the early 1920s was the so-called *Schutzpolizei Sonderwagen* (protection police special vehicle). It was intended for internal security duties and was a much more effective vehicle than the SdKfz 3. The state police numbered 150,000 men and the Versailles Treaty allowed one car for every thousand. However, only about 104 were actually made, three different types built by Daimler, Erhardt, or Benz. They were generally similar to the SdKfz 3 but had twin turrets with machine-gun, a cupola, and rear steering as well as front steering. They were built and placed in service between 1920 and 1923, and in the 1930s some were taken over by the *Reichswehr*. Later, a few were converted as armoured transport vehicles for high ranking Nazis and used as such in the Second World War. In the 1920s however, they served the police, 72 with the Prussian state police and 32 with other states.

Gathering momentum
In the meantime the Army was becoming more ambitious, and in the latter half of the 1920s, under the leadership of General Hans von Seeckt, development of armoured vehicles for the *Reichswehr* took on a new momentum. A secret agreement was signed with Soviet Russia (the Rapallo Agreement) in 1926, under the terms of which the Soviets agreed to make available testing facilities for the Germans at Kazan on the Volga. This meant that some of the Versailles Treaty limitations preventing testing of prototypes on German soil would be circumvented.

The various existing armoured cars did not satisfy von Seeckt. He wanted the army to have vehicles suitable for reconnaissance and with a good cross-country performance under all conditions. The designation *Gepanzerter Mannschaftstransportwagen* was retained, however, since 'armoured troop carriers' were still within the Treaty limitations.

In 1926, *Heereswaffenamt Waffenprufamt* 6 (Army Armaments Weapons Testing Dept 6) responsible for the supply of vehicles, issued a new specification and invited manufacturers to submit prototypes for testing. These vehicles were to be certainly the most technically advanced wheeled cars ever developed at that time.

The specifications were extremely exacting, and included:
Six or more road wheels with multi-axle drive
Top road speed of 65 kmph and minimum road speed of 5 kmph
A distance of 200 km to be attainable for three successive days at average speed of 32 kmph
Ability to climb gradients of 1 in 3
Ability to cross trenches 1·5 m wide unaided
Wading ability to a depth of 1 m
Front and rear steering with equal performance in each direction, the change-over not to take more than ten seconds
Maximum turning circle of four times wheel base
Minimum engine noise
Chassis weight of maximum 4 tonnes and overall combat weight of maximum 7·5 tonnes
Ground clearance of 0·3 m
The vehicle to be able to run on standard gauge railway tracks without preparation
The ability to float without preparation, and a swimming speed of 5 kmph (later dropped)
A crew of five: commander, driver, first gunner, second gunner, radio-operator/rear driver.

The technical requirements also included very stiff specifications for engine, transmission and armour application.

So sophisticated were all these features that it was not possible to adapt any existing design. New and radical prototypes were needed and Daimler-Benz, Magirus and Büssing-NAG were finally selected to submit designs.

The Daimler-Benz model was an eight-wheeler, the ARW/MTM 1 (*Achtradwagen/mannschaftstransportwagen* 1, or eight wheel vehicle/personnel carrier). This was one of the first designs to exercise the technical talents of Dr Ferdinand Porsche, whose many original ideas were later to make him famous both as a tank and car designer. All eight wheels on the Daimler were driven and it was of monocoque construction – a fairly novel idea for its day – with no conventional chassis. The front and rear pairs of wheels were steered and the four centre wheels turned as a group according to which end was steering. Water propulsion was achieved by a propeller and the hull was somewhat boat-shaped to facilitate the amphibious requirement. Two prototypes were built.

The Magirus design was an eight-wheeler similar to the Daimler but it does not appear to have been completed or delivered.

Büssing-NAG's design was even more complex, a ten-wheeler with all wheels independently suspended and with the two outer groups arranged in fours as steering bogies and the centre pair fixed. The steering wheel within the body was set on a vertical column and could be mounted into the steering box serving either end of the vehicle, depending on the direction of travel. Known as the ZRW (*Zehnradwagen* or ten-wheel vehicle) it had slab-sided additions to the basic boat-shaped superstructure.

The ZRW was not so successful as the Daimler eight-wheeler, being more unwieldy and – because of its central steering – having a less satisfactory degree of control. On amphibious trials the vehicle sank.

Officers consult their maps by an SdKfz 232 eight-wheeled radio-equipped armoured car

SIX-WHEELERS

The great financial slump of 1929–30, which affected Germany even more savagely than other major nations, saw the premature end of the ambitious and complex multi-wheel drive designs. It was clear that the high cost of such sophisticated vehicles would take up too much of the already limited defence budget and, at a conference held at the *Waffenprufamt* 6 offices in March 1930, the projects were reluctantly abandoned 'since the present financial status of the Reich makes vehicles of this size and type far too expensive'.

However, there was a fortuitous substitute, for in 1929 *Waffenprufamt* 6 had issued a requirement for a six-wheel military truck based on the production commercial 6×4 chassis built by the major truck factories. To meet this requirement Magirus had produced the M206, Büssing-NAG the G31 and Daimler-Benz the G3. A further suggestion put to the three manufacturers was the possibility of adding an armoured body to make a substitute armoured car to replace the abandoned eight- and ten-wheelers.

Daimler-Benz and Magirus were in the van of development – Daimler in particular being very successful with their G3 chassis. In 1929 an improved chassis, the G3a, was built. Daimler-Benz built the prototype armoured car in late 1929, using the G3 chassis, now known as the G-3(p) – for *panzerte* or armoured – with a faceted armoured body made by DeutschenWerke of Kiel. *Waffenprufamt* 6 had insisted on dual steering front and rear, but apart from this the layout of the vehicle betrayed its commercial truck origins. The water-cooled gasoline engine was at the front with drive to the rear axles, each of which had its own differential. The front axle steered from a normally positioned driving wheel, but there was an element of complication in providing the rear steering position, just forward of the rear axle, to steer the front

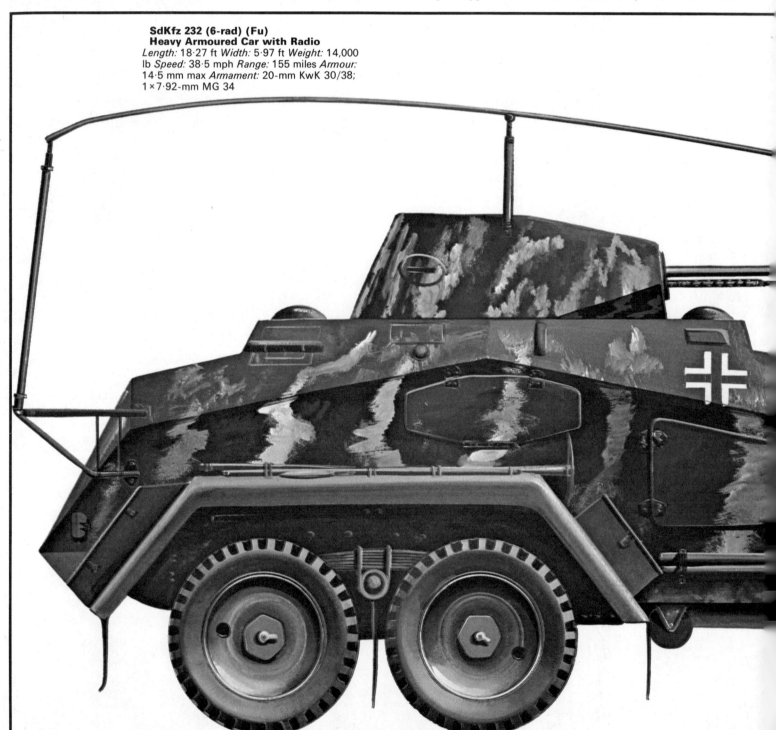

SdKfz 232 (6-rad) (Fu)
Heavy Armoured Car with Radio
Length: 18·27 ft *Width:* 5·97 ft *Weight:* 14,000 lb *Speed:* 38·5 mph *Range:* 155 miles *Armour:* 14·5 mm max *Armament:* 20-mm KwK 30/38; 1×7·92-mm MG 34

wheels. The multi-angled armoured super-structure followed the outline of a car, with a bonnet covering the front. There was a small turret on top with a 7·92-mm MG 13 machine-gun. The conventional chassis weighed 2·17 tons and the body weighed 2·26 tons, a total weight that made the fitting of large tyres essential. The gearbox was a commercial type with four forward and one reverse gear.

Trials of the prototype showed the need for a larger radiator and a stronger front axle, both of which were duly incorporated. One feature of the vehicle, which was to become a characteristic of later types also, was the canted steering wheel, slewed over on its column to reduce the overall height of the armoured superstructure by precious inches.

The first production vehicles, based on the Daimler-Benz G3a chassis, were delivered in 1932 and were used in the summer army exercises that year. Büssing-NAG

delivered their first 12 vehicles to an externally similar design the following year. The full army designation for this armoured car was *Schwerer Panzerspähwagen* (sPzspäh) SdKfz 231 *mit Fahrgestell das Leichter Gelandeganghger Lastkraftwagen (o)*, or, in English, heavy armoured reconnaissance vehicle type 231 on the chassis of the light cross-country truck (commercial). The short designation was SdKfz 231 (6-rad), the 6-rad (six-wheel) being added when eight-wheel vehicles appeared later. Basically, the vehicles had a simple girder type chassis, whose steering front axle had an additional linkage allowing them to be steered from a rear position. Semi-elliptic springs provided the suspension and a rubber loop type track or chains could be fitted round the rear pairs of wheels to give extra traction.

In 1934, Magirus-built vehicles of the same type appeared, but these differed in having side-mounted spare wheels which

were free to revolve, thus aiding cross-country traction if the vehicle became bogged down.

Production of these six-wheelers ran to 1000 units by 1936 when they were superseded by a new eight-wheel design. The six-wheelers remained in service, however, and were still first-line equipment at the time of the invasions of Poland in September 1939 and of France in May 1940. Thereafter they disappeared rapidly and were used for training and internal security duties up to the end of the war.

There were actually three variants of the six-wheelers. The first, sPzspähw SdKfz 231 (6-rad), was the basic reconnaissance vehicle used by the heavy platoons of motorised reconnaissance battalions, and was popularly known as the *Waffenwagen*. The Daimler-built vehicle had a single 7·92-mm MG 34 machine-gun in its turret, but the Büssing and Magirus vehicle had a co-axial mount with a 20-mm KwK 30 or 38 plus one MG 34. These latter vehicles also had a mount on the turret roof for carrying another MG 34 for anti-aircraft defence.

Commanders' vehicles

The sPzspähw (Fu) – for *Funkwagen* or radio vehicle – SdKfz 232 was the model used by unit commanders. It was based on either the Magirus or Büssing chassis and had the co-axial turret mount with 20-mm gun. A curved 'bedstead' frame aerial on the same axis as the body and attached by poles to the rear superstructure was the characteristic feature of this vehicle. An ingenious front support in the shape of a shallow 'U' was pivoted on the turret top, thus allowing the turret to traverse without hindrance from the top hamper of the aerial. A 100-watt radio for communications with the rear was fitted in this vehicle.

The sPzFuWg (*Schwerer Panzerfunkwagen*) SdKfz 263 was superficially similar to the SdKfz 232, but actually it differed in many ways. It was intended in the first place as an armoured command vehicle for a higher formation commander or his staff, and the turret was fixed in the forward position. The frame aerial was larger than that of the SdKfz 232 and was fixed, but had provision for being lowered. The vehicle lacked the 20-mm gun and had only an MG 34 in the turret front, with no AA machine-gun. When used as a mobile headquarters, a separate pole radio mast which was carried in sections in the vehicle could be erected on the ground. Lacking turret traverse gear, the vehicle had more internal space, allowing room for a staff officer's map board and a radio-operator and wireless gear to be carried.

There was only one other contemporary six-wheel armoured car in German service at this period. This was the SdKfz 247 (6-rad), which in German military terminology was the *Schwerer Gepanzerter Personenkraftwagen auf Fahgestell des Leichter Gelandegangiger Lastkraftwagen (o)* (heavy armoured personnel carrier on the chassis of the light cross-country lorry). This vehicle was of little significance and was essentially the chassis of the Krupp L2H 143 6×4 1·5-ton truck with an armoured body. It had no turret or armament, and the few built between 1936 and 1938 were used as command or observation vehicles for high-ranking officers in the early parts of the Second World War. The sloped armour body was similar in style to that of the SdKfz 231 (6-rad) six-wheeler.

An SdKfz 231 six-wheeled heavy armoured car, used during the invasions of Poland and France

Bundesarchiv

FOUR-WHEELERS

In the meantime, the limitations imposed by the defence budget and the Versailles Treaty had combined to push armoured car development in Germany in a new and more realistic direction. In order to train an army officially deprived of tanks in tank warfare, it was necessary to find some substitute for actual tanks.

In a 1928 military exercise a commercial truck was disguised to look like a tank, and the success of this idea led *Waffenprufamt* 6 to have some *Panzernachbildung* (simulated armoured fighting vehicles) made up for use on a more extensive scale. Made of wood or card outline, the earliest of these 'paper panzers' were simply sectional dummy tank bodies fitted round the Hanomag or BMW Dixi light cars then in military service.

In 1930 a standardised design was introduced, this time for a simulated armoured car, which featured an aluminium body on an Adler Standard 6 car chassis. Later, thin sheet steel was used. The Adler had a dummy rotating turret and carried a full crew of commander, gunner, radio-operator, and driver. (This idea proved so successful that dummy armoured cars and tanks were used for training up to the end of the Second World War, thus releasing valuable fighting vehicles for service.)

Having built a dummy armoured car on the Adler Standard 6 chassis, the next logical step was to use this readily available 4 × 2 chassis as the basis for an actual armoured car, resulting in the Kfz 13 series which were designated as *Mittlerer Gepanzerter Personenkraftwagen* (medium armoured passenger car). The requirement for this vehicle was issued in 1932, and it entered service in 1934, remaining in use into the early 1940s. Daimler headed the production consortium. The vehicle betrayed its motor car ancestry in its front engine layout and conventional leaf spring suspension. A radio/command version was also produced, the Kfz 14.

Basic weapons carrier

The Kfz 13 was a *Waffenwagen* (weapons carrier) for reconnaissance units, open-topped and very basic. The vehicle had a 3 litre 6-cylinder engine with four-speed gearbox, and the lightly armoured superstructure (5 mm to 8 mm) was welded. The semi-elliptic spring suspension was of the motor car type. The armament was an MG 34 on a pedestal mount with a light front shield. Internally, the vehicle retained a normal motor car style of layout. The Kfz 14 differed in carrying a radio set with a frame aerial which could be lowered when not required. Kfz 14s had no gun and always accompanied Kfz 13s since the latter lacked any kind of signalling facilities.

While excellent for training, both the Kfz 13 and Kfz 14 were patently below any minimum combat standard. They were too lightly armoured to withstand even small arms fire, the cross-country performance was severely limited, and they were rather unstable due to the high centre of gravity imposed by the armoured body. Even so, Kfz 13s were still in service as late as 1943 on the Russian front because of shortages of later

types. Thus, though to a large extent the Kfz 13 and 14 were expedients, they were important in enabling troops to familiarise themselves with the use and deployment of armoured cars. They saw their widest use in the Polish campaign of 1939, the occupation of Czechoslovakia earlier that year and the invasion of France in 1940.

In 1936, when re-armament was fully under way in Germany, there came the adoption of a number of *Einheits* (standard) chassis which were to serve as the basis for all new wheeled vehicle designs. Among them was the *Einheitsfahrgestell I für Schwerer Personenkraftwagen* (standard chassis model 1 for heavy passenger cars) which had a rear-mounted engine and was intended specifically for armoured car use. (*Einheitsfahrgestell II* was similar but had a front-mounted engine and was used for conventional military vehicles of the personnel carrier or *Kübelwagen* type.)

Most of the requirements originally laid down for armoured cars were built into the specification, albeit in a less exacting form, and there was an additional demand for maximum standardisation of components throughout the *Einheitsfahrgestell* range. The motor industry had to design new chassis to meet all these requirements as existing ones were not robust enough. Chassis prototypes were displayed at the 1936 Berlin Motor Show, and the first armoured cars on this chassis, intended to replace the Kfz 13 and 14, appeared in service the following year.

The full designation was *Leichter Panzerspähwagen mit Einheitsfahrgestell I fur Schwerer Personenkraftwagen* (light armoured reconnaissance car with the standard chassis model 1 for the heavy passenger car). The original requirement was for a *Waffenwagen* with a single machine-gun, but this was later changed to a 20-mm tank gun, and a radio car for command and communication use. In fact there were several variations, the most important models being:

Leichter Panzerspähwagen(MG) (SdKfz 221) *mit 7·92-mm MG 34.*

Leichter Panzerspähwagen (SdKfz 221) *mit 28-mm sPzD 41.* (This was a Second World War period conversion of the original model with a tapered bore anti-tank rifle.)

SdKfz 221 Light Armoured Car (opposite)
Length: 15·7 ft *Width:* 6·4 ft *Weight:* 8820 lb
Speed: 50 mph *Range:* 185 miles *Armour:*
14·5 mm max *Armament* 1 x 28-mm

Kfz 13 Adler Armoured Car
Length: 13·8 ft *Width:* 5·6 ft *Weight:* 4850 lb
Speed: 37 mph *Range:* 200 miles *Armour:* 8 mm
max *Armament:* 1 × 7·92-mm MG 34

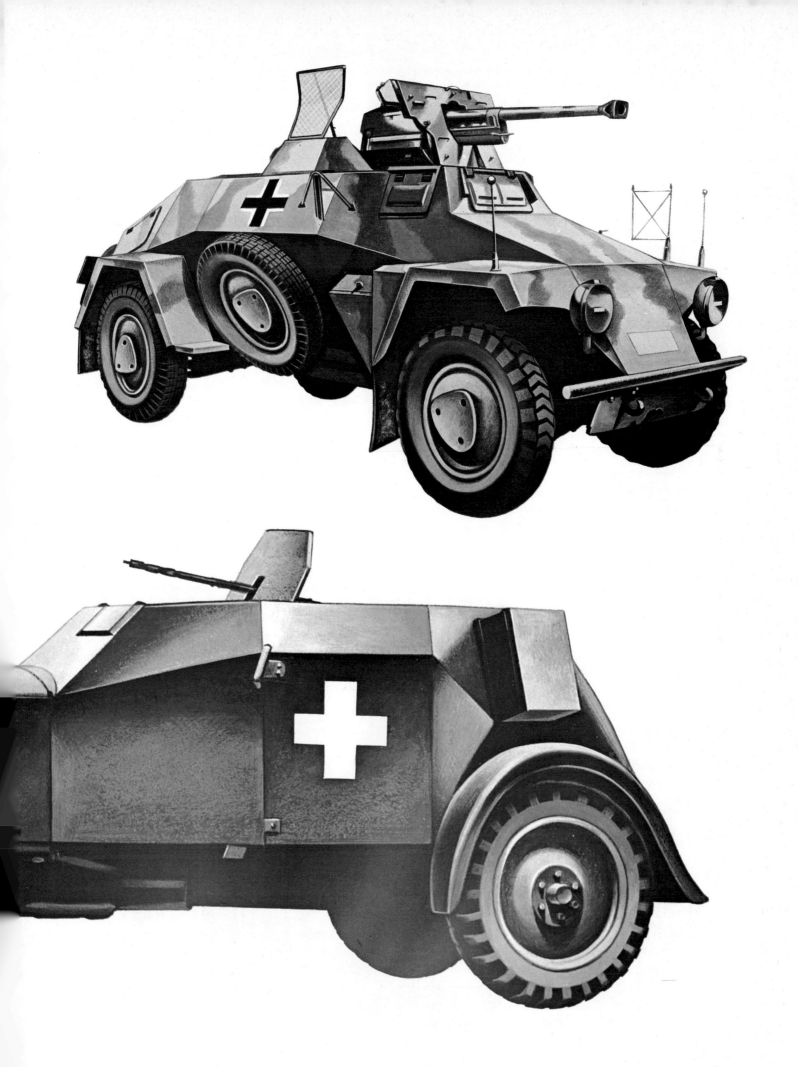

Leichter Panzerspähwagen (20-mm) (SdKfz 222). (This model had the 20-mm KwK tank gun to meet the revised *Waffenwagen* requirement.)

Leichter Panzerspähwagen (Fu) (SdKfz 223). (This model was radio equipped to meet the original *Funkwagen* requirement.)

Kleiner Panzerfunkwagen (SdKfz 260). (This was a specialised radio car development with more extensive equipment, intended for use by HQ units for communication on divisional or regimental networks. Its turret was set further back than on the 223 and it rarely carried a machine-gun.)

Kleiner Panzerfunkwagen (SdKfz 261). (This model was a later development of SdKfz 260 and was externally similar, but mechanical changes gave it a slightly better cross-country performance.)

All models in the SdKfz 221/222 series were based on a Horch/Auto-Union chassis (built by Horch-Werke of the Auto-Union combine) and Eisenwerk Westerhutte was the parent firm of the production contractors. Production in 1936–38 was on the Ausf (model) A chassis which had a 3·5 litre 75-hp engine, and 1939–42 production was on the Ausf B chassis with a 3·8 litre 81-hp engine. Production ceased in 1942 but these useful little vehicles remained in service for the duration of the war and the layout and style had some influence on British, French and Russian light armoured car design from 1940 onwards.

The chassis had four-wheel drive, fully independent suspension, optional four wheel steering (later this was dispensed with), and a self-locking differential which ensured both wheels on each side received power whatever the ground conditions. The hull was of welded armour plate, had a door each side, hinged visors front and rear, and a hand-turned turret traversed from the gun mounting. The turret was open-topped in the case of the SdKfz 221, seven-sided in

the SdKfz 221 and ten-sided in the SdKfz 222. Hinged wire-mesh screens on top were intended to deflect grenades.

The chassis proved to be very complex to build and hard to maintain, and this led to the early demise of the *Einheitsfahrgestell I*, for once the conditions of war had shown up the problems of both maintaining an increased production rate and keeping vehicles in service, there was a move to further rationalise chassis production. Since production of the *Einheitsfahrgestell I* chassis ceased in 1942, output of SdKfz 221 series armoured cars ended accordingly. Under the Schell programme, which sought to reduce and simplify the available range of military chassis from 1942, four-wheel chassis were discontinued, and all new armoured cars were big eight-wheelers.

The hulls were all welded as were the turrets when fitted. Cross-country performance of the vehicle was good, and armour protection was 14·5 mm at the front and 5–8 mm elsewhere. Weight was between 3·8 and 4·8 tons according to type.

In general the SdKfz 221 replaced the Kfz 13 in motorised reconnaissance units from 1936. Various other types of unit, including tank battalions, also procured these vehicles, sometimes for liaison work.

SdKfz 222 Light Armoured Car

Length: 15·7 ft *Width:* 6·4 ft *Weight:* 10,580 lb *Speed:* 50 mph *Range:* 185 miles *Armour:* 14·5 mm max *Armament:* 1 × 20-mm; 7·92-mm MG 34

Artillery units used them as mobile observation posts, but the major users were the reconnaissance companies of armoured reconnaissance battalions.

The SdKfz 221 was armed with only an MG 34 in the front of its small turret. A number of them were refitted with 20-mm tapered bore anti-tank guns, set high in the turret, the front of which was cut down, but this was not a widespread conversion as the importance of these small armoured cars decreased as the war progressed – most 20-mm tapered bore guns were fitted to other types, like half-tracks.

The SdKfz 222 with its 20-mm KwK 30 or 38 gun in the turret came into service from 1938. It differed from the SdKfz 221 in that it had a limited range radio set (the 221 had no communications gear), and a more prominently cut-away rear superstructure to give the driver a better rear view when reversing. The turret of the 222 was larger than that of the 221, having ten sides, and accommodated the 20-mm automatic gun adapted from an aircraft gun. It had a ten-shot magazine and could fire either armour-piercing or high explosive rounds.

The gun had a high rate of fire (1280 rpm in the KwK 30 or 480 rpm in the KwK 38). The high angle of elevation (87°) enabled the gun to be used against aircraft. A hand-wheel controlled both elevation and traverse, with a linking arm to traverse the turret with the mount. Two smoke projectors on each side of the turret supplemented the gun, and an MG 34 was coaxial with the 20-mm weapon.

The turret was fairly constricted, despite its increased size; the commander was also the gunner and the other crew members were the driver and the radio-operator, who also acted as loader. The KwK 38 gun had increased elevation and was fitted to the later production vehicles. The SdKfz 222, with its heavy armament, mainly equipped

the divisional reconnaissance units. The top speed was 50 mph and the radius of action was 110 miles in cross-country conditions.

The SdKfz 222 had a shorter run of service than most. Because of the very severe terrain on the Russian Front in 1941–42, these small four-wheel vehicles proved not entirely suitable for the divisional reconnaissance role, and they were largely replaced from 1942 by the SdKfz 250/9 half-track which was fitted with the turret from SdKfz 222.

The SdKfz 223 radio car came into service in 1938 and was a replacement for the Kfz 14. It had a small turret similar to that of the SdKfz 221, and a prominent frame-type radio aerial mounted on the hull. This could be collapsed on its pylons if desired to reduce the overall height and aid concealment. This aerial was rather cumbersome, and later models simply had a sectionalised pole aerial which was easier to handle. The SdKfz 223 was a companion vehicle to the 221 in the reconnaissance companies of armoured reconnaissance battalions.

The *Kleiner Panzerfunkwagen* (small armoured radio vehicle) SdKfz 260 and 261 were expressly designed for use by unit and formation commands, with radio suitable for maintaining divisional and regimental links. They had a more cut-away superstructure at the rear as in the 222, and the small turret was set further back to allow more room inside the hull for radio equipment. Though both vehicles in theory carried MG 34, in practice the armament was usually omitted due to the great encumbrance of the signals equipment.

The different designations for the SdKfz 260 and 261 were mainly to account for their different radio outfits. The 260 had only sectionalised pole aerials, usually carried with the lower sections rigged. Early models of the SdKfz 261 had the big frame aerial, but this was later replaced by pole aerials.

An SdKfz 223 armoured reconnaissance vehicle of Guderian's Panzergruppe *in France, August 1940*

SdKfz 261 Light Armoured Car
Length: 15·58 ft *Width:* 6·5 ft *Weight:* 9500 lb
Speed: 50 mph *Range:* 200 miles *Armour:* 8 mm
Armament: nil

One last light armoured car model in the same 'family' of four-wheelers was the SdKfz 247, produced in very small numbers on the *Einheitsfahrgestell II* chassis, which was the alternative passenger car chassis for the *Einheits* series, differing from the *Einheitsfahrgestell I* by having front engine and transmission rather than a rear engine. The manufacturer was again Auto-Union/Horch. This chassis was used for the big series of standard cross-country cars produced by the Germans and the SdKfz 247 – not to be confused with the different SdKfz 247 (6-rad) already described – had an armoured body in place of an open car body.

This vehicle was classed as an armoured personnel carrier, and some were issued to reconnaissance units. In the main they went to elite fighting units like Panzer Division Gross Deutschland or were used as personnel transport by formation staffs and commanders. There were actually two models: the SdKfz 247/I was the original version, while the other, SdKfz 247/II, had a more powerful radio set and a thicker front armour shield. A prominent 'star' aerial was a feature of this later model.

Light armoured cars were proved early on to be highly suited to the reconnaissance role, hence the subsequent designation of *Panzerspähwagen* (armoured reconnaissance car) for this type of vehicle. By the time the first Panzer Divisions were formed in 1938, three armoured recce companies with motor cycles and light armoured cars were organic to each division. In a Panzer Division there were 42 cars of various types with the three companies, and a further eight cars with the recce company of the division's mechanised infantry regiment. The reconnaissance company of an Infantry Division had three light armoured cars, intended for command and communication use. The recce company of a Motorised Division included 18 light armoured cars.

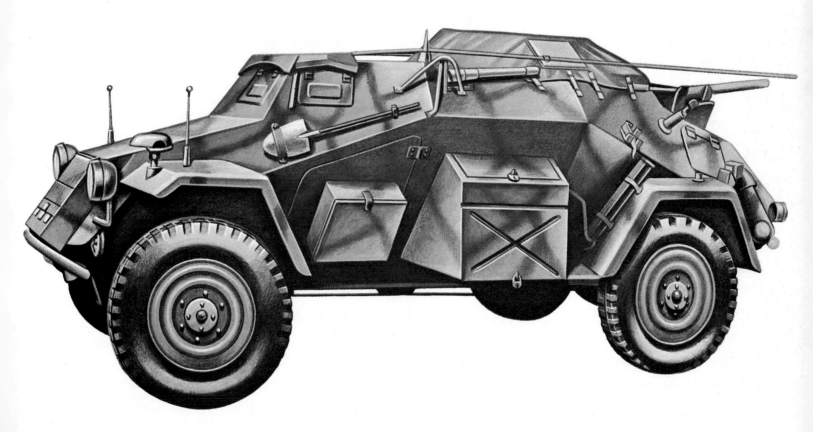

EIGHT-WHEELERS

An SdKfz 231 (8-rad) heavy armoured car

By far the most important of all German armoured cars in the Second World War were the impressive eight-wheelers which were originally conceived as heavy cars to complement the light four-wheelers. The origin of the eight-wheelers goes back to 1934 when *Waffenprufamt* 6 asked Büssing-NAG to develop an experimental eight-wheel chassis with all wheels driven and with steering on all axles. This was at the time when the *Einheits* series of cross-country chassis was being evolved to cover all weight chassis. It was the 'light' *Einheitsfahrgestell* which formed the basis for both the standard light cross-country personnel carrier and the light armoured cars of the SdKfz 221 series. The eight-wheel chassis design was very complex and the idea of a cross-country truck version was never pursued. However, it was decided to develop an armoured car on this chassis, and in 1935 the Ministry of War funded a prototype, known as VsKfz (*Versuchskraftfahrzeuge* or experimental vehicle) 623.

A production design was finalised and a preliminary order placed, the intention being to replace the rather unsatisfactory six-wheelers then in service. First deliveries took place in 1937 and by 1940 the various types of eight-wheeler were in widespread service. The initial types were intended as direct replacements for the existing six-wheelers and for this reason they took exactly the same ordnance numbers, but to differentiate between the old and new the number of wheels was indicated with the designation. Thus the old model SdKfz 231 (6 rad) was replaced by the SdKfz 231 (8 rad). Both the old and new models were in service concurrently, moreover, especially between 1938 and 1940. The basic eight-wheeler chassis was of sturdy two-section girder construction with tubular cross members. It was actually light for the size of the vehicle, but the long armoured body gave extra rigidity. A Büssing-NAG V8 gasoline engine of 155 bhp was mounted at the rear of the vehicle which was actually the front of the chassis – the chassis and body were reversed in relation to each other and there was no front and rear in the conventionally accepted sense. The gearbox was mounted centrally on the chassis, the wheels were mounted in fours on two bogies, all independently sprung, and also linked in pairs with semi-elliptic springs. All axles steered and there was a reduction gear box for each bogie. There were six gears available in either direction, an auxiliary high/low ratio being available with its own gear lever. Steering and driving controls were duplicated for driving in either direction. A special differential in each of the gear reduction boxes compensated for the differing turning radii of the inner and outer pairs of wheels. Simple pressed-steel disc wheels were fitted, with low-pressure self-sealing cross-country tyres. The entire design was well thought out and extremely advanced for its time.

The first vehicle type produced on the eight-wheel chassis, the SdKfz 231 (8-rad), followed the pattern of previous output in being a *Waffenwagen*. It had a turret mounting the 20-mm KwK 30 or 38 automatic cannon, with a co-axial MG 34, both fired by foot pedal. The four man crew consisted of a gunner, commander, driver and second driver/radio-operator.

Unlike the earlier types, the eight-wheeler was quite a roomy vehicle with good stowage arrangements. The welded hull was of face-hardened armour, 15 mm thick on the turret front, 8–10 mm elsewhere on vertical or near vertical faces, and 5 mm on sloped faces such as the glacis, although later production vehicles, built after combat experience, had the armour maximum increased to 30 mm. The very earliest vehicles had a front stowage basket on the nose with a forward face 10 mm thick, which acted effectively as spaced armour, giving extra frontal protection. Late production vehicles with 30 mm of armour were of course slightly heavier, and the engine was bored out to give 180 bhp, and thus an increased performance to compensate for the added weight.

Spacious turret

The turret of the SdKfz 231 was a roomy hexagonal structure. The commander was provided with a seat on the left-hand side, with a folding periscope which had provision for a camera attachment to photograph the terrain on reconnaissance work. There was a ball race for turret traverse with a combined traverse and gun elevating wheel with the gunner on the right and a more highly geared over-ride control for the commander. Seats and other fittings were attached to the turret sides and there was thus no turret floor. Vision ports in the sides and an opening roof hatch were other features of the turret.

The SdKfz 231 formed the backbone of the divisional armoured reconnaissance battalions in increasing numbers from 1937–38, and by late 1940 had completely supplanted the SdKfz 231 (6-rad) in first line units. As with the earlier types of car, there was a companion, radio-fitted car for the use of commanders (though the SdKfz 231 (8-rad) also had a wireless set). The special radio-fitted vehicle was the SdKfz 232 (Fu) (8-rad). This *Funkwagen* variant was in most respects similar to the SdKfz 231 (8-rad) except that it had a large frame aerial above the turret, as on the 6-rad version. A shaped pivoting front support allowed the turret to traverse without disturbing the aerial frame. The aerial was cumbersome and vulnerable, so on later production models the frame aerial was replaced by a sectionalised rod aerial which could be erected on the turret roof, and a star-shaped aerial which fitted on the rear decking. The SdKfz 232 (Fu) (8-rad) again was armed with the MG 34 and 20-mm gun in the turret.

Once again there was a special command vehicle for signals staff and headquarters use which in this case dispensed with the turret and was given special bodywork. It had a full-width superstructure replacing the turret and carried 100-watt or 80-watt medium wave radio equipment to maintain battalion, regimental and divisional links. Designated SdKfz 263 (8-rad), this was known as a *Schwerer Panzerfunkwagen*

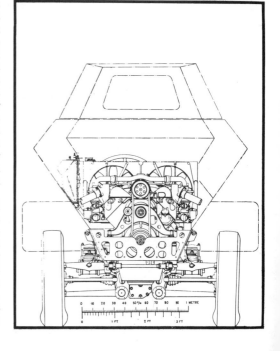

Warpics Ltd

SdKfz 231 (8-rad) Heavy Armoured Car
Chassis arrangement in side elevation and, opposite, rear elevation
 Length: 19·19 ft *Width:* 7·22 ft *Weight:* 18,300 lb *Speed:* 53 mph *Range:* 165 miles *Armour:* 10 mm max *Armament:* 20-mm KwK 30/38; 1×7·92-mm MG 34

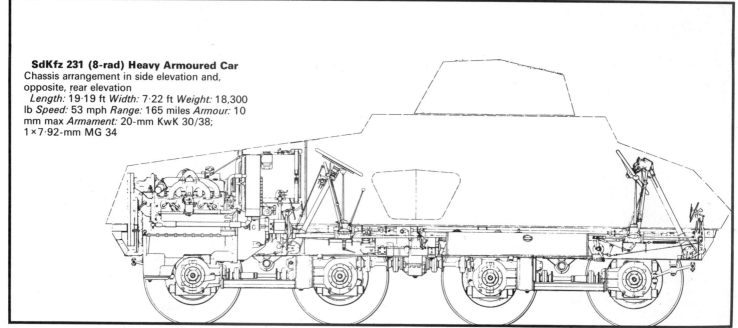

rather than a *Panzerspähwagen*. The super-structure followed the angles of the lower hull sides and there were access hatches on the roof. The only armament was a ball-mounted MG 34 in the hull front. A large frame bedstead aerial was mounted above the superstructure but, as with other radio vehicles, it was replaced by a sectionalised rod aerial on later vehicles. From the end of 1941 this type of vehicle went out of production, its function being undertaken by radio-equipped half-tracks. The SdKfz 263 (8-rad) served with signals platoons of armoured reconnaissance battalions as well as some formation commands. It weighed 8·5 tons fully laden, rather more than the standard *Waffenwagen*.

These were the main armoured cars in service when the great German campaigns of 1939–40 took place. At this period the armoured reconnaissance companies and battalions, equipped with motor cycle and armoured car platoons, were in the fore-front of the attack, playing an important part. From late 1941 onwards, however, the importance of armoured cars and the German reconnaissance troops declined. This was mainly because the German forces were forced largely on to the defensive in Europe, while in the Western Desert and on the Russian Front the half-track could undertake most of the functions of the armoured car, and other roles besides, particularly those of a defensive nature. Conversely, the Allies, who went on to the offensive, began to turn out more and more armoured cars, using them up to 1944 and 1945 – by which time they had almost disappeared from German service.

The changing needs of the German army are reflected in the first armoured car to depart from the established order. This was the *Schwerer Panzerspähwagen* 75-mm SdKfz 233, introduced for armoured recon-naissance units, which needed to be able to provide their own mobile fire support.

The gun was the well-tested 75-mm low velocity KwK 37 L/24 which had been used on the earlier PzKpfw IV tanks and various types of assault gun. It was mounted on the vehicle in the front of the superstructure, and both the turret and hull top as fitted to the SdKfz 231 (8-rad) were omitted, and replaced by low superstructure side additions, leaving the vehicle open-topped. The gun could fire smoke, high explosive and two types of armour-piercing round, a total of 55 rounds of all types being carried. The crew was reduced to three men – driver, rear driver/radio-operator and gunner/commander. The gun had very limited traverse and was aimed via a simple dial sight.

Protracted development

The status of armoured cars in the German forces is demonstrated by the time taken to get the final generation of eight-wheelers into service. A start was made in August 1940 to replace the eight-wheeler with an improved version. Externally the new vehicle appeared similar to the original Büssing-NAG design, but one major difference was that it had a monocoque hull instead of a separate chassis. The same type of eight-wheel steering was employed, but the bogies were attached directly to the lower hull. Büssing-NAG undertook design and construction, while Deutschen produced the hull, which was similar to that of the preceding series of eight-wheelers, but was much simplified. The two separate groups of mudguards that covered the wheels on the earlier models were replaced by new one-piece mudguards with integral stowage boxes.

A prime requirement for the new design was that it should operate well at extremes of climate, particularly in tropical climates. The Czech firm of Tatra was given the task of developing a high-powered engine to replace the gasoline engine, and a compact V-12 14·8 litre unit was evolved.

The prototype did not appear until July 1941, by which time German troops were fighting in North Africa, and the complicated Tatra engine proved too noisy for desert operations. It was rejected and an improved engine designed with the major aim of muffling the noise. By this time (late 1942), operations in North Africa were not going well, and priority was being given to tanks and assault guns, so progress on the new armoured car was slow.

In the event it did not enter production until 1943, and did not see service in any numbers until 1944. Armour thickness of 30 mm at the front was standard with 14 mm and 10 mm on other faces. Increased fuel capacity was a major feature of this vehicle, giving a range of 370 miles against the 170 miles of the earlier eight-wheelers. Later vehicles had even greater capacity, increasing the range to 620 miles. Around 2300 of this new series were built from 1944 up to March 1945, and the model, designated SdKfz 234, received an unexpected new lease of life when it was selected for increased production (100 vehicles a month) in the closing weeks of the war. This was largely because it was relatively simple for the hard pressed bomb-damaged factories to build as the military situation deteriorated. The scheme was something of a last ditch stand which never really materialised. Designated ARK by the manufacturers, the new eight-wheeler was built in several forms.

The basic *Waffenwagen* version had the usual 20-mm KwK 38 automatic gun and was designated *Schwerer Panzerspähwagen* (20-mm) SdKfz 234/1. The KwK 38 was arranged in an open-top six-sided turret, with high elevation (75°) making it suitable for engaging air targets. Folding mesh panels on top of the turret prevented

SdKfz 232 (8-rad) (Fu)
Heavy Armoured Car with Radio
Length: 19·19 ft *Width:* 7·22 ft *Weight:* 19,400 lb *Speed:* 53 mph *Range:* 165 miles *Armour:* 10 mm max *Armament:* 20-mm KwK 30/38; 1 x 7·92-mm MG 34

grenades from being hurled inside. There was a co-axial 7·92-mm MG 42 in the mantlet. Some 20 spare ammunition magazines were carried and radio-telephone and wireless equipment fitted.

The excellence of the eight-wheel suspension on these cars made their cross-country performance almost equal to that of a tank and the road speed was, of course, much superior. The Soviet Army still used many light tanks (for example the T-70) in their reconnaissance units and the German Army decided that an armoured car capable of tackling a light tank at close range was a desirable asset, since early contact with enemy attacks was most often at reconnaissance unit level.

Thus was evolved possibly the best-known of all German armoured cars, the Puma, which was more fully designated *Schwerer Panzerspähwagen* (50-mm) SdKfz 234/2. This vehicle retained the basic ARK series eight-wheel chassis but had a new oval turret of very fine ballistic shape, with a 50-mm KwK 39/1 L/60 tank gun in a streamlined *Saukopf* (pig's head) mantlet. The turret had originally been developed for the Leopard light tank which had been

projected for the reconnaissance role in 1943–44 but had been abandoned at prototype stage. The 50-mm gun was very effective for its calibre, having semi-automatic action at a muzzle velocity of 2700 fps when firing armour-piercing ammunition. The compact mantlet included a recoil mechanism mounted above the gun, a telescopic sight, and a co-axial MG 42. Three smoke projectors were mounted as standard on each side of the turret.

The weight of the vehicle was of course increased by the addition of the turret, to 11·5 tons in combat order and 11·3 tons unladen. The mantlet armour was up to 100 mm thick, but armour elsewhere was as for the other eight-wheelers of the ARK type. The turret had full 360° traverse and the vehicle was fitted with radio-telephones as standard but could also carry long range Fu 12 radio equipment. The crew of four men remained as in the standard car and the commander had a periscope in the turret roof.

To supplement the Puma a more heavily armed support version of the armoured car was produced as a result of Hitler's personal intervention. Soviet armour packed a

powerful punch, and the idea of the additional vehicle, designated *Schwerer Panzerspähwagen* (75-mm) SdKfz 234/3, was to give support fire to the Puma. Basically the SdKfz 234/3 was an updated version of the SdKfz 233, with similar placing of the 75-mm KwK 51 L/24 low velocity gun. The vehicle was open-topped, the gun had a limited traverse and 55 rounds of ammunition were carried. The combat weight of this vehicle was 9·8 tons.

The SdKfz 234/3 was, of course, something of an expedient in common with a large number of late-war period German self-propelled guns. In this case the L/24 guns were plentifully available, having been removed from early versions of the PzKpfw IVs.

Still more illustrative of the expedient nature of the later German equipment was the final vehicle in the SdKfz 234 series, the *Schwerer Panzerspähwagen* (75-mm lang) SdKfz 234/4. This was yet another of the types suggested by Hitler and was potentially an excellent type, reflecting the desperate situation on Germany's fighting fronts in late 1944/early 1945.

The SdKfz 234/4 was in essence the same vehicle as the SdKfz 234/3, but instead of the short 75-mm gun of limited effect, the SdKfz 234/4 had the powerful long 75-mm Pak 40 anti-tank gun. This was a weapon of proven effectiveness as a ground mount, the standard anti-tank gun with infantry units. The SdKfz 234/4 simply took the complete gun and carriage, and placed it straight into the open fighting compartment of the SdKfz 234/3. The result was a crude but potent wheeled tank destroyer, with its traverse limited to that of the Pak 40 carriage. The wheels and trail were, of course, removed from the carriage but the gun shield was retained.

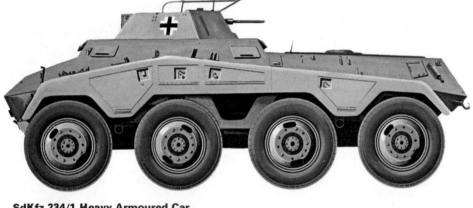

SdKfz 234/1 Heavy Armoured Car
Length: 19·72 ft *Width:* 7·74 ft *Weight:* 13,150 lb *Speed:* 53 mph *Range:* 370–620 miles *Armour:* 30 mm max *Armament:* 20-mm KwK 30/38; 1×7·92-mm MG 42

SdKfz 234/2 'Puma'
Heavy Armoured Car
Length: 22·3 ft (inc gun) *Width:* 7·64 ft *Weight:* 25,880 lb *Speed:* 53 mph *Range:* 500–620 miles *Armour:* 40 mm max (100 mm on mantlet) *Armament:* 50-mm KwK 39/1; 7·92-mm MG 42

The particular merit of the vehicle was its speed, low profile, and fast drive in either direction. Known to the troops as the *Pak-wagen* (anti-tank vehicle), it was intended to be built at a rate of 100 a month as a standard type. The chassis was well-proven, and relatively cheap, the gun was readily available and easy to fit. Its crudity did nothing to diminish its effectiveness – it could be turned out quickly and did all it was called upon to do. However, only a few were built, for in the closing months of the Second World War the entire system of production and procurement was in disarray and the promising *Pak-wagen* hardly got a chance to show its mettle.

Needless to say, these eight-wheel vehicles, well ahead of their time, have had much influence on subsequent armoured car development, and Britain, France, Holland and the Soviet Union, among other nations, have all produced multi-wheel drive vehicles which clearly owe something to the German designs that proved their worth in battle over several years of war.

The eight-wheelers did not exhaust German efforts at armoured car design, even though they represented the bulk of the output. Efforts were made continually during the war years to produce simpler four-wheel designs incorporating the lessons of battle.

Experimental vehicles

The *Mittlerer Panzerspähwagen* (medium armoured car) VsKfz 231 (4-rad) was an interesting vehicle which did not see production. Conceived in 1941, it was intended to be a four-wheel version of the eight-wheel SdKfz 231. The design also involved elements of the SdKfz 234, in that a six-cylinder Tatra air-cooled diesel engine was to be used. The VsKfz 231 had a shortened SdKfz 231 chassis, four wheels all driven, face-hardened armour of 30 mm frontal thickness and 8–14·5 mm elsewhere. The original idea was to have a 20-mm gun in a rotating turret, but this was changed to a 50-mm KwK 39. There was to be a co-axial MG 42 and a crew of four.

Büssing-NAG did the design work and Auto-Union/Horch were contracted to act as manufacturers. Orders for 1000 vehicles were placed, with production scheduled to start in October 1943. By the time plans had advanced this far, however, practical battle experience on the Russian Front had shown that half-tracks were more useful for reconnaissance work under the prevailing conditions and all work and production schedules on the car were cancelled.

Outside these activities came a whole series of prototypes for amphibious light armoured cars, built and designed by Hans Trippel, whose Trippelwerke factory had specialised in amphibious vehicle development before the war. Trippel made his prototypes as private ventures and never succeeded in having any designs adopted by the Wehrmacht, though they were tested and demonstrated. One major difficulty under war conditions was obtaining the necessary permission to carry out private venture work. However, in 1941 Trippel built an amphibious light personnel carrier similar in appearance to the famous Volks-wagen *Schwimmwagen* – almost like a bathtub on wheels. Using this vehicle, the SG 6, as a basis, he built a light amphibious armoured car called the *Schildkrote* (Turtle), a prototype of which was ordered by the

German Army in January 1942. It had an Opel 2·5 litre motor car engine, was armed with an MG 81 and had a turret searchlight.

The demonstration was not successful when it took place in March 1942, but the following month Trippel secured orders for two prototypes with improved steering which were designated *Schildkrote II*. Tests were carried out using vanes fitted to the wheels to give propulsion in water, but eventually propeller drive via a power link to the engine was adopted. These new prototypes lacked turrets but were intended to have octagonal open-topped turrets rather like those of the SdKfz 222 series vehicles, and 20-mm cannon or 7·92-mm machine gun armament was proposed. Demonstrations of these new vehicles in the summer of 1942 were again unsuccessful. A third prototype, *Schildkrote III*, was then built, with the Tatra diesel engine developed for the VsKfz 231 (4-rad). Otherwise similar to the *Schildkrote II*, the *III* was also considered unsuitable.

However, all this led to Trippelwerke being engaged to undertake development work on a standard type of amphibious chassis with rear mounted Tatra engine, the E3 light armoured car and E3M munition carrier. These were built in 1944 and prototypes were demonstrated in October of that year. Due to the deteriorating war situation, however, there was no production order. The E3 was one of a projected series of *Einheits* vehicles which were part of a long term plan to produce a number of basic standardised vehicle types and thus reduce the dozens of different vehicles on the ordnance inventory.

The Germans supplemented their own vehicles with a number of captured types, though these saw service only in limited numbers. For instance, some armoured cars of French and Austrian origin were taken into service. The only type to see wide-scale service, however, was the French Panhard 178, which was taken into German service as the *Panzerspähwagen* P 204 (f), of which well over 150 were employed. This vehicle was conventional in appearance: a four-wheeler with central turret and a rear-mounted 105-hp two-stroke engine. Some were used on the Russian Front. They had a 25-mm Hotchkiss gun and co-axial machine-gun, with a maximum armour of 20 mm and a top speed of 50 mph.

One special adaptation of these cars was as railway patrol cars for the Russian Front. Fitted with flanged wheels, they were used to escort armoured trains and for fast patrol duties along the many miles of captured railway line used by the Wehrmacht for bringing supplies up to the fighting fronts. About 40 Panhards were so converted in early 1942.

One of the Austrian cars which saw service (as the SdKfz 254) was the Saurer RR7, designated by the Germans *Mittlerer Gepanzerter Beobachtungskraflwagen* (medium armoured observation vehicle). This was actually a wheel-cum-track vehicle that could change from wheels to track while in motion. It had a six-cylinder diesel engine and was intended for the Austrian Army as a light gun tractor. When Germany annexed Austria in 1938, the Germans ordered an armoured version and a small number were built and entered service in 1942. Another variation, with different fittings, saw even more limited service as a repair and maintenance vehicle. The SdKfz 254 was armed with only an MG 34.

A captured Panhard armoured car with German police badge takes part in an anti-partisan action with the Wehrmacht forces in Russia in 1941

SdKfz 234/3 Heavy Armoured Car (left)
Length: 19·68 ft (inc gun) *Width:* 7·64 ft *Weight:* 22,050 lb *Speed:* 53 mph *Range:* 370–620 miles *Armour:* 30 mm max *Armament:* 75-mm K 51

ARMOURED HALF-TRACKS
CROSS-COUNTRY TRANSPORT

Armoured cars played a relatively minor part in German military activities during the Second World War, and a much more significant vehicle was the armoured half-track, which eventually took over many of the armoured car's roles. The half-track established a style for the use of armoured personnel carriers which is now an integral part of all major armies.

In fact, the idea of an armoured troop carrier was first exploited by Britain, for the thinking that led eventually to the development of tanks in Britain in 1915–16 stemmed from the need to carry infantry into action immune from machine-gun fire. The tank became an instrument of attack in its own right, but by 1918 the British had converted some to troop carriers – with sliding doors replacing the gun sponsons of Mk V tanks, or in the case of Mks V* and V** with lengthened hulls specially to carry troops. They were used as troop carriers not altogether successfully, at the Battle of Amiens in July 1918. The big Mk IX tank, known as the 'Pig' because of its ugly shape and bulk, was built, too late to see service, with ports for infantrymen to fire through from inside. It could also carry stores and equipment.

These efforts provided the British Army on the Western Front with a means of moving up infantry with a tank attack, and actually pre-dated by far similar German ideas. But with the coming of peace in 1918, these lessons were largely forgotten by the British who quickly reverted to what were virtually pre-war ideas. The poor economic condition of Britain was used, as always, as a convenient excuse to run down the Army and pack up development work.

Britain was not short of ideas, however, and an Experimental Mechanised Force was set up in the 1920s to study on a small scale the experiences of the last years of the Great War; they took one step further the idea of co-ordinated tank, infantry, assault engineers and artillery attack under one command – the basic principle of the Panzer Division in fact.

The leading British armoured warfare theorists were men who had been concerned with, or keen observers of, the development of armour in the First World War, among them Captain Liddell Hart and Generals Fuller and Swinton. Their writings on the subject were keenly followed in other nations, particularly in Germany. Liddell Hart, in particular, was a prolific writer and one of his prophetic statements in the 1920s was that 'motorised armed forces have to be able to perform similar feats to the marauding hordes of the Mongols'. Tactics demonstrated during exercises with Britain's Experimental Mechanised Force showed that it was possible to act as a 'marauding force' with fully mobile tanks, artillery and infantry breaking through conventional front lines, attacking or straddling lines of communication, operating against the enemy's undefended rear and so on.

But much of this work went unheeded in Britain, for by the 1930s the Experimental Mechanised Force has been disbanded and policy was switching back to older ideas: infantry walking into the attack behind slow-moving, machine-gun armed 'infantry tanks', barrages from prepared positions against fixed front lines and so on. The last vestiges of the 'marauding forces' idea were seen in 'cruiser tanks' and 'light tanks', the former to exploit the breakthrough made by the infantry, but acting largely independently. The nautical analogy in the classification did owe something to the writings of the pundits, but the essential ingredient of specialised infantry and artillery support was missing from British armoured doctrine in the 1930s.

Not so in Germany, however, where the German Army had the advantage of starting with a clean sheet, without preconceived ideas to obstruct innovations. As we have already seen, the idea of mechanised infantry was borne in mind from the start when the new *Reichswehr* formed seven motor battalions each with 15 armoured troop carriers. The seeds of the armoured half-track already existed in Germany.

French inspiration
The half-track idea was not, of course, German – it had been developed by Adolphe Kégresse, a French engineer who managed the Czar of Russia's personal motor fleet in the early 1900s. Kégresse had made a bogie assembly with rubber tracks to replace the back wheels of one of the Czar's cars, thus giving it superior traction in winter snow and ice. After the Russian Revolution of 1917, Kégresse returned to his native France and Citroën took up his half-track ideas with commercially successful results.

The French, British and American armies, among others, all had Citroën-Kégresse half-tracks in the 1920s and 1930s, but failed to exploit their full military potential. The British, in particular, had dropped half-tracks altogether by the late 1930s, though the US Army, under the impetus of war, went on to develop their own successful line of M3 series armoured half-tracks, derived directly from the simple Kégresse ideas. These American half-tracks proved to be among the most durable of military vehicles and many hundreds were still in service with the world's smaller armies (notably Israel) well into the 1970s, none of the surviving vehicles having been built later than 1945.

The first German half-track for military use was the Daimler-built Bremer Marienwagen. This appeared in 1917–18, and was basically a Daimler lorry with the body of an Erhardt armoured car. The back wheels were replaced by simple rubber-band type track units – and later one vehicle was built with track units replacing the front wheels also. Only four vehicles were built, and for various reasons, notably poor steering, uneasy control characteristics and track-shedding, the Bremer Marienwagen was abandoned. It was however designed as an armoured troop transport and would have been the German equivalent of the British troop-carrying tank. Various less significant prototypes had been built towards the end of the First World War, among them the Mannesman-Mulag *Panzerkraftwagen* an armoured lorry type of troop carrier with rifle ports, and the Daimler DZVR.

Half-track development took its most significant step forward in 1926, when the German War Ministry tested a number of available trucks and half-tracks to deter-

SdKfz 251/1 (opposite)
Medium Armoured Personnel Carrier
Length: 19·02 ft *Width:* 6·89 ft *Weight:* 8430 lb *Speed:* 34 mph *Range:* 185 miles *Armour:* 12 mm max *Armament:* 2×7·92-mm MG 34

'Mongol hordes': Panzer Grenadiers move in to attack a village from their SdKfz 251s

mine future procurement and operational policy. At the time there were several four-wheel drive tractors in production, built by firms like Kraus-Maffei, and some of these had 'add-on' half-track units similar to the basic Kégresse type (several manufacturers came up with their own half-track units, in the 1920s). A number of such vehicles were purchased, mainly as artillery tractors, and it was decided that for cross-country use efforts should be concentrated on this type of vehicle in future. Commercial output could not keep up with demand, however, and some rationalisation was needed. The other important point was that Hans von Seeckt, the Inspector of Mechanised Vehicles, realised that all arms would need to use half-track vehicles for cross-country transportation, not just the artillery.

By this time, moreover, the future structure of the German Army was being planned, and the deliberations led eventually to a requirement for prototypes in six weight classes. These, it was envisaged, would satisfy all the requirements of the various fighting arms. In 1932 contracts were placed with various automotive firms for trial vehicles.

All six prototypes were built and tested, and in time they led to the famous series of half-track gun tractors used extensively by the Germans in all theatres of the Second World War. They fall outside the armoured half-track story except for a few late-war extemporised conversions. However, the half-tracks in the 1-tonne and 3-tonne classes were developed further into armoured vehicles as well, and the two types formed the basis for the standardised infantry carriers which equipped the Panzer Grenadiers in many forms in the Second World War.

It was in 1935 that the idea of adapting the new artillery tractors took root. By this time the plans for the new Panzer Divisions were well in hand and the first tanks (PzKpfw I and II) were in production. To carry a complete infantry section or squad of ten men the 3-tonne tractor chassis was of suitable size. The choice of this chassis may also have been influenced by the construction by Rheinmetall in 1935 of an experimental armoured self-propelled anti-tank gun, the 37-mm *Selbstfahrlafette* L/70, which had a 37-mm anti-tank gun in a fully traversing turret. It was tested by *Heereswaffenamt*, but never put into production.

The armoured troop carrier version was started in 1937 and only minimal changes were needed – such as canting back the steering wheel – to suit the basic chassis to the armoured superstructure's shape. A faceted, well-sloped armoured body was designed with a strong family resemblance to that used on the armoured cars. Development vehicles included some with rear engines, but it was decided that simple open bodies allowing troops to disembark from the back doors or over the sides were more practical, so the armoured personnel carriers retained the same front engined layout as the artillery tractors. The prototype, designated *Gepanzerter Mannschrafts-transportwagen* (Gp MTW) was ready in 1938 and was rushed into production after

SdKfz 251/9
Medium Half-track with 75-mm Gun
Length: 19·02 ft *Width:* 6·89 ft *Weight:* 18,800 lb *Speed:* 34 mph *Range:* 185 miles *Armour:* 12 mm max *Armament:* 75-mm StuK 37; 1×7·92-mm MG 42 (some models only)

successful trials. At that time the Panzer Divisions' motorised infantry was still lorry-borne and there was an urgent need to give them armoured vehicles. The current (1938) production version of the 3-tonne tractor was used, the H kl 6, and this basic design was 'frozen' and used until 1945, though many details on the vehicle were changed as time went by.

The Hanomag-built chassis had a body by Büssing-NAG. There was a frontal armour of 14·5 mm, with 8 mm on the sides. The vehicle was given the ordnance designation SdKfz 251, the appellation by which it is best known, and was called the *Mittlerer Schutzenpanzerwagen* (medium infantry armoured vehicle). The first production vehicles were ready in the spring of 1939 and went to equip an infantry company in I Panzer Division for troop trials. General Heinz Guderian, Inspector General of Armoured Troops, among others, was highly impressed and called for further developments to make the vehicle of universal use within the armoured division.

By the time the Germans invaded Poland in September 1939 more SdKfz 251s were in service (now also with II Panzer Division) and they spearheaded the attack along with the tanks. This was the first full-scale vindication of the ideas first put forward by Liddell Hart and other protagonists of armoured warfare.

To a great extent the SdKfz 251 was a compromise. It was not as heavily armoured as might have been desired and it was in essence no more than an armoured taxi, for the idea then was simply that the infantry should be carried in the protected vehicle where they would de-bus to fight on foot. However, for its period, it was revolutionary, and directly spurred on the Americans to develop their own similar armoured half-track, the M3.

In an SdKfz 251/6, Luftwaffe officers co-ordinate an attack while a panzer commander looks on

While other nations' half-tracks mostly used the simple rubber band and spring bogie half-track unit of the type produced by Kégresse and his imitators, the German half-tracks were radically different and highly sophisticated for the period.

Strictly speaking the chassis was a three-quarter track: normal front wheels, steered conventionally, supported the front end and the actual drive was taken via the transmission to the front sprocket wheels of the long track units. These supported the weight of the vehicle, and a Cletrac-type steering unit (a controlled differential transmission with steering brakes on the shaft) was fitted within the drive system on the front axle of the suspension. The sprocket wheels had rollers on the wheel perimeter rather than actual sprocket teeth to engage the track.

The Cletrac steering unit acted as the front wheel was steered, thus braking the tracks as required to assist the vehicle round the radius of the turn; brake drums were an integral part of the drive wheels. The tracks themselves were of highly sophisticated design, with sealed, lubricated needle roller bearings on the track pins, and detachable rubber track pads on the inside to cushion the wheel paths and engage the sprocket rollers positively. These features gave long track life and excellent traction, but they were of high quality and thus expensive to produce. Later production vehicles had sprocket teeth of conventional type and dry pin tracks to simplify production and reduce costs. The suspension was by sprung torsion bars and the wheels were interleaved and of the perforated disc type with solid rubber tyres. The interleaved suspension was designed to give excellent flotation, though as with all other examples of this German interleaved suspension they were vulnerable to frozen snow if parked overnight.

The gearbox was a four-speed type with two-speed auxiliary boxes for cross-country driving. There were eight forward and two reverse gears. Steering was of the Ackermann type and the controlled differential steering came into operation automatically after the front wheels were turned more than a certain amount. The hand brake and foot brake both operated on the track brakes.

The chassis was of conventional girder frame type of welded girder construction with cross-members. There were armoured belly plates under the chassis and the hull itself was in two sections bolted together, the front one consisting of the engine and driving compartments, the other containing the passenger and fighting compartment. In most cases the hull was of welded construction, but there was an alternative riveted body since some firms in the SdKfz 251 construction group had facilities for riveting but not for welding. The SdKfz 251 became a major production type and by 1944 nearly 16,000 had been built by a consortium of companies which included Skoda, Adler, and Auto-Union with many small sub-contractors. Production figures of 348 in 1940 increased to 7800 by 1944. The engine was a Maybach HL 42 6-cylinder, 100-hp water-cooled unit of 4 litres.

Simplifying production

There were four basic production models, each a further simplification of its predecessor with a view to reducing production time and costs. Mechanically they were all similar but there were external detail differences. Ausf A, which was the first production type in 1939, had three prominent vision ports in each side of the superstructure. The radio aerial was fitted on the right front fender and a simple swivel bracket without a shield of any kind was provided at the front and rear ends of the fighting compartment. The Ausf A was soon succeeded by the Ausf B (the major type in service in 1940) which incorporated all the improvements suggested by early experience. The side vision ports were omitted, leaving only those in the front superstructure for the driver and commander. Tools and equipment were re-arranged and the characteristic shield for the forward MG 34 mount was added. Stowage lockers were fitted each side between the mudguards and the superstructure.

The Ausf C went into production in mid-1940, though production of the Ausf B continued until the end of that year. The Ausf C again featured all the improvements resulting from further battle experience. A single front plate was fitted on the nose, replacing an angled two-piece plate on earlier models. The radiator was now exposed to the bottom of the vehicle but armoured cooling intakes were fitted prominently on the sides of the engine compartment (some late Ausf Bs also had these). The radio aerial was resited on the superstructure in both the Ausf B and C.

By 1942 the German economy was suffering from the war effort, and losses on the Russian Front were making swift replacement of material imperative. To speed up production and cut costs many sorts of German AFV were simplified as far as possible. In the case of the SdKfz 251 this involved greatly simplifying the armoured superstructure to eliminate all unnecessary machining and fabrication time. Faceted areas, such as at the back and the engine compartment sides, were replaced by single large plates. The engine air intakes were abandoned and the stowage boxes at the sides, originally detachable, were built in as part of the structure, while the vision ports were replaced by simple vision slits. The all-welded Ausf D was in production up to the end of the war.

The basic vehicle had a simple interior with padded bench seats along each side to hold nine or ten men. The complete infantry squad with their MG 34 rode the vehicle, four vehicles carrying a complete infantry platoon and ten vehicles carrying a complete company – with the company commander riding in the tenth vehicle. In fact there was a lot of variety in actual establishment due to the fact that supply could never keep up with demand. In theory, all four battalions of a Panzer Grenadier Brigade would be wholly equipped with half-tracks but in reality only one battalion, or at best two, would have them, the remaining battalions being equipped only with trucks. Thus the half-tracks tended to go to the best battalions, and elite divisions like Panzer Division Gross-Deutschland or Waffen-SS Panzer divisions would generally have the most generous allocations.

Within a Panzer Division the task of an armoured battalion of Panzer Grenadiers was to follow closely on the heels of the attacking tanks, co-operating with them as the situation demanded. The usual German method of tank attack was to concentrate first on any enemy artillery with the first wave of tanks and either eliminate the enemy, force them to withdraw, or throw them into chaos. The second wave was usually accompanied by the Panzer Grenadiers in their half-tracks and the task of this wave was to engage enemy infantry and anti-tank guns. Assault guns were also generally employed at this stage, possibly substituting for the second wave of tanks. A third wave of tanks or assault guns with an accompanying wave of Panzer Grenadiers would concentrate on mopping up or engaging any remaining pockets of resistance. The motorised infantry battalion carried in ordinary trucks (the second battalion of the regiment) would then relieve the armoured battalion, so that the armoured element was freed if necessary to carry on the forward push. This, at least, was the theory, and while there were textbook attacks carried out this way there were many variations on the theme.

For example, in an attack on a major target, dive-bombing by Stukas or a big artillery bombardment would be carried out rapidly and with great precision. Tanks and Panzer Grenadiers would encircle the target area while the bombardment took place, then half the force would turn inwards to make an attack from the rear, while the other half maintained the circle, partly to stave off any attempted counter-attack but also to catch any enemy forces attempting to flee the area.

Panzer Grenadiers were also used as an actual spearhead attack force in some cases where the defending forces were only lightly equipped. Here tanks engaged the enemy forces with the maximum amount of noise or fire, then drew the enemy fire while a Panzer Grenadier company in half-tracks made a swift flank attack from another direction. This was a favourite tactic for capturing defended points and bridges. In defence, a typical tactic was the reverse of this. The Panzer Grenadiers would engage the enemy frontally while tanks attempted to work their way round to attack the flanks.

The need for Panzer Grenadiers to spearhead their own attacks led to the development of many of the variants of the SdKfz 251, for the armed versions which appeared gave further strength to their own firepower. In effect the family of SdKfz 251 variants gave the Panzer Grenadiers their own range of armed vehicles to back up the requirements of the armoured infantry. There were 23 official variants, most of them used in armoured divisions, and there were also unofficial variations.

One variation of the basic vehicle, SdKfz 251/1, carried a quadrant and mount for the 'heavy' MG 34 machine-gun, ie, on its adjustable tripod mount with optical sight. There was no armoured shield and a second MG 34 with heavy mount was carried in the vehicle. This variation was used by the heavy machine-gun squads of the heavy or support platoons. The standard SdKfz 251/1 carried MG 34s in their 'light' form only.

One of the most spectacular variations of this vehicle was also designated SdKfz 251/1 but it was fitted with six frames for launching 280-mm or 320-mm *Wurfkorper* rockets. Three launching frames were carried on each side, and the rockets had jellied petrol warheads. Basically the idea was to provide heavy fire support cheaply, using existing types of rocket. The frames were attached at about 45° elevation and the entire vehicle was lined up for firing. There was a sight vane on the engine cover for the driver's use, but sometimes the commander used an artillery type periscope for more accurate sighting. The rockets – also known as *Wurfrahmen* – were fired in succession, not together. The whole set-up was crude, but proved quite effective, specially in street fighting. In Stalingrad and Warsaw these weapons were used with spectacular and terrifying results. They were easy to load and made the vehicle into a useful type of terror weapon.

Mortar carrier

Just as the SdKfz 251/1 served the needs of the rifle squads, so the vehicle was adapted to carry the 80-mm mortar of the heavy support company. It could be fired from the vehicle if necessary, but was normally fired from cover on the ground. Some of the seating was taken out of this variant to allow stowage space for the mortar ammunition. Later in the war the light half-track, SdKfz 250, was more commonly used as a mortar carrier, since the SdKfz 251 was larger than was really necessary and half-tracks were in short supply.

For unit commanders, the SdKfz 251/3 was equipped with a medium-wave radio set, for maintaining regimental and divisional links, fitted with the characteristic bedstead aerial. The vehicle took over the functions of the radio-equipped armoured cars in the early years of the war. After about 1942 a sectional aerial mast replaced the bedstead aerial. A second version of the 251/3 was fitted with extra radio equipment for communicating with tanks when operating with tank and other AFV formations. Later models had pole aerials in the usual style and differed little externally from the standard model. Yet further variations of the SdKfz 251/3 were built for other special communications roles, some for large formation command use and others for air control and liaison.

The SdKfz 251/4 was a special infantry gun tractor for hauling the 105-mm guns in

SdKfz 251/16
Medium Half-track with Flame-thrower

Length: 19·03 ft *Width:* 6·89 ft *Weight:*
19,180 lb *Speed:* 34 mph *Range:* 185 miles
Armour: 12 mm max *Armament:* 2 x 14-mm
flame projectors; 1 x 7·92-mm MG 34

*An SdKfz 251/1 follows an Sdkfz 251/10 (with
37-mm gun) across a frozen Russian landscape*

Army Group, and Divisional commanders.

The SdKfz 251/8 was an armoured ambulance – *Krankenpanzerwagen* – with racks for two stretchers. It could carry two litters and four seated wounded, or the stretcher racks could be folded back to provide for eight seated wounded. This variant carried no armament.

The first half-track to have a gun fitted was the SdKfz 251/10 which was given a 37-mm Pak 36 mounted on the forward superstructure. The idea was postulated as early as the original troop trials using the vehicle in 1939, and by 1940 the first SdKfz 251/10s were in service. The Pak 36 was simply the complete weapon less its wheels and field carriage. The platoon leader's vehicle was usually an SdKfz 251/10 and its function was to provide covering fire for the other vehicles of the platoon.

In later vehicles the original high gun shield was considerably reduced, and some had no shield at all or a shield on one side only. These were common variations of the half-track. The second gun-armed variant to be produced in quantity was the SdKfz 251/9 which entered service in 1942. It carried the familiar 75-mm KwK 37 L/24 short low velocity gun of the type originally carried by the PzKpfw IV.

When the Panzer IV began to be equipped with the long 75-mm gun there was a gap in the armoury of an armoured division in that the usefulness of the short gun as a support weapon was missed. Assault guns filled this gap to some extent, but they were not always av...able. Moreover, it was useful for the Panzer Grenadiers to have their own fire support capacity, and most battalions had a gun company of six of these vehicles. The L/24 gun was mounted above the front of the superstructure. Later conversions had raised armour plates on the sides to protect the crew. This was a popular and valuable vehicle known to the men as *Stummel* (Stump). The SdKfz 251/11 was a telephone line layer, equipped with cables and apparatus for rigging field telephones. SdKfz 251/12, 251/13, 251/14 and 251/15 were all specialist artillery vehicles for such functions as artillery survey, sound ranging, and flash spotting. They were used by assault gun units and artillery units within the armoured division.

Flame-throwing half-track

An interesting and potent variant was SdKfz 215/16, the *Flammpanzerwagen* which carried two 700-litre flame-fuel tanks and two 14-mm projectors, one on each side of the superstructure. Some models had a third projector on a long hose extension for use outside the vehicle. The rear doors were welded shut as the flame fuel tanks were in the hull rear. Up to 80 two-second bursts of flame were possible with the equipment and the range was up to 35 metres.

SdKfz 251/17 was another type introduced in 1942, with a 20-mm Flak 30 or Flak 38 mounted in the rear compartment. There were several variations in the mounting, and those with the Flak 38 sometimes had bulged superstructure sides that let down to give better traverse and more room for the crew in action. These vehicles were intended to give anti-aircraft support for Panzer and Panzer Grenadier units, but the guns could be (and often were) used against ground targets.

Also designated SdKfz 251/17 was an elaborate design with enclosed remote-control turret which entered service in

armoured divisions. It was also used to tow other types of gun such as the 50-mm Pak 38, the 75-mm Pak 40, and the 105-mm *Leichte Feldhaubitze* 18, among others. Externally the vehicle resembled the SdKfz 251/1 but without the machine-gun mounts.

The SdKfz 251/5 was yet another variant for carrying the various assault engineers of an armoured battalion with their inflatable boats and assault bridges. Another variant – the *Pioneerpanzerwagen* – was the SdKfz 251/7 which had special fittings on the superstructure sides to carry assault bridge ramps. Assault boats and demolition stores were carried.

The SdKfz 251/6 was an important command version elaborately equipped for the use of senior formation commanders. It had comprehensive office facilities with mapboards, cypher and encoding machines, and multi-wave radio equipment in a variety of alternative combinations. The vehicle was popularly known as a *Kommandopanzerwagen* and was used by virtually all Army,

very small numbers in 1944. There was a covered front end where the small turret was located. Armament consisted of the Flak 38 20-mm gun.

SdKfz 251/18 was an armoured observation vehicle for artillery use, and 251/19 was a telephone exchange vehicle used by communications units. Of greater interest was SdKfz 251/20 *Uhu* (owl) which gave the Army 'eyes' for night fighting. Developed in 1944, when special night attack units were formed, each with six Panther tanks, it carried an infrared searchlight (*Beobachtungs Gerät* 251) on a turntable mount in the rear compartment. A special sighting telescope was also carried in the vehicle. In action the searchlight operator picked out suitable targets with the searchlight beam – which was of course invisible to the human eye – then called up the Panther tank to engage the target. The tank also had its own, less powerful, night sights, and could spot targets unaided by *Uhu* at up to 500 metres. This was the start of what is now a standard aspect of armoured warfare – night fighting – and almost all modern AFVs carry infrared searchlights.

Allied air superiority was a constant problem for ground forces by 1944, and one

Bundesarchiv

expedient produced at this time was SdKfz 251/21, a *Flakpanzerwagen* fitted with three 15-mm Drilling MG 151 machine-guns in a triple shielded mount. The guns were obsolete Luftwaffe pieces and the conversion added a cheap but effective Flak vehicle to the inventory.

More firepower against the ever-growing number of tanks led to the 1944 conversion of the SdKfz 251/22 which originated as a last ditch SP vehicle exactly like the SdKfz 234/4 armoured car. Hitler ordered

maximum self-propelled anti-tank firepower and the Pak 40 75-mm gun was simply fitted into the fighting compartment of the half-track, just as in the armoured car. The wheels and trails were removed and the roof of the driving compartment was cut out to facilitate gun traverse. Like the SdKfz 234/4, this was a most effective expedient, and although the 251/22 was not as mobile, it was built in larger numbers.

This was the last official version of the ubiquitous 3-tonne half-track, though there

were numerous prototypes carrying such pieces as 88-mm Pak guns, *Flakvierling* quad 20-mm mounts, and a variety of old tank turrets and odd weapons. It had proved to be a remarkable and immortal vehicle, and the Skoda-built model of the SdKfz 251 emerged again after the Second World War. Basically similar to the Ausf D version of the original design, it was designated OT-810 and remained as one of the standard troop carriers of the Czechoslovakian Army well into the 1970s.

LIGHT HALF-TRACKS

Second only to the 3-tonne armoured half-track in importance was the 1-tonne model or *Leichter Schutzenpanzerwagen* which complemented the heavier vehicle and, in some forms, took over the functions of armoured cars. The original 1-tonne unarmoured half-track appeared with the other series of prototype machines in 1934–35, built in this case by Demag.

It was intended at the time specifically for infantry roles such as towing the small 37-mm anti-tank gun, the 75-mm infantry howitzer, or ammunition trailers. With a simple truck-like engine cover and an open body it was taken into service as the SdKfz 10 *Leichter Zugkraftwagen* (light prime mover). Pre-production models culminated in the definitive Demag D7 of 1939, which became the major service version and remained in production until 1944. Mechanically this was a most efficient little vehicle, and as with the Hanomag, there was a big group of contractors and subcontractors involved in production work.

By 1939, when the prototype Hanomag 3-tonne half-track SdKfz 251 had proved successful, the Army's thoughts turned to utilising the 1-tonne half-track in a similar fashion by giving it an armoured body. The only snag was that the vehicle was very small (15·3 ft long, 6 ft high, and 5·75 ft wide) and of limited power, having a Maybach HL 42 100 hp engine. With the added weight of an armoured body, it was feared that performance would be reduced, but this problem was overcome by shortening the suspension so that the vehicle had one less road wheel on each side.

Büssing-NAG accordingly designed a scaled down version of the armoured body used on the 3-tonne vehicle and fitted it to a prototype shortened chassis supplied to them by Demag. The resulting prototype performed well and the type was ordered into instant production. It was not so successful as the 3-tonne model, carrying only six men and having inferior performance. However, in certain functions it proved itself. It was ideal as a platoon or company commander's vehicle, as an 80-mm mortar carrier, and as an observation vehicle. Where it could do the same job as a 3-tonne and thus release the 3-tonner for more urgent functions, the small half-track was a valuable asset.

In its armoured form the 1-tonne vehicle was designated SdKfz 250 *Leichter Schutzen-panzerwagen*. Some 14 variants were produced, several of them parallel in function to the SdKfz 251 variants, and in this respect the vehicles were interchangeable. The basic SdKfz 250 was 15 ft long, 6·2 ft wide, and 6·4 ft high. Its top speed was 37 mph and the Maybach Variorex gearbox gave a choice of seven forward and three reverse gears. Mechanically the SdKfz 250 was similar to the SdKfz 251; wheels, suspension and track arrangements were all basically the same except in physical disposition. The combat weight of the vehicle was 5·3 tons (1 tonne was the payload) and the armour was 14·5 mm thick at the front and 8 mm elsewhere. The early vehicles entered service in 1940 and were

first used in the invasion of France. Production continued until 1943 in the vehicle's original form, after which the body was slightly simplified (as with the 3-tonne vehicles) to expedite production.

Because of its small size, stowage arrangements in the SdKfz 250 were very compact. The original superstructure was quite complicated in shape, being 'bowed' outward in the middle to give maximum passenger space. Like that of the 3-tonner, the body was faceted with multi-angled plates. The modifications made in 1943 were considerable: the number of panels used in constructing the body was reduced by nearly half; integral side stowage lockers were built in; front and rear plates were reduced to single pieces; vision flaps were replaced by slits; a larger rear door was fitted; hoops were fitted to support a canvas cover over the passenger compartment; and headlights were eliminated in favour of a standard *Notek* night driving light.

The basic model, essentially a plain armoured troop carrier, was designated SdKfz 250/1 and was principally used by platoon and company commanders. Like the 3-tonne model it normally had two MG 34s on pivot mounts, but sometimes had only one, or carried the heavy tripod-mounted version. Apart from the basic radio-telephone set the vehicle could carry another set if necessary for a particular command function.

Communications equipment

The second model produced was the SdKfz 250/2 telephone line layer which was also used as an observation vehicle. Carrier frames for field telephone lines were supplied and could be fitted on the mudguards or in the fighting compartment. Up to three cables could be laid at a time. A radio set (*Funksprechgerät F,* the R/T set in all vehicles) was standard equipment (in the front superstructure ahead of the passenger seat alongside the driver) throughout the SdKfz 250 series.

The SdKfz 250/3 was the *Leichter Funk-panzerwagen* and like other radio vehicles had the inevitable bedstead frame aerial. As with the 3-tonne equivalent, several sub-variants of this vehicle had different radio equipment according to their function, although externally they were all similar. One of the most famous wartime vehicles was of this type – Rommel's personal command and liaison vehicle, *Greif*, which he used while commanding the Afrika Korps in the Western Desert in 1942. Later examples of the SdKfz 250/3 had the bedstead aerials replaced by pole and star aerials as required and externally were similar to the basic SdKfz 250/1 when their aerials were dismantled.

The SdKfz 250/4, an air liaison or observation post vehicle, was specifically intended to carry the staff of assault gun batteries. A 1943 production vehicle on the simplified chassis, it replaced an earlier type, the SdKfz 253. The SdKfz 250/5 *Leichter Beobachtungspanzerwagen* was a similar vehicle intended for the same role. Externally it resembled the SdKfz 250/1, but it

had the necessary radio equipment for its observation post role.

Next in the series, SdKfz 250/6, was a specialised ammunition carrier for assault guns. There were two variations, one with racks for 70 rounds for the short 75-mm gun and another which carried 60 rounds for the long 75-mm gun.

SdKfz 250/7 was a specialist 80-mm mortar carrier, which largely replaced the 3-tonner in this role. Another version of the 250/7 had racks for carrying 80-mm mortar ammunition, and was often used to transport the commander of the heavy platoon of a Panzer Grenadier battalion. The mortar of an SdKfz 250/7 could be fired from the vehicle but was fired from the ground whenever possible.

A 1-tonne version of the 3-tonne half-track with 75-mm short gun was an obvious development, and the SdKfz 250/8 came into service in 1943 to equip the heavy gun platoon of a Panzer Grenadier battalion, (with six vehicles to a platoon). In all respects, it was similar to the 3-tonne vehicle. An MG 42 was mounted above the gun to act as a ranging piece and to give local defence fire.

By 1942 the armoured car had been proved unsuitable for the Russian Front, and half-tracks had a better survival and maintenance record than wheeled vehicles. In severe snow and mud the four wheeled armoured cars were very limited in operation. It was therefore decided to replace armoured cars as far as possible by half-tracks. This decision resulted in the SdKfz 250/9 *Leichter Schutzenpanzerwagen* (20-mm) being built to replace the SdKfz 222 armoured cars. The SdKfz 250/9 simply had the complete turret assembly of the SdKfz 222 armoured car placed atop the superstructure, which was suitably roofed in. In effect a half-track armoured car was the result. The six-sided turret retained the mesh anti-grenade covers. Later vehicles had a simpler turret.

Further self-propelled gun vehicles were also produced. Just as the 37-mm Pak gun was fitted to the SdKfz 251 half-track (resulting in the SdKfz 251/10 platoon commander's vehicle), an identical operation converted the 1-tonne vehicle to the SdKfz 250/10. This was the first gun-armed variant to appear in service, during the 1940 campaigns. A similar vehicle, the SdKfz 250/11, replaced the SdKfz 250/10 from 1942. This had the tapered bore 20-mm *Schwerer Panzerbusche* 41 mounted on the superstructure front.

The remaining variants of the 1-tonne half-track were all ancillary vehicles for artillery use. SdKfz 250/12 was a *Leichter Messtruppanzerwagen* (light artillery survey vehicle) which was equipped with range-finding periscopes and levelling and signalling apparatus for the control of artillery batteries.

A very early variant of the SdKfz 250 was a light ammunition carrier designed specifically to support the new family of assault guns. It had a different ordnance inventory number, SdKfz 252, and was a good example of the thoroughness with which the Panzer

Into action on the Russian Front: Panzer Grenadiers leap from their half-track

Divisions were planned. Purpose-built for its single support role, its body was distinctively cut away to keep weight to a minimum. It was fully enclosed, with hatches for the crew members and double doors at the rear to give access to the ammunition compartment. A two-wheel trailer carrying additional ammunition was towed.

During the invasion of France SdKfz 252s acted as limber vehicles for the new StuG IIIs, but it was quickly realised that such a purpose-built vehicle was a luxury, even in 1940, and production soon ceased. Thereafter the basic vehicle was adapted for use as a munitions carrier and redesigned SdKfz 250/6 as previously described.

Also dating from the early period was the SdKfz 253 *Leichter Gepanzerter Beobachtungskraftwagen* (light armoured observation vehicle), another fully enclosed type which was intended as a command vehicle for the assault gun batteries. As with the munition carrier, however, it proved an elaborate luxury and was an early victim of rationalisation. Subsequently the basic SdKfz 250 vehicle was adapted for the observation post role as the 250/4 and 250/5.

As these early changes indicate, it was apparent almost from the first that having armoured half-tracks in both the 1-tonne and 3-tonne classes was not altogether satisfactory. There was much duplication of production effort, only a limited number of common components for maintenance purposes, while the 1-tonne vehicle was too small for some roles (eg troop carrying) and the 3-tonner was over-large for some of its roles (eg mortar carrying). This conclusion led to an attempt to rationalise the armoured half-track into one single design.

Stemming directly from the existing vehicles, the HK 600 series of prototypes were built by Demag and Hanomag in 1942, with a strong family resemblance to the SdKfz 251 half-track. The old division into 1-tonne and 3-tonne classes was retained, the HKp 606 being a 1-tonner and the HKp 603 a 3-tonner. They were greatly simplified in comparison with the SdKfz 250 and 251, in order to reduce costs and production time. They were not so radically improved, however, as to justify disrupting production of the SdKfz 250 and 251 at a time when these vehicles were needed more urgently than ever. Consequently, the HK 600 prototypes were abandoned in 1943, but the work was not wasted: many of the simplified features of the superstructures were used in the 1943 rationalisation programmes of the SdKfz 250 and 251.

Hitler's promptings led *Heereswaffenamt* to seek an altogether more radical solution. While some of Hitler's ideas proved wildly impractical or fanciful, many of his fundamental notions were proved right. Indeed, in his desire to standardise effort where possible, and to go for maximum armour and firepower, he was usually some way ahead of his AFV advisers.

In May 1942, he suggested the development of new 'locust-like' vehicles to overcome all the difficulties of traction encountered during the first severe winter on the Russian Front. They were to be both simple and easily adaptable to all functions. Hitler wanted them to be in production within a year and the Adler firm was given the task of making two prototypes. These were ready at the end of 1942 and were known as LeWS (*Leichter Wehrmachtschlepper* or light military tractor).

The LeWS was radically different from

Rommel's personal command car, Grief (opposite), an SdKfz 250/3 light armoured half-track

**SdKfz 250/5
Light Half-track with Radio**

the SdKfz 250 and emphasised the passing of the leisurely days of complicated automotive engineering. The LeWS had a simple chassis, pressed steel dics road wheels, low pressure front tyres, steel dry pin semi-tracks, and used tank type sprockets and idlers. The armoured front cab and engine compartment followed the layout of a conventional truck and the engine was an ordinary Maybach HL 30 four-cylinder unit, restricting maximum speed to only 17 mph. The body floor took the form of a flat pan so that various ordnance mountings or self-contained superstructures could be simply bolted on. All fabrication was of flat plates and simple pressings. Provision was made for production without the cab so that a completely integral cab and super-structure could be fitted to certain versions.

However, the promising LeWS was never proceeded with, probably because it still did not overcome the problem of two different weight classes. In any case, a full-tracked truck, the *Raupenschlepper Ost* (tracked tractor Eastfront) was being pro-duced to meet many of the same require-ments. It was even cheaper and simpler to build so that production of LeWS would have involved precisely the sort of duplica-tion the design had set out to avoid. A third LeWS prototype was built in 1944 and there were plans for a fourth with Maybach-Olvar tank type transmission, but the *Leichter Werhmachtschlepper* was a lost cause and was never ordered.

A companion design, however, the *Schwerer Wehrmachtschlepper* (SWS – heavy military tractor), had a more success-ful history and was the nearest the Wehrmacht came to a standardised half-track replacement for the earlier designs. It was actually intended to replace the 5-tonne (unarmoured) series of artillery tractor half-tracks and was produced with a cargo body and unarmoured truck type front end for this role. Like the LeWS, the SWS was produced with an armoured cab and engine cover, and a flat body floor to which various combinations of ordnance and superstructure could be fitted. The prototype was built in 1942, and although production did not start until late 1943, the type saw wide and useful service in the last 18 months of the war.

The basic vehicle was produced as an armoured supply truck or munitions carrier with plain wire-mesh drop sides. The most important variants, however, were those fitted with the 150-mm *Panzerwerfer* 42 rocket launcher and the 37-mm Flak43 AA gun. The former had a low armoured super-structure in which the rounds were carried with the ten-barrel launcher in a traversing mount on top of the superstructure. The Flak gun version retained the wire mesh drop sides of the supply carrier with the tra-versing gun mount welded to the body floor.

Captured half-tracks

The SWS was the final development of the German armoured half-tracks, but in addition to their own vehicles, the conquests of 1939–41 brought them a number of foreign half-tracks which were added to the stock of German built vehicles.

France was the major European builder of half-tracks before the war and the Unic and Somua-Kégresse were the standard French types. The Somua was the most widely used, and fitting armoured bodies to it produced an armoured personnel carrier similar to the SdKfz 251. This

SdKfz 250/7
Light Half-track with 80-mm Mortar

SdKfz 250/8
Light Half-track with 75-mm Gun
Length 14·96 ft *Width:* 6·4 ft *Weight:*
12,565 lb *Speed:* 40 mph *Range:* 200 miles
Armour: 12 mm max *Armament:* 75-mm KwK 37;
1 x 7·92-mm MG 34

*September 1941: German troops with their SdKfz
253 light observation vehicle watch smoke rising
from the city of Novgorod, set on fire by the retreating Russian troops*

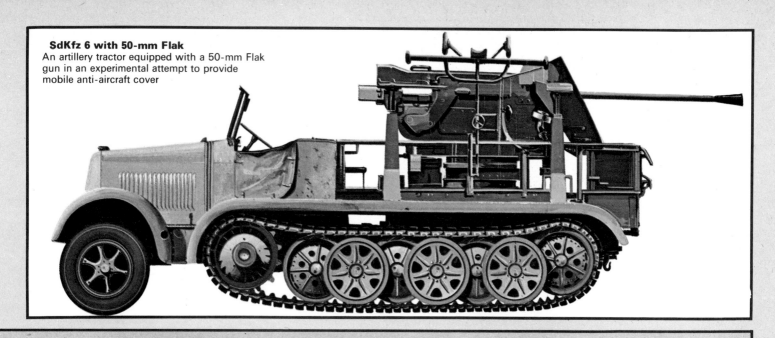

SdKfz 6 with 50-mm Flak
An artillery tractor equipped with a 50-mm Flak gun in an experimental attempt to provide mobile anti-aircraft cover

vehicle was designated *Mittlerer Schutzenpanzerwagen S 307 (f)* and was used as a troop carrier and 105-mm howitzer tractor in some armoured divisions. A further armoured conversion was built in occupied France and consisted of 16 captured French 80-mm mortars mounted on a traversing mount in a twin bank on the rear chassis. Another variant carried a 150-mm *Panzerwerfer 42* launcher (mounted as in the SWS), while a 1944 conversion mounted the 75-mm Pak 40.

The Unic-Kégresse half-track was also given an armoured superstructure. Designated *Leichter Schutzenpanzerwagen U 304 (f)*, it was used as a substitute for the SdKfz 250 by German units based in France.

Conscripted vehicles

As the war situation deteriorated, all sorts of vehicles not intended for offensive use were given armoured bodywork and pressed into service with first line units. An early example was the experimental fitting of a captured Russian 76·2-mm anti-tank gun to an SdKfz 6 5-ton half-track tractor. Only nine such vehicles (popularly known as Diana) were converted and some were sent to Libya in 1942. This opened up the idea of further conversion and many unarmoured half-tracks were given armoured cabs and superstructures.

This was particularly true of vehicles with anti-aircraft mounts. Allied air superiority on all fronts meant the Germans had to divert a lot of resources into providing air defence for their ground troops – unlike the Allies, who were much less frequently bothered by the Luftwaffe.

Some SdKfz 9 18-ton half-tracks – the largest of all – were given armoured cabs and engine covers and fitted with the 8·8-mm Flak 37 gun. The SdKfz 7/1, SdKfz 10/5 and SdKfz 7/2 all had anti-aircraft mounts and varying degrees of improvised armour protection.

Finally, there was the Maultier truck, a 3-tonne capacity lorry converted to a half-track. Initially the idea stemmed from a field conversion in Russia, where some troops placed Panzer I track units on a truck in place of the rear wheels. Subsequently Fords, Opels and Daimlers were given factory conversions to compensate for the lateness of the *Schwerer Werhmachtschlepper*. Even some 4·5-tonne Mercedes trucks were converted with PzKpfw II track units. The Opel model was the most numerous of these, and to help overcome the constant shortage of armoured half-tracks it was decided to design an armoured superstructure for it. Some 3000 armoured versions of the Opel Maultier were ordered and delivered in 1944–45 as armoured gun tractors/munition carriers. They were used for towing wheeled *Nebelwerfer* mounts, with the rockets carried in the armoured body.

A further 300 Maultiers were ordered to be fitted with the 150-mm *Panzerwerfer 42*. The ten-barrel launcher was fitted on a low superstructure as in the SWS, and the Maultier and SWS were used interchangeably as the standard mobile rocket launcher for specialist rocket troop regiments in armoured divisions. Again, it must be stressed that this was an expedient design – no more than a lightly armoured tracked lorry – but it was quite an effective design.

The last half-track worthy of mention was not an armoured vehicle at all – the tiny NSU *Kettenrad*, a half-track motor-cycle. Produced as part of the Wehrmacht's half-track series, it was designated HK 100. The front forks and suspension were of motorcycle type and the rear end was made up of a punt type chassis with miniature half-tracks. The engine was a water-cooled 1·5 litre Opel motor car unit of 36 hp. Top speed was 50 mph.

The original idea was to provide a small tractor for the use of airborne troops, and the design work was conditioned by the need for a vehicle light enough and small enough to be carried by the standard Junkers Ju 52 transport. The *Kettenrad* could be slung between the aircraft's undercarriage or manhandled (just) through the fuselage cargo door. It could tow a light ammunition trailer, a 28-mm or 37-mm anti-tank gun, or a 75-mm light infantry gun. It was designated SdKfz 2 *Kleine Kettenrad*. The first examples were issued for service in June 1941 and the type became immensely popular with the troops specially on the Russian Front. A couple of variants carried reels or spools in the rear compartment for telephone or cable laying. An enlarged version, the *Grosse Kettenrad*, was projected but resources were not available for such luxuries and it never went into production.

End of the line

The half-track had almost had its day by 1945. It was complicated to build and a compromise in performance terms. The open top, while considered satisfactory in the late 1930s, left much to be desired when the occupants came up against aircraft strafing and shell splinters.

This is not to say that armoured half-tracks were not valuable. The Panzer Divisions with their combination of tanks and half-tracks and the tactics that went with them were almost unbeatable until they were overwhelmed by sheer weight of numbers. By 1942 there were sufficient half-tracks available for all Panzer Divisions to have one or more battalions of Panzer Grenadiers equipped with half-tracks and over 15,000 of the 3-tonne model alone were built. Increased firepower was provided by adding the self-propelled gun variants to the armoured infantry battalions, but the grave shortage of vehicles was inevitably revealed.

A replacement vehicle for the armoured half-track was planned, to apply all the lessons of five years of war. It was intended to base a new fully-tracked armoured troop carrier on the chassis of the small but highly efficient Czech PzKpfw 38(t), and to use the German adaptation of the chassis as the basis for a whole series of vehicles.

The armoured personnel carrier variant, *Schutzenpanzerwagen* Ausf 38(t), would have been fully enclosed and would have had rear doors and a 20-mm gun in a traversing turret. However, by the closing months of the war there was little chance of the project proceeding beyond the drawing board stage and no prototype ever materialised. While no new half-track designs have appeared since the German and American types of the Second World War, virtually all tracked armoured personnel carriers built by the major powers since then have been of the full-track type similar to that planned by the Wehrmacht in 1945.

Panzerwerfer *42 ten-barrel 150-mm rocket launcher mounted on the back of a converted Opel Maultier truck. More than 3000 of these trucks were converted to half-tracks, 300 of them with the* Panzerwerfer *mountings*

Opel Maultier Panzerwerfer Half-track
Conversion

MOBILE SUPPORT

Self-propelled guns on tracked chassis date back almost to the invention of the tank. The very first self-propelled gun was British, the Gun Carrier Mk I of 1916, brainchild of the ingenious men of the Landships Committee who sponsored Britain's first tanks in 1915–16. The Gun Carrier Mk I was the progenitor of all self-propelled guns, and by carrying a complete 60-pounder gun and its carriage in demountable form, it was also the direct ancestor of the *Waffentrager* (weapons carrier) type of vehicle which the Germans also developed during the Second World War.

While some 48 vehicles of this type were built by the British, they rarely used them for their intended role; instead, they were relegated to supply carrying duties. But although the British did not persevere with them, the French Army took up the idea and made all the running in 1916–18. By the time of the Armistice there were about eight self-propelled gun types in French service, usually with a chassis based on the Holt Caterpillar derived type (St Chamond) which had inspired the early French tank efforts. Calibres of the French self-propelled pieces ranged from 280-mm down to 75-mm. The 75-mm version was built on the Renault FT 17 chassis, while larger pieces were on Holt-type chassis. As it happened, two of the largest equipments developed in 1917, the 280-mm and 194-mm guns, remained in French service throughout the inter-war period and were still in use in 1940. The French, however, had made no design progress with self-propelled guns, so there were no successors to these weapons.

American developments

However, the US Army took over where the French left off. When the Americans entered the Great War in 1917, training and supply of the American Expeditionary Force were jointly undertaken by the British and French. The Americans used a number of self-propelled guns (or 'gun motor carriages' as they called them) of the French type, and built others of their own based closely on the French ideas. The resulting pieces were essentially big guns in open mounts on tracked chassis.

Various prototypes were produced until 1922, when development ceased as a result of major cuts in defence expenditure. Around a dozen self-propelled pieces of various calibres were built, from a 240-mm howitzer down to a 75-mm gun. Conclusions drawn by the American Caliber Board – set up to investigate future artillery policy in the 1920s – was that self-propelled guns had a major drawback: the vulnerability of the motorised carriage. If this broke down or was damaged in action then the gun itself was immobilised. The Caliber Board suggested all future artillery should be tractor-hauled and for the rest of the inter-war period only nominal experimental work in self-propelled guns took place. During the Second World War, however, the Americans revived gun motor carriages in a big way.

In Britain in the 1920s there was a short

period of revival for self-propelled guns which coincided with the forward-looking period of the Experimental Mechanised Force. Some efficient and compact self-propelled 18-pounders were built on the Vickers chassis of the medium Mk I tank, then in British Army service. The idea for the prototypes was encouraged by the Director of Artillery and, indeed, the main prototype was called the 'Birch Gun' after the then Master-General of the Ordnance. The very first self-propelled guns in the modern sense, the prototypes had an up-to-date purpose-built chassis comparable in performance and standards to contemporary tanks.

The new guns (only three were built) operated experimentally as an integral part of the Mechanised Force on exercises and received world-wide publicity. The Birch Gun looked dramatic, purposeful and efficient. However, although well ahead of any previous thinking, and greatly in advance of the rather crude French and American motor carriages, it was not well received in the British Army. The Royal Artillery were suspicious, and the Tank Corps saw them as usurping tanks on the battlefield (similar differences of opinion held up British self-propelled gun development again during the war), so these promising designs were not pursued. The best self-propelled guns in the British Army in 1939 were a few 'support tanks' which could fire smoke shells by replacing the 2-pounder gun with a 3-in howitzer, and no British self-propelled artillery was even under development in the early stages of the war.

While the German Panzer Divisions were being built up in the 1930s the problem of divisional artillery was not overlooked. But priority was given to calibre and efficiency rather than mobility. To a large extent the problem was one of economics. Tanks or any other type of specialised tracked vehicle were extremely costly and priority was given to tank production. It was much cheaper to use wheeled vehicles to tow artillery pieces than to provide tracked carriages or tractors.

However, prewar German tacticians had not discounted the idea of self-propelled guns, and various prototypes were produced in the 1930s, including an armoured half-track tank destroyer in 1935. None of these was put into production, but the desire for mobile anti-tank guns within Panzer Divisions led to the very swift introduction of a new class of vehicle in 1940, the *Panzerjäger* I (tank hunter) which was based on the Panzer I chassis with an anti-tank gun replacing the original tank turret. This was the first of an increasing number of *Panzerjäger* vehicles which were turned out in rapid succession over the next few years. These were the most important of the early German self-propelled guns. While the *Panzerjäger* were a valued additional weapon in the French campaign in 1940, they became essential after the start of the Russian campaign in 1941. Here thousands of Russian tanks were encountered; the attrition rate on both sides was enormous,

and numbers of fighting vehicles were improvised by fitting guns to obsolescent or captured tank chassis.

The most powerful anti-tank gun available at this time was the Russian 76·2-mm weapon, large numbers of which were captured by the Germans in the early stages of the invasion of the Soviet Union. This was adapted to fit old PzKpfw II and PzKpfw 38(t) chassis. Work started on the new vehicles in December 1941, Alkett converting the PzKpfw IIs and Bohmisch-Mahrische of Prague the PzKpfw 38(t)s. Guns and mounts were adapted by Rheinmetall-Borsig. These vehicles, SdKfz 131 and 132 respectively, were subsequently better known as the *Marder* (marten) II and III.

Plans were made to employ the German 50-mm Pak 38 gun in a similar manner, but this was dropped in favour of the more powerful new 75-mm Pak 40 which was comparable to the Russian gun. In mid-1942 this gun appeared in a more elaborate conversion (also SdKfz 131) on the later model Panzer II chassis. The capture of much French equipment in 1940 led to a similar conversion for the so-called (by the Germans) *Lorraine Schlepper* in 1942, the firm of Becker adding Pak 40 guns and superstructure to mak the SdKfz 135, better known as the *Marder* I.

Makeshift conversions

Of necessity these *Panzerjäger* conversions were hasty, crude, and makeshift and from the outset they were regarded as stopgaps while purpose-built vehicles were developed. High silhouettes, thin armour, poor crew protection, low speed, and instability were recognised defects. Improved models based on the PzKpfw 38(t) chassis were the SdKfz 138 and 139, the *Marder* III. They had better armour disposition but still used the tank chassis virtually unchanged. Meanwhile work was undertaken on the basic PzKpfw 38(t) chassis to make it more suitable for the SP role; the rear engine was moved to the middle allowing a larger fighting compartment to be built at the rear and a lower silhouette. Vehicles using this chassis, also designated *Marder* III, appeared in 1943. This was the last of the important *Marder* family, for in 1944 it was succeeded by the *Hetzer*, a fully armoured low silhouette vehicle purpose-built for the anti-tank role.

The new type of anti-tank vehicle introduced the final classification, the *Jagdpanzer* (hunting tank), a type intended to supersede the *Panzerjäger*, even though this aim was never realised. Tank destroyer versions of the Panther and Tiger fell into this category. At this stage the classification of anti-tank vehicles by type became more complicated and terminology was changed by the Germans. Some types originally designated *Panzerjäger* (eg the *Elefant*) were already fully armoured and were redesignated *Jagdpanzer*, into which category the fully armoured assault guns (*Sturmgeschütze*) were also subsequently placed.

Then, at the beginning of 1945, there was

SdKfz 139 'Marder' III
The first *Panzerjäger* 38(t) mounted the Russian
76·2-mm Pak 36(r) on the PzKpfw 38(t) chassis

a further change. All the fully armoured vehicles with a tank destroyer capability were now redesignated as *Panzerjäger* while the erstwhile (and by now obsolescent) *Panzerjäger* of the *Marder* type were recategorised as Pak (Sf) or 'anti-tank carriages'. There were also a number of less well-armed vehicles, mainly based on captured infantry carriers, which had lighter guns. These were mainly used for police work, home defence, or patrol work in occupied countries. A small number of *Panzerjäger* with heavier calibre guns were produced (notably the *Nashorn* [Rhinoceros] and *Elefant*) as well as a new generation of *Panzerjägerkanone* under development in 1945, based on improved versions of the PzKpfw 38 chassis.

The basic unit for *Panzerjäger* deployment was the company, which was subdivided into platoons. Due to their vulnerability it was usual to protect the flanks or rear of a *Panzerjäger* company with infantry, tanks, or towed anti-tank guns. On

the offensive, *Panzerjäger* were used to follow up an attack rather than lead, and they protected against breakthroughs by enemy tanks or picked off retreating stragglers. Platoons or companies usually fired en masse for maximum effect as directed by radio by the company commander, individual platoons engaging different targets as necessary.

On the defensive *Panzerjäger* were kept concealed (but not dug in) as much as possible and were held as mobile reserve to guard against breakthroughs by enemy tanks. File, line, or arrowhead deployment was used. In the last year of the war attacks from the air became a major limitation on the deployment of *Panzerjäger* in the field, these open vehicles being particularly vulnerable to strafing. Infantry, motorised infantry, and later *Volksgrenadier* divisions all had *Panzerjäger* companies, and they were also found in armoured divisions before being displaced by *Jagdpanzer*. Heavier

vehicles like the *Nashorn* and *Elefant* were found in GHQ companies. Despite their tactical limitations, *Panzerjäger* provided fire power and movement in the anti-tank role and played a major part in German armoured warfare operations throughout the Second World War.

While the *Panzerjäger* class of vehicle was the most important, and most numerous, there was parallel development in mobile support artillery, furnished principally for Panzer and Motorised Divisions but later found in nearly all types of fighting division. With tanks and tank hunters in service, the towed artillery now lacked comparable mobility and just as the *Panzerjäger* was quickly improvised by mounting an anti-tank gun on an old tank chassis, so assault howitzers (*Sturmhaubitze*) and infantry guns (*Infanterie Geschütze*) vehicles were produced by placing existing types of howitzer or gun on modified tank chassis.

The first to see service was the 150-mm

L/12 mounted on a Panzer I chassis. This was converted and in service with the Panzer Divisions by the time of the invasion of France in 1940, and enabled the infantry gun companies of Panzer Divisions (six guns to a company) to exchange their tractor-drawn 150-mm infantry guns for the same weapon on a highly mobile chassis. It was a very top heavy and unsatisfactory conversion but it proved the point and led to better things. By 1942 production of self-propelled infantry guns or howitzers was in full swing, the most famous and widely used being the *Wespe* (Wasp) on the PzKpfw II chassis, and the *Hummel* (Bumble Bee) on the PzKpfw III/IV chassis. These self-propelled infantry guns took second place to the tank hunters in priority of output and allocation, but nonetheless usually at least one artillery battalion in a Panzer Division was fully equipped with self-propelled guns.

A later, but similar, type was the self-propelled field howitzer or gun (*Feld-haubitze* or *Feldkanone*), which featured a lighter field piece, also on an obsolete tank chassis. Collectively these SP infantry support weapons were called *Panzer-artillerie* (armoured artillery) and the type of vehicle so produced was known as a *Geschützwagen* (motor-gun), GW for short.

The other major category of self-propelled gun was the *Sturmgeschütze* (assault gun), the one type of tracked artillery vehicle to be purpose-built. Development of assault guns started before the outbreak of war as an integral part of the Panzer Division idea, and a type expressly intended to supplement the towed infantry gun. True mobile support vehicles, with a gun in a tank type mount in a low-profile superstructure, they had a common chassis with the battle tanks – the *Sturmgeschütz* III, (StuG III) was mounted on the PzKpfw III chassis, and the *Sturmgeschütz* IV (StuG IV) on the PzKpfw IV chassis. Indeed, had the German arms industry been geared up for a long war in 1939–40, there would not have been such a vast output of crude improvised types.

It was originally intended that the *Sturmgeschütz* type should supply all the needs of mobile support artillery. They were therefore organised in batteries for allocation by local group commanders to support infantry battalions in the attack. The first two *Sturmgeschütz* battalions were in service for the invasion of France in 1940, and thereafter they became of increasing importance. Assault guns gave the infantry fast-moving firepower, leaving tanks to engage other tanks or spearhead fast-moving thrusts into the enemy flanks.

The assault gun, with a low velocity 75-mm gun, inevitably grew in power; later the high velocity long 75-mm gun was fitted, giving them an effective anti-tank capacity as well, so that there was a merging of the functions of the *Sturmgeschütz* and the *Panzerjäger*. By the latter part of the war there was much intermixing of types and functions, and the *Sturmgeschütz* found itself increasingly replacing tanks.

In fact, the Wehrmacht found the merging of capabilities very useful, for the *Sturmge-schütze* could carry a heavier gun in its

Wespe in action. This self-propelled infantry gun mounted a 105-mm gun on a PzKpfw II chassis

JgPz Tiger (P) 'Elefant'
The most famous of Germany's 'Hunting tanks',
Dr Porsche's 'Elefant' was a formidable weapon
Length: 26·71 ft *Width:* 11·25 ft *Weight:*
158,000 lb *Speed:* 12·5 mph *Range:* 95 miles
Armour: 200 mm max *Armament:* 88-mm Pak 43/2

limited traverse mount than a tank of corresponding size, it was of lower silhouette, had better armour, and was less expensive and quicker to build – all important factors as time and money began to run out. In the latter part of the war *Sturmgeschütze* sometimes partially replaced tanks in tank battalions, and were used to counter enemy tanks. The *Sturmgeschütz* was, of course, greatly superior to the improvised types, and the lessons were quickly learned by the Soviets, who directly copied the German ideas to produce their famous series of SUs (*Samochodnaya Ustanovka* or self-propelled gun) such as the SU-85 and SU-100 (on the T-34 tank chassis) and the SU-122 and SU-152 (on the KV and Josef Stalin tank chassis).

These Russian assault guns, mounting even heavier weapons than the German *Sturmgeschütz*, proved formidable weapons, able to combine all the functions of several types of SP gun in just a few basic types. Even the British Army toyed with the assault gun idea, and a very effectively-armed, though slow, tank destroyer/assault gun version of the Churchill tank was built

(the Churchill 3-in gun carrier); this, however, never saw service, mainly due to argument as to whether it should be a Royal Artillery or Tank Corps responsibility.

Closely allied to the *Sturmgeschütze* were one or two less important classes of vehicle which in most cases were built in quite small numbers. These included the *Sturmpanzer* (assault tank) and *Sturmmörser* (assault mortar), both evolved for street fighting on the Eastern Front where the Russians stubbornly held towns and cities street by street – the experience at Stalingrad in 1942 leading directly to their development.

Lastly, and most importantly, came the anti-aircraft tank or *Flakpanzer*, which was developed in the latter part of the war. As we have seen, there were several types of AA mounting on half-tracks, armoured, semi-armoured or unarmoured. As Allied air strength increased from 1943 onwards, these half-track AA vehicles were not enough and a series of *Flakpanzer*, mostly based on the PzKpfw IV, were produced. Towards the end of the war some heavier AA pieces on the Panther tank chassis

were planned, but none was produced.

The full inventory of self-propelled gun types in the Wehrmacht runs into three figures and it is not possible to describe every single type; moreover, some of them were single prototypes or built only in handfuls. Here, then, we look at the types of major importance only since these were the types in large scale service.

Waffenträger

Weapons carriers were unique to the German Army during the Second World War, and characteristic of the very advanced thinking which the German design engineers applied to military problems. The *Waffenträger* was a class not developed for the armoured forces, but evolved in the first place for the field artillery. Had developments been carried through to their logical conclusion and full scale production started, then almost the entire field army, as far as 'teeth' arms were concerned, would have been fully motorised and to a large degree fully armoured. The development history is quite distinct from that of self-propelled artillery and the functions of *Waffenträger*

SiG 33 SP infantry guns in an armoured column move up through a devastated Russian landscape

PzJg RSO
This *Panzerjäger* mounted a 75-mm Pak 40 on the chassis of the *Osttractor* artillery tractor

as originally foreseen were simply to facilitate the mobility of the artillery, not to fight in the armoured role.

In 1942, the Army's Artillery Branch, no doubt inspired by developments in the self-propelled gun field, put up proposals and requirements for a form of carriage which enabled the field artillery to carry out its traditional role, harnessing a tracked vehicle to replace the horse team or gun tractor. There was some evidence here of inter-branch controversy, for the artillerymen did not consider assault guns and similar self-propelled guns with limited traverse mounts to be fully suitable for a divisional artillery role, though they liked the mobility which these weapons enjoyed.

Detailed specifications were issued, and subsequently a large number of firms built prototypes. In fact, many *Waffenträger* types were projected but relatively few were actually made. The PzKpfw IV chassis was used for the 1942–43 designs, and two early projects which did not see fruition featured either a demountable 150-mm FH 18 or a 128-mm FH 81. These were very heavy pieces and the designs were probably dropped because of weight problems. The very first *Waffenträger* built was the *Heuschrecke* (Locust) IVb which, as the designation implies, was also on the Panzer IV chassis. This was fully designated GW IVb *für* 105-mm le FH 18/1 (chassis for the 105-mm light field howitzer 18/1).

A development of the *Heuschrecke* was the *Grille* (Cricket). Fully designated 105-mm le FH 18/40 *auf Fahrgestell* GW/III/IV (105-mm light field howitzer 18/40 on the chassis of gun carrier III/IV), this vehicle was on the running gear of the proposed PzKpfw III/IV.

What finally evolved were two standard chassis – a light weapons carrier (*leichter Waffenträger*) and a medium weapons carrier (*mittlerer Waffenträger*). Mechanically these two vehicles featured the running gear, steering transmission, and final drive of the original PzKpfw 38 (t) design.

Production of these standard *Waffenträger* was scheduled to begin in March 1945 and reach an output of 350 a month by late that year. Firms which had been involved in PzKpfw IV production were to switch over to building the new *Waffenträger* vehicles. Adding to these the *Hetzer* and the proposed *Schutzenpanzerwagen* auf 38(t) it can be seen that a fully rationalised output of light armoured vehicles would be produced starting in 1945, which would replace dozens of older types of self-propelled weapon and the various half-tracks. Taking into account also the E series tanks and the projected Panther II, it is apparent that a completely new look Panzer Division was not far off – lean, streamlined and fully rationalised.

What was projected on paper, however, was far too late to stand a chance of realisation for while the plans were sound, the German war machine and the economy were fast collapsing in the early spring of 1945. By the time Germany surrendered, the only real progress with this new-look equipment was the prototypes, some jigs, and a lot of drawings. It is of interest to note, however, that what had started out as a rather esoteric artillery dream of fanciful vehicles which could emplace their own guns had turned by degrees into a highly sophisticated and ruthlessly standardised type of vehicle.

STURMGESCHÜTZE

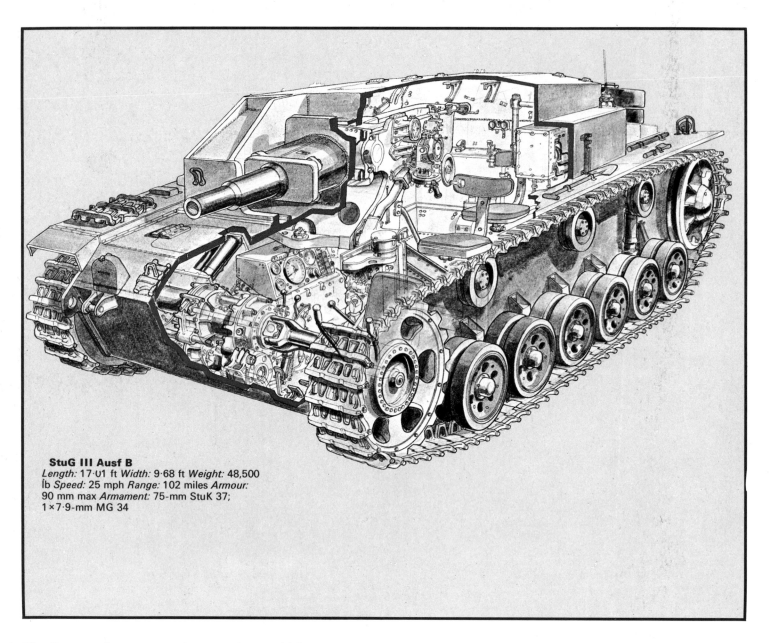

StuG III Ausf B
Length: 17·01 ft *Width:* 9·68 ft *Weight:* 48,500
lb *Speed:* 25 mph *Range:* 102 miles *Armour:*
90 mm max *Armament:* 75-mm StuK 37;
1×7·9-mm MG 34

The *Sturmgeschütz*, or assault gun, takes pride of place in the story of German self-propelled guns. The first steps to procure an assault gun go back to mid-1936, when *Heereswaffenamt* agreed to a request by the Inspector of Infantry's department for an armoured vehicle to provide supporting fire for the infantry.

Daimler were asked to prepare designs based on the Panzer III which was already going into production, and Krupp were asked to design the superstructure and armament. As finally drawn up, the vehicle, designated *Sturmgeschütz* III, was exactly like the current PzKpfw III A Ausf E mechanically and structurally up to track cover level, above which it had a low enveloping armoured superstructure in place of the usual turret, since a low silhouette was specially requested. Armament was the familiar 75-mm L/24 low velocity gun as fitted to the PzKpfw IV tank. Armour protection was 50 mm at the front, reducing to a minimum of 10 mm elsewhere. A pre-production batch of 30

vehicles was ordered in 1939 and the first of these were in service for the invasion of France, where they proved successful.

Lagging production

The original designation was *Gerpanzerte Selbstfahrlafette für Sturmgeschütz* 75-mm KwK (armoured carriage for 75-mm assault gun). A full production order was placed for the vehicle, deliveries to start in September 1940 and to proceed at the rate of 50 a month. Built on the by then current PzKpfw III Ausf F turret they were little different from the pre-production type, being designated SdKfz 142. The Maybach HL 120 TR engine with Maybach Variorex gearbox gave performance similar to that of the PzKpfw III tank, with ten forward speeds and one reverse. The empty weight was 20·2 tonnes and combat weight 22 tonnes. There was a crew of four and 44 rounds were carried. Built during the winter of 1940–41, some 184 vehicles of this type were completed. By this time of course production was already lagging behind

requirements and this marked the start of production of the improvised *Panzerjäger* and *Sturmhaubitze* types to keep up the numbers of vehicles in the field.

An improved engine and only detail changes marked the Ausf B, the third type to appear. Improved frontal armour was featured and 548 were built by the firm of Alkett. All these three variants were externally similar and are easily distinguished by the short L/24 gun.

However, the shape of things to come was shown in a new interim production model designated *Sturmgeschütz Lange* 75-mm *Kanone* L/33 (assault gun with long 75-mm gun L/33). This recognised the need for an anti-tank capability and featured a lengthened higher velocity version of the 75-mm weapon. Krupp and Alkett built these vehicles early in 1942, largely as a result of early experience on the Russian Front. The vehicle was unchanged from the Ausf B except for the new gun.

By now the Wehrmacht was fully involved in the Russian campaign and the vast

StuG III Ausf G, with 105-mm howitzer in place of the 75-mm gun, Zimmerit anti-magnetic paste and rails for Schürzen side plates. Inset, left: General Heinz Guderian, mastermind of the Panzer Divisions. Sacked after the German defeat at the battle of Moscow, he was later appointed Inspector of Armoured Forces, though a clerical error left 90% of self-propelled guns outside his terms of reference

numerical superiority of Soviet tanks made it imperative that the assault gun must now be fully capable of taking on tanks. As a result the long 75-mm StuK 40 L/43, as then being fitted to PzKpfw IVs was fitted in the mantlet of the *Sturmgeschütz* from February 1942. Some early vehicles had no muzzle brakes. Only 119 of these vehicles were built before production changed to a newer improved model. All these early vehicles had quite simple armour disposition with plain hatches, sighting window in the front superstructure, and an armoured radio box each side.

However, from June 1942, a much revised model was produced, resulting from an order given by Hitler at the end of 1941 after the Russian T-34 tank had been encountered for the first time. The new model was designated 75-mm *Sturmgeschütz* 40 Ausf G, having the even longer StuK 40 L/48 gun as standard and an improved superstructure with frontal armour increased to 80-mm thick. Built on the PzKpfw III Ausf G chassis, it continued to be produced unchanged until late in 1943. In later models the Ausf J, L and M chassis were used with consequent changes in appearance of sprocket wheel, idler and track width. This vehicle had an MG 34 with 600 rounds as standard, and an empty weight of 21·6 tonnes. An armoured commander's hatchway and revised superstructure were new features.

In 1943 the model was again improved, to be more commonly known as the *Sturmgeschütz* III Ausf G (StuG III Ausf G). Spaced frontal armour, a commander's cupola, and side skirts were featured. Later vehicles had a much improved cast *Saukopf* mantlet replacing the fabricated mantlet of earlier vehicles. Many were coated with the Zimmerit anti-magnetic paste, and a layer of concrete up to six inches deep was added to the armoured roof of the driving compartment. The final production vehicles were fitted with the *Nalwerteidgungswaffe* (close defence weapon), an internally loaded bomb thrower which projected anti-personnel and smoke charges against attacking infantry. A remote-control MG 34 was also by now standardised. This late production Ausf G had been built on the chassis of the late (10/ZW) PzKpfw III tank, and when Panzer III production ceased completely in August 1943, the chassis was produced solely for assault gun use. Some detail changes, eliminating non-essential fittings, were made to the chassis at this time. Over 9000 of the various production versions of the StuG III Ausf G had been built by the time the factories were captured by the Allies.

Support StuGs

Giving the StuG III an anti-tank capability detracted from its performance in the infantry support role, so an additional series of vehicles was built to support the StuG IIIs. These were simply StuG III Ausf G models with a 105-mm *Sturmhaubitze* 42 howitzer; consequently the type was designated 105-mm *Sturmhaubitze* 42 Ausf G. Some had the *Saukopf* mantlet and others the cast type; some lacked a muzzle brake. A few vehicles were unarmed and were used as munitions carriers for the assault guns, while others were used by engineers of tank battalions to carry bridging equipment. Racks were fitted on top of the superstructure to hold assault bridge sections. Over 1000 105-mm *Sturmhaubitze* 42 vehicles were built in 1943–45.

In 1942 one other type of assault gun on the PzKpfw III chassis was produced. This was the *Sturm-infanteriegeschütz* 33 *auf* PzKpfw III, which featured the 150-mm L/11 howitzer in a high armoured superstructure. A crew of five was provided and the vehicle weighed 22 tonnes. The idea of this vehicle was for use in close street fighting, and 12 were built for combat trials. Subsequently it was decided that heavier armour was required but the Panzer III chassis was not powerful enough to allow the extra weight needed, and the project was abandoned in favour of the *Brummbar* (Grizzly Bear) on the PzKpfw IV chassis. The 12 SiG 33s built saw service on the Russian Front.

Yet another version of the *Sturmgeschütz* saw service in larger numbers, however, the StuG IV, basically the same conversion as the StuG III, but on a Panzer IV chassis, with the difference in overall length made up by fabricating a new roof section. Some 632 StuG IVs were built, starting in mid-1943.

StuG III Ausf G with bolted box mantlet and Schürzen *advances through a maize field*

StuG III Ausf G
Length: 20·14 ft *Width:* 9·71 ft *Weight:*
52,690 lb *Speed:* 25 mph *Range:* 105 miles
Armour: 90 mm max *Armament:* 105-mm StuK 42;
1 x 7·92-mm MG 34

PANZERJAGER

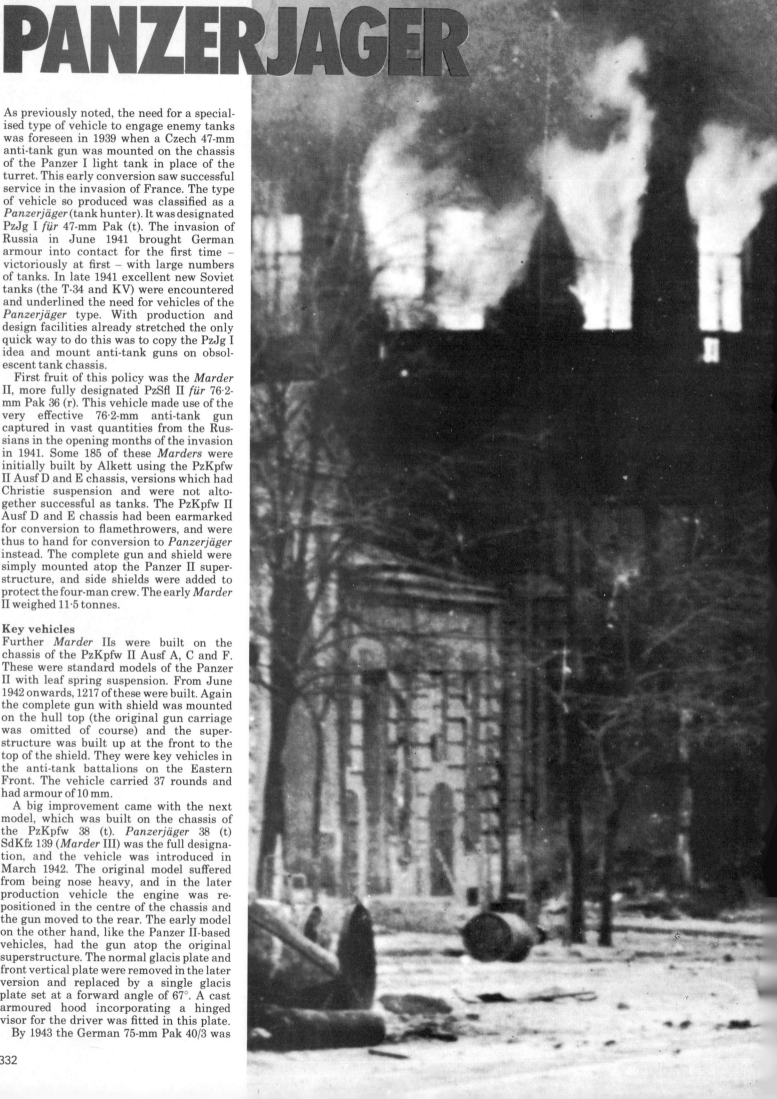

As previously noted, the need for a specialised type of vehicle to engage enemy tanks was foreseen in 1939 when a Czech 47-mm anti-tank gun was mounted on the chassis of the Panzer I light tank in place of the turret. This early conversion saw successful service in the invasion of France. The type of vehicle so produced was classified as a *Panzerjäger* (tank hunter). It was designated PzJg I *für* 47-mm Pak (t). The invasion of Russia in June 1941 brought German armour into contact for the first time – victoriously at first – with large numbers of tanks. In late 1941 excellent new Soviet tanks (the T-34 and KV) were encountered and underlined the need for vehicles of the *Panzerjäger* type. With production and design facilities already stretched the only quick way to do this was to copy the PzJg I idea and mount anti-tank guns on obsolescent tank chassis.

First fruit of this policy was the *Marder* II, more fully designated PzSfl II *für* 76·2-mm Pak 36 (r). This vehicle made use of the very effective 76·2-mm anti-tank gun captured in vast quantities from the Russians in the opening months of the invasion in 1941. Some 185 of these *Marders* were initially built by Alkett using the PzKpfw II Ausf D and E chassis, versions which had Christie suspension and were not altogether successful as tanks. The PzKpfw II Ausf D and E chassis had been earmarked for conversion to flamethrowers, and were thus to hand for conversion to *Panzerjäger* instead. The complete gun and shield were simply mounted atop the Panzer II superstructure, and side shields were added to protect the four-man crew. The early *Marder* II weighed 11·5 tonnes.

Key vehicles

Further *Marder* IIs were built on the chassis of the PzKpfw II Ausf A, C and F. These were standard models of the Panzer II with leaf spring suspension. From June 1942 onwards, 1217 of these were built. Again the complete gun with shield was mounted on the hull top (the original gun carriage was omitted of course) and the superstructure was built up at the front to the top of the shield. They were key vehicles in the anti-tank battalions on the Eastern Front. The vehicle carried 37 rounds and had armour of 10 mm.

A big improvement came with the next model, which was built on the chassis of the PzKpfw 38 (t). *Panzerjäger* 38 (t) SdKfz 139 (*Marder* III) was the full designation, and the vehicle was introduced in March 1942. The original model suffered from being nose heavy, and in the later production vehicle the engine was repositioned in the centre of the chassis and the gun moved to the rear. The early model on the other hand, like the Panzer II-based vehicles, had the gun atop the original superstructure. The normal glacis plate and front vertical plate were removed in the later version and replaced by a single glacis plate set at a forward angle of 67°. A cast armoured hood incorporating a hinged visor for the driver was fitted in this plate.

By 1943 the German 75-mm Pak 40/3 was

Panzerjäger I
First of the long line of German specialised anti-tank SP mountings
Length: 14·51 ft *Width:* 5·07 ft *Weight:* 14,110 lb *Speed:* 25 mph *Range:* 87 miles *Armour:* 14 mm max *Armament:* 47-mm Pak(t)

A Panzerjäger *I enters the blazing city of Rostov towards the end of the first phase of the battle for Russia, in November 1941*

SdKfz 132 'Marder' II
A potent tank destroyer, the first 'marten'
mounted the captured Russian 76·2-mm anti-
tank gun on PzKpfw II Ausf D or E chassis

A heavily camouflaged Marder II *on the move*

Bundesarchiv

available, and in the late *Marder* III it was mounted on a platform at pannier height and was shielded by a three-sided superstructure of 10-mm plate that extended over the tracks and to the extreme rear of the hull. This arrangement extended the fighting compartment more conveniently to the back of the hull. A support for the gun when in the travelling position was mounted on the front horizontal plate. About 750 early *Marder* III and 800 late *Marder* III were built. The gun could be traversed 30° left and right and elevated −10° to +25°, common for all various *Marder* models.

Marder I was entirely different in appearance, built on the chassis of the captured French Lorraine carrier which was taken into German service as a standard type, and used as the basis for several SP vehicles. *Marder* I, full designation *Panzerjäger für 75-mm Pak 40/1 (Sf) Lorraine Schlepper (f)*, ran to only 184 vehicles converted by the firm of Becker in 1942–43. These were all used in France, where they first saw action in the battle of Normandy in 1944. The Lorraine carrier was a very stable tracked munitions and personnel carrier taken over in large numbers when France capitulated. It had a flat load space which lent itself admirably to conversion to a gun carriage and formed a basis for several improvised types.

While the *Marder* series were the most standardised and numerous of the improvised *Panzerjager* types, there were many others, large and small. Of major importance was the *Nashorn* (Rhinoceros) or *Hornisse* (Hornet) which was fully designated 88-mm Pak 43/1 *auf Fgst* PzKpfw III/IV, later *Panzerjäger* III/IV. As its designation implies, its chassis was a combination of Panzer III and Panzer IV components. The powerful 88-mm Pak 43 was put on a mobile mount as soon as production guns were available in 1943. Some 473 *Nashorns* were built. The open superstructure was only lightly armoured due to the weight of the gun, which rather overloaded the chassis, and production ceased as improved *Jagdpanzer* types became available.

Most of the other *Panzerjäger* types were small and relatively unimportant. Among them were the 37-mm Pak 35/6 on infantry

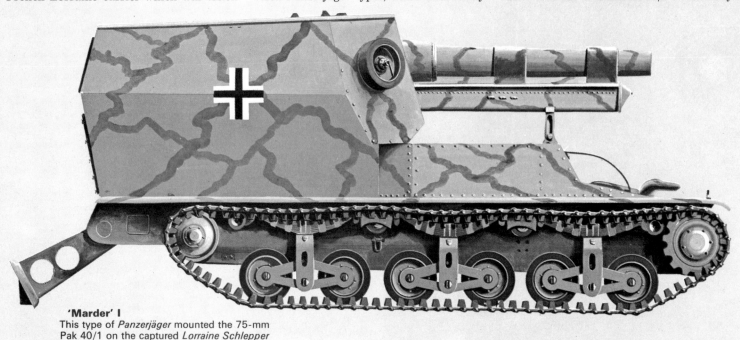

'Marder' I
This type of *Panzerjäger* mounted the 75-mm Pak 40/1 on the captured *Lorraine Schlepper*

SdKfz 138 'Marder' III
A late model *Marder* III, with the fighting compartment extended to the back of the hull
Length: 15·26 ft *Width:* 7·09 ft *Weight:* 23,150 lb *Speed:* 26 mph *Range:* 115 miles *Armour:* 25 mm max *Armament:* 75-mm Pak 40/3

carrier chassis such as the French VE and the ex-British Bren or Universal carrier (these were captured at Dunkirk). The ex-French Renault R-35 and Hotchkiss H-39 tanks were fitted with the 47-mm Czech gun and the Pak 40 to give *Marder*-like vehicles; these were used mainly by the occupying forces in France and fought in the Normandy campaign. An unusual conversion in the same vein featured a Hotchkiss 20-mm anti-tank gun mounted in a shield on a captured British Matilda tank, and some of these were used by occupying troops in Denmark, France and the Low Countries.

When the *Raupenschlepper* (Eastfront tracked tractor) appeared in 1943, 83 of them were converted to unarmoured *Panzerjäger* and entered service in 1944 as the 75-mm

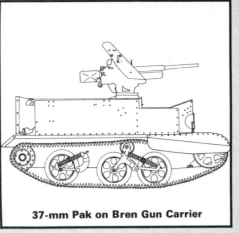

37-mm Pak on Bren Gun Carrier

Pak 40/1 auf RSO (Sf1). The Pak 40 gun was a field anti-tank piece less carriage mounted on a turntable on the load compartment. A canvas tilt was provided which served to disguise the vehicles true nature when it was rigged. Biggest of all these *Panzerjäger* types was the *Panzerjäger* 128-mm *Kanone* Sf VK 3001 (H) which was built on the chassis of one of the Henschel 30-ton prototypes which appeared during the development of the Tiger tank. This prototype tank was discarded in 1941 when 45-ton prototypes were called for. Two of the four VK 3001 (H) chassis completed were converted to very heavy *Panzerjäger* with the massive 128-mm K 40 as an anti-tank weapon against heavy tanks. An open top superstructure with rear door was fitted.

SdKfz 164 'Hornisse'
Also known as 'rhinoceros' the 'hornet' mounted the powerful 88-mm Pak 43 on the PzKpfw III/IV chassis

47-mm Pak 40 on Renault R-35
One of many minor *Panzerjäger* conversions on captured foreign vehicles

JAGDPANZER

The *Jagdpanzer* (hunting tank) was a natural development from both the *Sturmgeschütz* and *Panzerjäger* types and exhibited characteristics of both. The interim designs of 1943 mostly carried two designations, one as a *Panzerjäger* and another as a *Jagdpanzer*.

The classic example of this was the massive *Elefant* (originally *Ferdinand*, after its designer, Ferdinand Porsche), which was one of the most famous of all German AFVs. The original designation was PzJg Tiger (P) *Ferdinand für 88-mm Pak 43/2* and this was subsequently changed to *Jagdpanzer Elefant für 88-mm Pak 43/2 L/71 (SdKfz 184)*. It was decided to alter the uncompleted Porsche Tigers rejected in favour of the Henschel models, to make a self-propelled mount for the 88-mm Pak 43.

The *Elefant* was a complete re-design of the original Porsche project – virtually a new vehicle. Only the suspension and hull shape of the original Porsche Tiger remained. The original petrol-electric drive was retained, but was modified in that the Porsche air-cooled drive motors were replaced by two Maybach 300 hp HL 120 engines. These were centrally sited instead of being located at the rear, leaving the rear hull available for the fighting compartment. Drive was to the rear sprocket.

The driver and radio-operator sat in the hull front forward of the engines, the driver having a hydropneumatic steering system. Fuel tanks flanked the central engine compartment and the rear was given over to a full-width, slope-sided fighting compartment, which housed the commander, gunner and two loaders.

The *Elefant's* punch

The punch of the *Elefant* was its 88-mm Pak 43/2 L/71, a later development of the 88-mm Flak 36 which had been adapted as a weapon for the Tiger tanks. This was a more powerful (and longer) gun than the earlier 88-mm weapons. Originally the 88-mm gun was the only armament, but early combat experience showed that close range armament was necessary to prevent infantrymen from attacking the vehicle. Hence, a machine-gun was added in the front hull.

Vision from the *Elefant* was poor – forward only – so cupolas were added, and appliqué armour was extensively added in bolt-on form, with 100 mm added to the nose and 200 mm on the superstructure front. Elsewhere the armour thickness was up to 80 mm. There was a large circular hatch in the superstructure rear for weapons maintenance and this featured a smaller hatch for ejecting spare cartridges. Other superstructure apertures were pistol ports and two roof hatches. The Porsche suspension was novel, consisting of three twin bogies on each side sprung by torsion bars. The wheels were all steel with resilient rims.

The prototype *Elefant* appeared in March 1943, and in July that year the *Elefants*, by now allocated to Panzer Regiment 654, were used for the first time in action with disastrous results. Still mechanically unreliable, many broke down, became bogged down or were over-run and captured by

infantry at close quarters. Their use was then restricted and most of the survivors found their way to Italy in 1944, where many were lost to Allied tanks.

Developed from the StuG III and IV was a much refined design specially for tank destroying, and this vehicle, the *Jagdpanzer* IV, was the first true *Jagdpanzer* designed as such. Designated *Jagdpanzer IV Ausf F (75-mm Pak 39 L/48) SdKfz 162*, it was first in service late in 1943. It incorporated all the lessons learned from the use of the StuG III in the anti-tank role. Sloped armour with a very low silhouette was a distinctive feature, and the vehicle had 60 mm thick upper and lower frontal armour, well-sloped for optimum shot deflection. The sloping superstructure sides were carried over the full track width to give increased ammunition stowage, compared with the StuG III, of 79 rounds. The top of the superstructure was in one piece, with two hatches, and there was another small hatch through which a dial sight could be extended. Spaced armour was carried on each side of the superstructure.

The main armament of one 75-mm Pak 39 was mounted in the sloping superstructure front, the mount being of gimbal type protected by an external cast mantlet. Elevation limits were $-8°$ to $+10°$, with traverse 12° left and 10° right. The gun fired APBC (Armour-Piercing Ballistic Cap) ammunition at a muzzle velocity of 2300 fps, or HE shells at 1800 fps. Hollow charge rounds, AP 40 shot and smoke could also be fired, making the vehicle very versatile and effective. Early models had rounded front plates and late ones lacked a muzzle-brake on the gun. In late 1944 a revised model appeared with all-steel wheels (dispensing with rubber tyres) and an improved StuK 42 75-mm L/70, known as the *Jagdpanzer* IV/70. These vehicles were first-line equipment until the end of the war.

The finest *Jagdpanzer* of all, however, was the splendid and impressive *Jagdpanther*, the most important derivative of the Panther tank and another of the classic wartime AFVs. Both the *Elefant* and the *Nashorn* had been built as *Panzerjäger* types, as already described, but neither of these was satisfactory.

The need for a fast, up-to-date tank destroyer on a modern chassis was met by adapting the Panther as the previous attempts to produce a heavy tank destroyer had been so unsuccessful. The 88-mm Pak 43 had been mounted on the Porsche Tiger chassis (to make the *Ferdinand*) and on the PzKpfw III/IV chassis as the *Nashorn*, but both of these improvisations proved unsatisfactory: *Ferdinand* was too heavy and *Nashorn* too small and under-powered.

By 1943, however, there was an urgent need for tank destroyers in quantity so it was decided to utilise the best available chassis, that of the Panther. MIAG were asked to work out the design and the prototype was first demonstrated, in the presence of Hitler, on 20 October 1943. The Panther chassis was used unaltered, but the front and upper side plates were extended upwards to make a well-sloped enclosed

Jagdpanzer IV, mounting the 75-mm StuK 40 on the PzKpfw IV chassis, the first true Jagdpanzer

JgPz V Jagdpanther
Best of all the 'hunting tanks' the *Jagdpanther* was one of the all-time classic fighting vehicles
Length: 32·35 ft *Width:* 10·76 ft *Weight:* 95,900 lb *Speed:* 29 mph *Range:* 130 miles *Armour:* 80 mm max *Armament:* 88-mm Pak 43/3; 2×7·92-mm MG 34

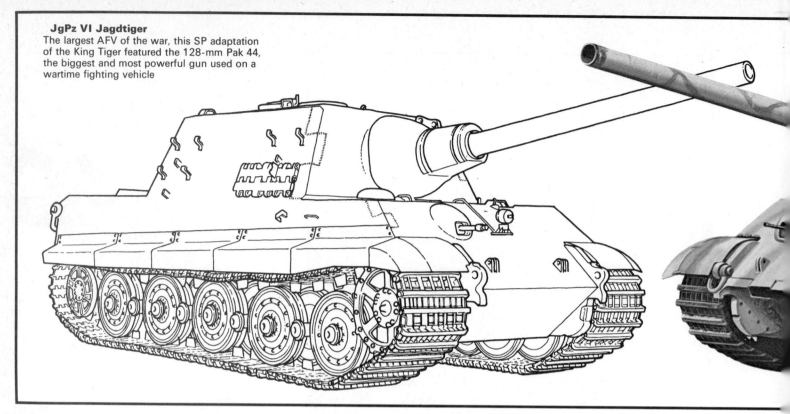

JgPz VI Jagdtiger
The largest AFV of the war, this SP adaptation
of the King Tiger featured the 128-mm Pak 44,
the biggest and most powerful gun used on a
wartime fighting vehicle

superstructure. The mantlet was fitted in
the centre of the hull front with a limited
traverse for the 88-mm Pak 43/3 L/71 gun of
11° each side. Armour was 80 mm in front
and 60 mm at the sides. A ball-mounted MG
34 was fitted in the right front of the hull and
the driver sat in the usual position in the
left front. Sighting equipment consisted of a
rangefinder and periscope telescope. The
telescope protruded through a slot in the
roof within an armoured quadrant arc
linked to the gun mount.

The new SP version of the Panther was at
first designated 88-mm Pak 43/3 *auf Panzerjäger* Panther (SdKfz 173) but at Hitler's
personal suggestion in February 1944 it was
redesignated simply as the *Jagdpanther*
(Hunting Panther).

MIAG commenced building *Jagdpanther*
in February 1944, using the Ausf G chassis
which had by then become the current
production type. By the war's end 382 had
been completed. First production *Jagdpanthers* had a one-piece barrel, but later a
two-piece barrel was used on the 88-mm
weapon to ease barrel changing, as the
barrel did not wear uniformly and it was
economical to make in it two parts. Later
Jagdpanthers had a simplified collar round
a thicker, bolted mantlet.

Crew of the *Jagdpanther* consisted of a
commander, gunner, two loaders, radio-
operator/machine-gunner and driver. The
vehicle carried 60 88-mm rounds. The
Jagdpanther was the best and most potent
of all the German tank destroyers. It was
well-shaped, low, fast and heavily armoured. It was intended to build *Jagdpanther*
at a rate of 150 per month, but disrupted
production facilities in the last year of the
war made this target quite impossible.

Following *Heereswaffenamt* policy, a
limited traverse tank hunter version of the
King Tiger tank was also produced. This
vehicle, originally designated *Panzerjäger*
Tiger Ausf B (SdKfz 186), consisted essentially of the King Tiger hull with a box-
like superstructure holding a 128-mm Pak
44 L/55 gun, the largest gun installed in any
wartime AFV. An unarmoured full-size

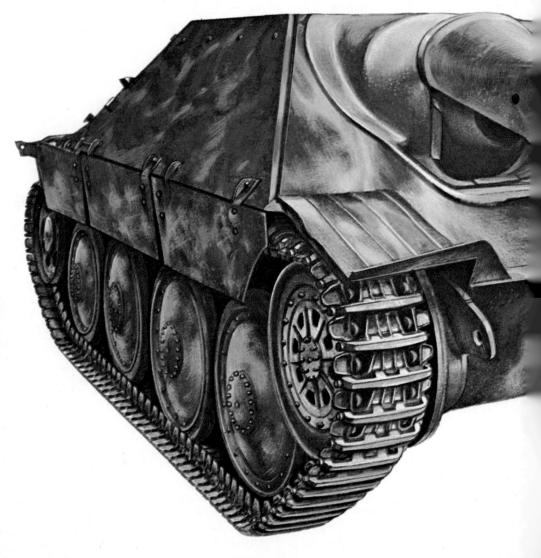

Length: 32·35 ft Width: 10·76 ft Weight: 158,000 lb Speed: 24 mph Range: 105 miles Armour: 250 mm max Armament: 128-mm Pak 44; 1×7·92-mm MG 34

JgPz 38(t) 'Hetzer'
The 'baiter', a light *Jagdpanzer* using the ubiquitous PzKpfw 38(t) chassis first appeared in mid-1941. Its armour disposition, concentrated around the hull and turret front, is shown

Length: 20·57 ft Width: 8·63 ft Weight: 35,275 lb Speed: 26 mph Range: 110 miles Armour: 60 mm max Armament: 75-mm Pak 39; 1×7·92-mm MG 42

mock-up of the new vehicle was completed in October 1942 at the same time as the prototype King Tiger. At 76 tons the *Jagdtiger* was the largest fighting vehicle to see action during the war.

Henschel co-operated with Krupp in the design of the gun mount, which featured the small *Saukopf* mantlet. To utilise internal stowage to the best advantage, separate ammunition was used with the 128-mm gun, and to assist with carrying the added weight the suspension was spaced out by an extra 260 mm. A total of 150 *Panzerjäger* Tiger Bs were ordered and they were built by Steyr-Daimler-Puch at St Valentin in Austria. In the event, shortages and disruptions meant that only 70 vehicles were built, 48 of them in 1944. Initially there was an interruption in the supply of the 128-mm gun and it was proposed to overcome this by installing the same 88-mm gun as in the *Jagdpanther*. In February 1944, on Hitler's orders, the name of the vehicle was simplified (in common with other types) and it was called the *Jagdtiger*, the designation by which it is best remembered.

Dr Porsche attempted to improve on the design by installing torsion bar bogie suspension, similar to that used on the prototype Tiger and the *Elefant*. One vehicle was converted to this standard, and Porsche claimed that production was greatly simplified. However, the urgent need for tanks did not allow time for changes to be made and any possible slight advantage of the Porsche suspension did not justify the added time lag. Externally this vehicle was distinguished by having one less road wheel each side.

Jagdtiger saw service from late 1944 until the end of the war, but although a most formidable weapon, it demonstrated that the tactical limitations of such large, heavy vehicles became a liability, even in the defensive type of warfare the panzers were then fighting. Even the lighter King Tiger strained logistic and tactical resources, and the lumbering *Jagdtiger*, a large, slow-moving target, subject to frequent breakdowns, was in no way a success. It remains an impressive engineering achievement, however, and is assured of its place in AFV history by virtue of its immense size and firepower.

Hetzer
In March 1943 it was also decided to build a light *Jagdpanzer*, and the ever reliable PzKpfw 38(t) chassis was called upon. What amounted to a miniature version of the *Jagdpanther* was produced. The vehicle was fitted with the 75-mm KwK 40 L/48 gun and designated *Jagdpanzer* 38(t) *Hetzer*. The first models appeared in May 1944 and all subsequent PzKpfw 38(t) production was diverted to build *Hetzer*. Over 1500 had been built by the end of the war and the vehicle was highly successful. A number were completed as flame-throwers, with a flame projector replacing the gun. Frontal armour was 60 mm, all plates were well sloped and the vehicle weighed 16 tonnes. A remote-control MG 42 was fitted in the roof for close-in defence.

The PzKpfw 38(t) chassis was so good that it was chosen to form the basis for a whole new series of light armoured vehicles. A *Panzerjäger Kanone* with 75-mm L/70 gun was planned in 1945, but never produced. The chassis for the new standard series was designated 38(d).

PANZERARTILLERIE

SiG 33s: 150-mm infantry assault guns on PzKpfw
II Ausf B chassis

150-mm SiG 33 auf PzKpfw II
Length: 15·58 ft *Width:* 7·34 ft *Weight:* 26,455
lb *Speed:* 25 mph *Range:* 124 miles *Armour:*
20 mm max *Armament:* 150-mm SiG 33

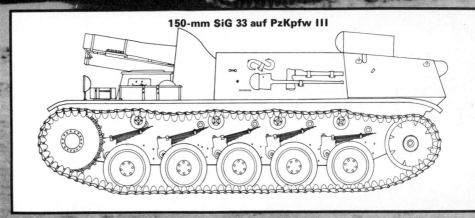

150-mm SiG 33 auf PzKpfw III

SdKfz 124 'Wespe'
The 'wasp' conversion of a PzKpfw II ausf H
Length: 15·72 ft *Width:* 7·35 ft *Weight:* 25,350
lb *Speed:* 25 mph *Range:* 87 miles *Armour:* 90
mm max *Armament:* 105-mm PzFH

Hummel *(bumble bee), featuring a 150-mm gun on the hybrid PzKpfw III/IV chassis*

The classification *Panzerartillerie* covers self-propelled equipment which gave fire support in the field but was not used in an anti-tank role: *Infanterie Geschütze* (infantry guns), *Sturmhaubitze* (assault howitzers), *Panzerhaubitze* (armoured howitzers) and so on. Many of these had foreign chassis and the types were diverse.

The first was the much over-loaded 150-mm SiG 33 4/12 auf PzKpfw I of 1940, which, like the PzJg I, proved the value of this type of mobile artillery. The Panzer II proved a more stable and suitable carriage for the 150-mm gun and many had entered service by 1942, serving with particular success in the Western Desert. Designated 150-mm SiG 33 L/12 auf Fgst PzKpfw II (sf), it came in two forms, one on the unaltered chassis and one on an extended chassis with an extra wheel each side. The superstructure was open with low sides; the vehicle weighed 10 tonnes.

Yet another infantry gun model was the 150-mm SiG 33 (Shl) auf PzKpfw 38 (t) *Bison*, an extremely crude 1942 adaptation which was replaced by a 1943–44 version in which the engine was moved forward into the original fighting compartment and the gun and superstructure were moved to the rear. This made for a better balanced vehicle. About 370 of both types were built.

Much more successful was the 150-mm *Panzerhaubitze* 18/1 auf PzKpfw III/IV *Hummel* (Bumble Bee) which was a *Panzer-*

artillerie equivalent to the *Nashorn.* Built on the PzKpfw III/IV chassis, this was produced from late 1942 until the end of the war, and over 660 were built. Weight was 26 tonnes and there was a five-man crew. The *Hummel* had the long 150-mm gun rather than the infantry gun.

One of the best conversions was the 150-mm SFH 13/1 auf GW *Lorraine Schlepper* (t). The gun was an obsolete Krupp piece of 1917 which was adapted and fitted to the Lorraine chassis by the firm of Becker of Krefeld. The mounting of this quite heavy weapon rather overloaded the suspension, but it was popular and effective, and the conversion was quickly carried out. This 8·3-tonne vehicle was used on all fronts and first served widely with the Afrika Korps in 1942. It was in service up to the end of the war.

The most important of all *Panzerartillerie* vehicles was probably the *Wespe* (Wasp), designated fully as le FH 18/2 *auf* Fgst PzKpfw II (sE) (light field howitzer 18/2 on carriage of the PzKpfw II tank) which went into production in 1942 and continued until the end of the war.

By the last quarter of 1942 the Russian Front situation was stable enough to allow the design of more suitable *Panzerartillerie* conversions. The PzKpfw II Ausf F was selected for carrying the 105-mm light field gun. The design was ordered and by December 1942 the first *Wespen* were leaving the Famo assembly plant at Warsaw.

In February 1943 Hitler ordered that the total production capacity of the Panzer II be diverted to this purpose and that of the PzKpfw 38(t) to carrying the anti-tank guns. His reasons appear to have been the popularity of the first vehicles in service and an early attempt at standardisation, but most of all, the fact that the newly introduced hollow charge ammunition for artillery weapons would in time supplant the conventional anti-tank guns. So, by April 1943, only *Wespen* were appearing on the PzKpfw II chassis. Due to the war situation, production ceased in mid-1944, but by then 682 had been built and a further 158 without main armament were supplied as *Munitions-Selbstfahrlafette auf* PzKpfw II. This was because the *Wespe* could carry only 32 rounds and a munition carrier was needed to supply each battery. Generally, Wasps served with the Panzer or Panzer Grenadier Divisions in batteries of six, with between two and five batteries to a division.

The 105-mm le FH 18 was similarly fitted to the *Lorraine Schlepper*, and 24 were converted under the designation 105-mm le FH 18 auf GW *Lorraine Schlepper* (t). Forty-eight Hotchkiss H-39 tank chassis, 24 FCM tank chassis and a few Char B2s were given a similar conversion, and all these French vehicles served the occupying troops in France and saw combat in the Normandy fighting. There were also several other field howitzer vehicles of a minor nature.

Yet another conversion of a foreign tank, the 105-mm light field howitzer on the Lorraine Schlepper

105-mm Light Field Howitzer on Char B1 bis
A conversion of the captured French tank
provided a makeshift SP gun

STURMPANZER

The last major types were the *Sturmpanzer* (assault tank) class, of which the most famous was the *Sturmpanzer* IV *Brummbar* (Grizzly Bear). Over 40 were built in 1943–44, specifically for street fighting on the Eastern Front. Very heavily armoured, the *Brummbar* weighed 28 tonnes and had 100 mm of front armour, and with a *Sturmhaubitze* 43 low velocity gun in an armoured mantlet it was a massive looking vehicle. Later examples had an added ball-mounted MG 34 in the hull front for self-defence. By the time the vehicle had been developed street fighting in Russian cities was a thing of the past, but *Brummbar* was a spectacular and formidable vehicle, difficult to knock out and living up to its name.

Even more formidable, however, was the heftiest of all, the *Sturmtiger*. Also known as the *Sturmpanzer* VI or *Sturmmörser*, this weapon was developed to requirements from the German Army engaged in the heavy street fighting at Stalingrad and other similar places in Russia. The full designation was 380-mm RW 61 Auf StuMrs *Tiger* (380-mm rocket projector Type 61 on Tiger chassis). The design owed its origins to a request for a self-propelled 210-mm howitzer capable of following up the advancing German troops and able to engage difficult targets with high angle fire. Hitler was personally responsible for instigating the idea for the weapon.

When development work started on the project, it was decided that the then-new PzKpfw VI Tiger Ausf E chassis would be used, but it was found that no suitable 210-mm gun was available. It was finally proposed to use the *Raketenwerfer* 61 L/54, a weapon that had originally been developed by the firm of Rheinmetall-Borsig as an anti-submarine device for the German Navy.

A model of the *Sturmtiger* was first shown on 20 October 1943, and the type went into limited production in August 1944, when ten existing Tiger tanks were converted by the firm of Alkett. For their intended role as mobile assault howitzers against troop concentrations and fortifications, they were heavily armoured. The suspension, power train, engine and hull were those of the basic Tiger E, but the normal superstructure and turret of the tank was replaced by a heavy rectangular superstructure. The welded superstructure was made of rolled armour plates, and the side plates were interlocked with the front and rear plates. A heavy strip of armour reinforced the joint between the front plate and glacis plate on the outside.

The rocket projector, mounted offset to the right of centre in the front plate of the rectangular superstructure, consisted of a tubular casting and spaced rifled liner and cast mantlet The mantlet, an integral part of the tube, protected the joint of the tube and mount. Gases were deflected between the tube and liner and escaped through a perforated ring at the muzzle end. A rectangular loading hatch was located in the centre rear of the top plate and was closed by two doors, one forward and one to the rear. The rear door, spring balanced and hinged to open outwards, could be opened independently of the forward door and mounted a smoke projector. A small crane for loading the rocket projectiles was mounted on the superstructure.

Six ammunition racks on each side of the fighting compartment were provided within the vehicle to accommodate a total of 12 rounds, while an additional round could be carried in the projector tube. The rounds were of HE and Hollow Charge type. The 380-mm (15-in) HE projectile (380-mm R *Sprenggranat* 4581) had an overall length cf 56 in, a total weight of 761 lb and a maximum range of 6200 yards.

Sturmpanzer VI Sturmtiger
Length: 20·7 ft *Width:* 12·25 ft *Weight:* 156,800 lb *Speed:* 25 mph *Range:* 87 miles *Armour:* 6 in max *Armament:* 380-mm rocket projector; 1×7·92-mm MG 34

FLAKPANZER

Tanks specially adapted for the anti-aircraft role did not come into service until 1943 when the menace from Allied aircraft was becoming critical. The PzKpfw IV was the obvious choice for the major production models because of its stability as a gun platform, its size and its availability. The various models were produced quickly and the conversions were simple, an existing AA mount placed above the original turret space.

First off was the *Flakpanzer* IV (20-mm *Flakvierling* 38) *Möbelwagen* (furniture van). This simply had the *Flakvierling* 38 quadruple mount on its turntable, with hinged 10-mm armour sides which were lowered to give a 360° traverse for the mount. This was built on the PzKpfw IV Ausf H or S chassis, and over 200 were converted from standard tanks.

First of the Flakpanzer *was* Möbelwagen, *a quadruple 20-mm Flak 38 on PzKpfw chassis, with hinged armoured sides*

Flakpanzer IV 'Möbelwagen'
The second version of the 'furniture van' replaced the quadruple 20-mm mount with a single 37-mm weapon

A further variant was the *Flakpanzer* IV (37-mm Flak 43 L/60) *Möbelwagen* which was similar to the other *Möbelwagen* except that a single 37-mm mount replaced the *Flakvierling* 38.

Also developed late in 1943 was the *Flakpanzer* IV (20-mm) *mit PzFGst Panzer IV/3 Wirbelwind* (whirlwind) which was simply a basic PzKpfw IV Ausf J chassis with a 20-mm *Flakvierling* 38 quadruple mount in a multi-sided light armoured (10-mm) turret with 360° traverse. Over 340 were built during 1944 by the firm of Ostban. A few are believed to have been built on the PzKpfw III chassis.

Its counterpart with 37-mm single gun was *Flakpanzer* IV (37-mm) *Ostwind* (east wind) which was exactly like *Wirbelwind* except for the single 37-mm gun replacing

the quadruple mount. Deutsche Eisenwerke built 205 of these vehicles in 1944–45, converting them from PzKpfw IV Ausf J. An improved model with a later version of the 37-mm gun, in prototype stage at the end of the war was designated *Ostwind* II.

By 1945 a much improved *Flakpanzer*, designated *Leichter Flakpanzer* IV (30-mm) *Kugelblitz* (ball lightning) was at testing stage, but only five were built. The *Kugelblitz* had a fast-traversing power-operated turret with twin machine cannon MK 103/38 30-mm weapons. The guns had 80° maximum elevation. A further variant of this, *Zerstorer* 45 (destroyer 45) was at prototype stage by the end of the war. It had four 30-mm MK 103/38 guns and would have been a formidable machine had its development not been curtailed.

On the Russian Front a very simple vehicle, *Flakpanzer* 38 (t) was built to counter low flying Sturmovik aircraft. This was essentially the PzKpfw 38 (t) in its basic gun carriage form (as used for the *Marder* III) but with a single 20-mm Flak gun mounted. Over 160 of these were produced in the 1943–44 period.

The only other Flak vehicles akin to *Flakpanzer* were two experimental types on the *Grille* prototype *Waffentrager* chassis. They featured the 88-mm Flak 37 and 88-mm Flak 41 respectively in a mount with folding shields rather like the *Möbelwagen*. There were no production plans for these costly and heavy vehicles, however, and they appear to have been made by Krupp solely to make use of the available chassis which would otherwise have been discarded.

Flakpanzer IV 'Ostwind'

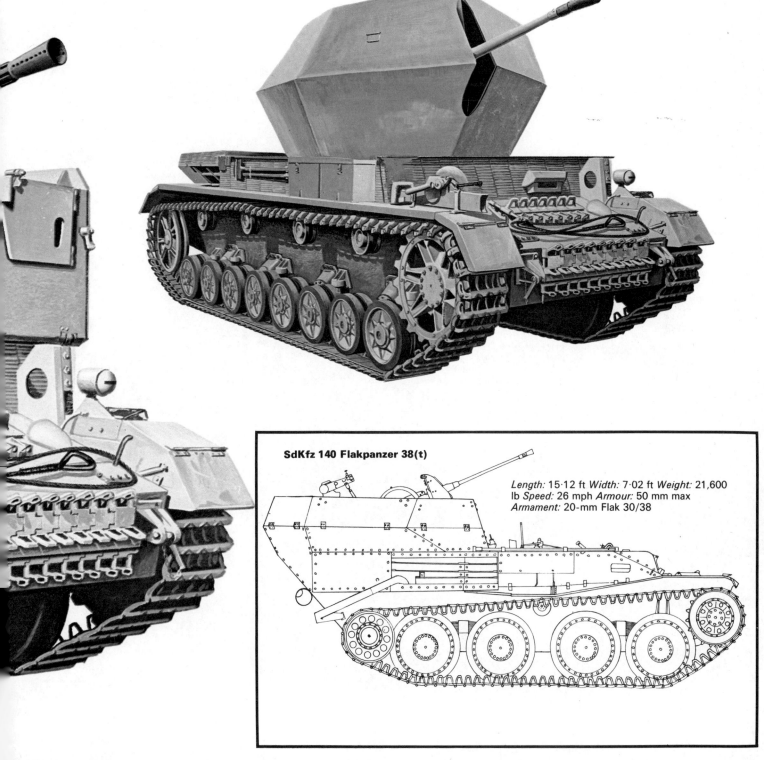

SdKfz 140 Flakpanzer 38(t)

Length: 15·12 ft *Width:* 7·02 ft *Weight:* 21,600 lb *Speed:* 26 mph *Armour:* 50 mm max *Armament:* 20-mm Flak 30/38

THE GIANT GUNS

600-mm Mortar 'Karl' (below and opposite)
Length: 36·58 ft *Width:* 10·33 ft *Weight:*
264,455 lb *Speed:* 6 mph *Armour:* 15 mm
Armament: 600-mm 040 mortar

Bundesarchiv

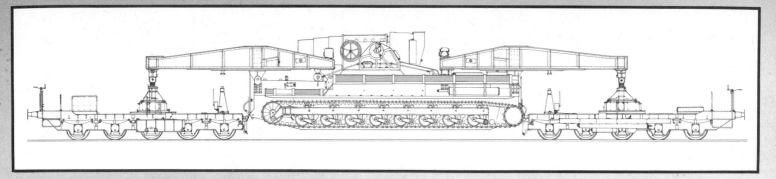

The largest tracked vehicles of all were the huge 600-mm mortars which were projected before the war. These might best be described as motorised tracked gun carriages. They were not Panzer Division vehicles and were in fact classified as siege artillery, but were actually self-propelled carriages with driving compartments. The first vehicle was built by Rheinmetall in 1939, under the designation 600-mm *Mörser Karl* (*Gerät* 040), and was intended to bombard the massively defended forts of the Maginot Line on the Franco-German border. The vehicle weighed 120 tonnes, was 11·15 metres long, and had 12-mm armour plate. The vehicle had a top speed of 6 mph. A second vehicle was built, designated 600-mm *Mörser Karl* II. By the time the two vehicles were in service France had fallen. Named *Thor* and *Eva*, the two pieces saw service on the Eastern Front and were particularly successful in the siege of Sevastopol in 1942, and also at Brest-Litovsk.

In 1943, as a result of these operations, six improved vehicles were ordered, designated 540-mm *Mörser Karl* (*Gerät* 041). The mortar was of different calibre, but it was generally similar to the original vehicle and weighed 130 tonnes.

To support these huge mortars, a special series of *Munitionspanzerwagen* IV was produced, based on the PzKpfw IV Ausf F. This type had a compartment for three rounds, and a 3·5-tonne crane lifted the round to mortar carriage. On the carriage itself the weapon detachment worked from platforms above the tracks. These platforms folded up alongside the carriage when the vehicle was on the move.

'Karl' being readied to hurl its 600-mm shell in a plunging arc towards Sevastopol. The piece could move short distances under its own power, but was slung between two rail bogies for longer journeys. Also shown is the Munitionenspanzerwagen *IV special ammunition carrier*

German troops training with a 3.7-cm Flak 18 mounted on an SdKfz 6/2 half-tracked self-propelled mounting

Ian Hogg